OCCUPATIONAL SAFETY AND HEALTH MANAGEMENT

OCCUPATIONAL SAFETY AND HEALTH MANAGEMENT

SECOND EDITION

Thomas J. Anton

Purdue University Calumet
Guest Lecturer and Supervisor

McGRAW-HILL BOOK COMPANY

New York St. Louis San Francisco Auckland Bogotá
Caracas Colorado Springs Hamburg Lisbon London Madrid
Mexico Milan Montreal New Delhi Oklahoma City Panama
Paris San Juan São Paulo Singapore Sidney Tokyo Toronto

This book was set in Times Roman by the College Composition Unit
in cooperation with Black Dot, Inc.
The editors were Anne T. Brown, Lyn Beamesderfer, and Bernadette Boylan;
the production supervisor was Denise L. Puryear.
The cover was designed by Warren Infield.
R. R. Donnelley & Sons Company was printer and binder.

OCCUPATIONAL SAFETY AND HEALTH MANAGEMENT

1 2 3 4 5 6 7 8 9 0 DOC DOC 8 9 4 3 2 1 0 9

ISBN 0-07-002108-2

Library of Congress Cataloging-in-Publication Data

Anton, Thomas, John
  Occupational safety and health management.
  Includes bibliographies and index.
  1. Industrial safety. 2. Industrial hygiene.
I. Title.
T55.A523 1989 620.8'6   88-38018
ISBN 0-07-002108-2

ABOUT
THE AUTHOR

THOMAS J. ANTON is a veteran of the U.S. Air Force (formerly the U.S. Air Corps) in the European-African Theater of Operations during World War II. He received his B.S. degree in education from Ball State University.

His industrial experience includes management positions in production for the Krick-Tyndall Company and safety engineering for the Budd Company over a 30-year career span.

For the past six years, he has been teaching safety engineering and supervision courses at Purdue University Calumet. He has also taught the Voluntary Compliance Safety Course to Federal Compliance Officers at the OSHA Training Center in Rosemont, Illinois. He has also held training seminars for Bethlehem Steel Corporation, Avery Graphics, and the Gary School System Maintenance Division. His other professional activities include industrial accident investigations for client lawyers, in the Chicago area, of Technical Services for Attorneys of Fort Washington, Pennsylvania.

Mr. Anton is the author of two widely used career textbooks for elementary schools: *Foreman Trainee,* 1975, and *Safety Engineer,* 1976, both published by the Educational Research Council of America. He is also the author of *Occupational Safety and Health Management* (1st edition, 1979) published by the McGraw-Hill Book Company.

He is a member of the Tri-State Chapter of the American Society of Safety Engineers and past member of the National Safety Council, where he served on the Power Press Safety Standards Committee. He is also a past member of the Indiana Manufacturers' Association, having served on its Workers' Compensation Committee.

Dedicated to my grandchildren,
Aaron and Elizabeth
—future members of the next century.

CONTENTS

PREFACE

Since the first edition of this book was published, events have occurred which have shown that the methods advocated have proven to be sound and that accident prevention can be a profitable combination of humanitarian motives and good business policy. The philosophy presented continues to be the basis of sound management practices combined with a moral obligation for the safety of the employee.

This book has been revised to continue to serve the basic needs of students, managers, and safety specialists whose professional work requires a sound knowledge of the fundamental principles of safety, particularly those relative to management and engineering.

This edition has been expanded from 11 to 20 chapters to provide for full coverage of occupational and health management. The phrase "safety in the workplace" has created a new awareness of safety requirements for workers. Under its impact, existing safety and health laws have been expanded, and new federal and state laws have exerted even more pressures on business.

Much of the work in Chapters 1 through 11, including some of the chapters themselves, have been rearranged into other parts of the text, while other chapters have been reorganized to include additional or updated materials.

The newer chapters added to the text include: Accident Investigation: A Management Function, Medical and Health Surveillance Systems, Psychological Aspects Related to Safe Performance, Construction Industry Safety and Health Standards, Radiation Safety, and Hazardous-Waste Management.

ACKNOWLEDGMENTS

Occupational Safety and Health Management is the product of the many people and organizations who have helped to coordinate my efforts in this endeavor—both directly and indirectly. It would be nearly impossible to list the names of all who have contributed to it, but it is most fitting to acknowledge those nearest to my endeavors in this project. They are as follows: Paul O. Sichert and David A. Verbeke; Susan-Marie Kelly, Managing Editor, *Safety and Health,* National Safety Council (formerly *National Safety News*); Professor Adam E. Darm, University of North Florida; Professor Carl Jenks, Purdue University Calumet; Peter J. Sheridan, Editor, *Occupational Hazards*; Ronan Engineering Company; Eileen Wolan, Occupational Safety and Health Administration of the United States Department of Labor; United States Environmental Protection Agency, Office of Solid Waste Management; and the United States Department of Education.

I would also like to thank the following reviewers for their many valuable comments and suggestions: Harvey E. Clearwater, University of Maryland; Linda Glazner, UCLA School of Nursing; and Carl G. Haselmaier, Colorado State University.

Last, and most important, a special mention of gratitude to my wife, Martha, for her utmost patience and encouragement in this endeavor.

Thomas J. Anton

BASIC PRINCIPLES OF ACCIDENT PREVENTION

Successful accident prevention programs depend on three essentials:

- Leadership by the employer
- Safe and healthful working conditions
- Safe work practices by employees

If any of these three essentials is missing, accidents on the job are likely to occur. These may result in injuries or even deaths, damage to property, loss of production, increased expenditure for insurance and compensation, and other costs.

Top management cannot delegate responsibility in this area. Employers, company presidents, and plant managers must be willing to accept the responsibility for occupational safety and health as an integral part of their jobs. They must establish safety policies, stimulate awareness of safety in others, and show their own interest if others are to cooperate in making working conditions safe and healthful.

Representatives of management, both line and staff, must reflect this interest in safety. Each department head must assume leadership for his or her own department and must be given the authority to fulfill responsibility for the safety and health program.

The most important person in the chain is the front-line supervisor, who deals most directly with the employee and thus bears the greatest responsibility for implementing the safety and health program. Front-line supervisors must be given the appropriate authority, assistance, and support. If they are to fulfill their responsibilities, they must be highly trained in safety and health—the most highly trained of any representative of the employer except for safety and health professionals such as safety engineers, and industrial hygienists.

DEFINITIONS

Accident

An "accident" may be defined as something that is unplanned, uncontrolled, and in some way undesirable; it disrupts the normal functions of a person or persons and causes injury or near-injury. During an accident, a person's body comes into contact with or is exposed to some object, other person, or substance, which is injurious; or the movement of a person causes injury or creates the probability of injury.

Injury

An "injury" may be defined as a harmful condition sustained by the body as the result of an accident; it can take the form of an abrasion, a bruise, a laceration, a fracture, a foreign object in the body, a puncture wound, a burn, or an electrical shock.

Serious Physical Harm

The Occupational Safety and Health Administration (OSHA) recognizes "serious physical harm" (a term devised by OSHA) as that type of harm which causes permanent or prolonged impairment of the body in that (1) a part of the body is permanently removed or rendered functionally useless or substantially reduced in efficiency on or off the job, or (2) a part of an internal bodily system is inhibited in its normal performance to such a degree as to shorten life or cause reduction in physical or mental efficiency.

REPORTING INJURIES

When an employee reports an injury to the supervisor, the supervisor is responsible for ensuring that the injured employee receives prompt medical attention. Company procedures should require that employees report all injuries to their supervisor. Furthermore, it should be company policy that if an employee does not report an injury, no matter how slight, to the supervisor, that is a violation of company safety rules.

A strict interpretation of this requirement means that there can be instances of employees reporting injuries incurred off the job, going to the plant dispensary for treatment, and thereby making the company liable. Investigation of minor injuries and follow-up action based on the findings are effective both in taking corrective measures and in showing good faith to employees and to OSHA. Management should watch closely the number of minor injuries because an increase in that number means an increase in the injurious contacts employees are having with objects in their working environment. Naturally, as the number of injuries increases, the probability of the injuries becoming more serious increases too.

INVESTIGATION OF ACCIDENTS

The occurrence of an accident proves that at some point prevention was inadequate; the failure may be by the management, the employee, or perhaps the builder of a machine, a process, or the plant itself. Something was not planned right, built right, or adhered to.

The purpose of accident investigation is to discover point of failure— the causative factors, the hazardous conditions or practices that brought about the accident—so that proper action can be taken to prevent a recurrence. There are many causes of accidents; and for any particular accident, there are usually several contributing factors. Learning where, why, when, how, and to whom accidents are happening means a great deal in learning how to teach employees to avoid them.

Finding the Facts

A "fact" is defined as something known to have happened or to be true. Fact-finding is, obviously, determining the facts. Every accident brings into play a system of fact-finding by many people. It can be started by management and union after an accident has occurred, continued before a board hearing an industrial case, and carried to a court of law hearing a third-party liability suit.

Every accident should be thoroughly investigated as soon as possible to find its major cause and contributing causes. All near-accidents should also be investigated; studying them can prevent recurrences.

It is important that the actual scene of the accident be inspected as soon as the occurrence of an accident is known of. Photographs are a necessary part of the investigation, both for the immediate description to top management and for future reference.

The supervisor is one of the most important people in an investigation, for several reasons. First, the supervisor is on the job almost all the time. Second, he or she is in constant contact with workers and is probably thoroughly familiar with all the hazards that could develop in the workplace. Third, it is most likely that if there were any witnesses to the accident, they may also have been employees under his or her jurisdiction.

FIGURE 1-1
Some large plants have their own ambulance equipment to get injured employees to the plant first-aid station as quickly as possible.

Don't underestimate the value of the supervisor on the line. The supervisor can prevent accidents and can tell you what has happened if an accident does take place.

When an accident is being investigated, the best results are obtained if the fact finders exercise good judgment, collect data ably, arrive at logical conclusions, and make sensible recommendations regarding prevention. It is neither sensible nor productive to recommend a guard after an accident has occurred but overlook an unsafe act, or to limit attention to the unsafe act if a better guard would suffice as a preventive measure.

A company's medical department, while separate from the safety program, is closely associated with it in accident prevention and therefore in fact-finding. The plant doctor should be as familiar as possible with the physical and environmental requirements of the plant in order to be able to analyze injuries. Information concerning the nature of each injury can help in planning to prevent recurrence. The doctor can also, of course, minimize the severity of injuries through prompt treatment.

Who Should Investigate Accidents?

Who should investigate accidents? First, the person (supervisor) who is directly in charge of the injured person's activities. The supervisor not only is the best qualified but has the best opportunity as well; therefore, it is the supervisor on whom management must rely to get all the facts as they occurred. The supervisor knows the nature of the work, how it should be done, and the people who do it. To repeat, then, without any question, the supervisor should be the first manager involved in the investigation of an accident.

Getting all the facts is no small responsibility. Collecting information from all witnesses to the accident is clearly very important in the investigation, and here again it is the supervisor who knows the workers and how best to approach them for facts.

ACCIDENT REPORTS

The supervisor must also be able to record the facts in an accident report in such a way that anyone who is remotely involved with his or her area of responsibility—or with the plant as a whole, for that matter—can understand what occurred and how it occurred. The report should also be understandable by an OSHA inspector. A poorly written report not only might show a poor attitude on the supervisor's part; it could also show a lack of interest by the company if the report were to be examined in a workers' compensation case or in a court of law. Once a supervisor completes an accident investigation and writes a report, he or she never knows whose desk it may cross within the company and outside the company. Therefore, the report must be as accurate as possible, and must include all the facts that the supervisor could gather.

Writing Up the Accident Report

An accident report, as has been said, should be written so clearly that it can be understood by anyone who reads it. Ideally, the report should be impartial and objective, clear and accurate in the presentation of facts, concise and unemotional. Descriptions of injuries, mechanisms, and processes and interpretations of facts should be stated as accurately and concisely as possible.

Supervisors in the course of their work must write a considerable variety of reports, including memoranda, letters, proposals, specifications, and production reports. No supervisor likes to spend hours doing this paperwork; but, on the other hand, no one on the receiving end likes to spend valuable time struggling to decipher the meaning of an incoherent report. In the most elementary terms, writing a report involves two aspects: (1) what caused the accident or property damage; (2) what recommendations or corrections the writer has made or is making.

Approach   The point of attack in writing the accident or property-damage report is the analysis of "what happened." An effective report, like a good product, must serve a particular purpose in a particular situation. In analyzing the report, you should try to keep in mind the following questions:

What is the purpose of the report?
Who will read it?
How will it be used?
What is wanted?
When is it wanted?
What discussions will be based on the report?

Too few supervisors realize that everything they write is used at some time by someone in making a decision.

Four Basic Principles   The following principles are fundamental to writing good reports:

1 Always have in mind a specific reader, real or imaginary; and always assume that this reader is intelligent but uninformed.

2 Before you start to write, always decide what the exact purpose of your report is, and make sure that every word, sentence, and paragraph makes a clear contribution to that purpose.

3 Use language that is simple, concrete, and familiar.

4 Check your writing against this maxim: "First you tell the reader what you're going to say; then you say it; then you tell the reader what you've said."

Communicating through the Report

As seen by you, the writer, the elements in the communications situation should be broken down into two categories, as follows:

1 Users of the report
  a Who they are
  b How much they know about the subject of the report
  c What their responsibility is concerning any action taken on the basis of the report
  d What their probable attitude is toward the conclusions and recommendations of the report
2 Uses of the report
  a As a guide for action
  b As a repository of information
  c As a long-term or short-term aid
  d As an aid under specified conditions

The "Big Six": Questions a Report Must Answer   There are six basic questions that must be answered when an accident is investigated, regardless of its nature or magnitude:

Who was injured?
Where did the accident happen?
When did the accident happen?
What was the immediate cause, and what were the contributing causes, of the accident?
Why was the unsafe act or unsafe condition permitted?
How can this type of an accident be prevented from happening again?

Basic Reasoning   We all know that the reason for investigating an accident is to find out what caused it and to prevent it from occurring again.

Without any doubt, an unsafe act or an unsafe condition will most generally be the cause of an accident. But the investigation of an accident should not stop after an unsafe act or condition has been identified. The investigator should delve much deeper into the causes of the accident. For example, how much knowledge did the injured worker have of the task being performed? How much training had the worker received in preparation for the task or job? Did the worker have problems with the machine, with co-workers, or with family? All these factors must be carefully and analytically considered when making the investigation. In addition, the investigation must be undertaken promptly; and the investigator must be open-minded. Without this type of an approach, the "big six" will not be scientifically put together in their proper perspective.

The investigation of an accident must not be considered as simply the accumulation of a mass of data; it must be approached as fact-finding with the purpose of preventing further injury. The ability to investigate and report accidents can be developed only by acquiring a basic knowledge of the causes of accidents and by experience in investigating them.

FIXING RESPONSIBILITY FOR SAFETY

Today, modern management insists on operating policies based on measured performance tasks, enforced standards, and appropriate action when performance is poor. Therefore, when a line supervisor is made responsible for safety—just the same as for production—the end result can only be for the good.

By their conduct and attitude during an accident investigation, supervisors can in almost all cases convince employees that the investigation is intended primarily to find the cause of the accident. If fair and impartial decisions are made after the investigation of an accident, employees will cooperate in the investigation and be willing to give accurate information to their supervisors.

If the supervisor has done something that contributed to an accident, such as knowingly allowing an unsafe act to occur or an unsafe condition to exist, he or she must share in the blame for it. For management to ignore such conduct by a supervisor would be hard to justify. It should be remembered that in fixing responsibility, the seriousness of the accident or the extent of the injury is not the factor to consider; what really counts is the conduct of the supervisor or the employee which contributed to the accident.

But if a supervisor is to be held accountable, he or she must have the authority to make procedural or physical improvements for the sake of safety. If management does not back up the supervisor, employees will notice this lack of coordination and will rightfully blame management for accidents.

BEHAVIOR MODIFICATION

This may come as a complete surprise to many readers, but safety is *not* a state of mind. It is a batch of behaviors which usually differ widely in different settings. There are generally two types of solutions to safety problems: (1) designing safe ways of doing things and (2) using personal protective equipment. Generally, either of these solutions, but especially the second, means one thing to the worker: more work, or less favorable conditions to work in. The behavioral term for this is "response cost." It costs sweat to do things the safe way. The worker is informed that the alternative to the increased response cost is increased danger. This leads us to an important behavioral principle.

Improbable future events (like a burned eye, a broken toe, a cut finger, or lung damage) are ineffective in controlling current behavior. Only immediate consequences are effective. This is why the warnings about cigarettes and cancer have been so ineffective. The immediate social or personal pleasures of smoking are far more powerful than the potential dangers of cancer. "It can't happen to me."

In the case of respirators, for example, the immediate cost of extra sweat and an unpleasant appearance or discomfort far outweigh the distant and improbable possibility that the employee might develop an occupational disease. The supervisor is therefore faced with a problem: "future dangers" are inadequate to sell the use of respirators.

What is the solution? Some behavioral scientists believe that the key is "incentives," and some companies are using various gimmicks as incentives in promoting safe work practices. These incentives are money, attractive job assignments, and holding up employees who have worked many years without accidents as "living proof." Since everyone is different, many different motives must be appealed to. The approach must be positive, so that employees will not feel that they look foolish to their co-workers. Don't forget: employees are not children. They are one of our most valuable tools in industry today—the workers, the people on the job, the people who hit the buttons. They are the keystone of our advanced technological society. Selling safety can be hard work, but it can be very satisfying when good results are achieved.

Understanding Behavior

As managers, supervisors, or whatever we want to call ourselves, we must learn some fundamental facts about behavior if we are going to learn to modify people's actions so they will not get themselves injured on the job.

Behavior is caused. Regardless of what people do or say, there's a reason for doing or saying it. Sometimes managers forget this, especially when someone does something that will cause problems; and sometimes managers are only fooling themselves if they think they know what causes employees to do certain things. For a variety of reasons, they often get into a pattern of "labeling" every employee. Labels are neither causes of behavior nor explanations for behavior.

If you truly want to understand behavior, you must forget about labeling and get to the true causes of employees' behavior patterns. You must put yourself in their place. If you can do this, you will note that in many respects we are all alike. All human beings are continuously striving to satisfy needs and desires. All behavior, whether good or bad, is directed toward this end. Our basic needs are simple and few; but once these are satisfied, we learn to want other things in life. There is no end to these other needs or desires; they stay with us for life. We not only desire material things of life, but we also desire intangibles, such as being liked, being loved, and being respected. And so it goes, from the cradle to the grave.

In this respect, the worker is no different from the supervisor or the manager. We all want the same things out of life. Managers want assurances that they are accepted by others; so do employees. We all want approval as people. But too often we fail to recognize these behavioral similarities in each other. Only when managers and supervisors understand these basic characteristics in employees do they realize that what employees do and say makes sense.

What happens when basic needs of employees go unsatisfied? Of course, different people will react differently. Some become apathetic and indifferent. Others become frustrated and resentful. Eventually, something has to happen. There has to be an outlet. That outlet takes various forms. It may lead to troublemaking, agitation, hostility toward the supervisor, suspicion about the company, and even trouble at home.

When an individual employee is trapped in such a situation, he or she can become erratic at work, and neither the worker nor the supervisor may be aware of this. Even if performance has not yet shown signs of deterioration, the worker will eventually be involved in an accident. Perhaps the accident will be slight, and the worker will fail to recognize it as a warning—and so may the supervisor who is not alert to the worker's attitudes and mental condition.

Developing Positive Attitudes toward Safety

The line supervisors, who are the direct link between employees and top management, can develop good attitudes toward safety in several ways. First, they can show that they accept the top management's safety program. Second, supervisors should never condone unsafe acts or themselves violate safety rules. Third, they must apply their own knowledge about safety in all situations where it is required. Finally, supervisors should never, through words or actions, show a disregard for the company's safety program in the presence of employees. If the supervisor disregards the safety program, both the supervisor and the company lose credibility in the eyes of the employees. The supervisor must continually show, through actions and attitude, that he or she is just as serious about the employees' safety as the company is. Furthermore, the supervisor must constantly remind employees of the benefits of working safely. Keeping employees convinced calls for continual persuasion on the part of the supervisor.

UNCOVERING UNSAFE ACTS AND CONDITIONS

When one is attempting to identify an unsafe act as the probable cause of an accident, there are several things that must be considered. For example, what did someone do or fail to do that caused an unexpected occurrence and thus an injury? What was it that made someone do something differently this time? Or, perhaps, what specific acts or movements should be eliminated from or added to the job procedure to prevent the same accident from recurring? Does the situation call for behavior modification, a change in instructions, a guard, or perhaps closer supervision? Sometimes the direct answers to such questions will identify the unsafe act. And, furthermore, they may also produce a change in behavior, supervisory surveillance, or job procedure.

An investigation may often determine that an unsafe condition contributed to the unsafe act which caused the accident. But there does not have to be an unsafe condition involved in every accident; experience has shown that the majority of accidents are caused by unsafe acts. Nevertheless, unsafe acts by employees are still a liability for management: management may not be responsible for them, but management pays for them. What can management do to stop unsafe acts? The answer to this question is that management must continuously, through its supervisors, drive home the fact of its concern for employees and its refusal to condone unsafe acts.

The great majority of unsafe conditions are created by human beings: someone has failed to modify, repair, or maintain; or someone has carelessly created a condition without thinking that it could cause an injury.

Citations or reports of unsafe acts and unsafe conditions should say specifically what caused or contributed to an accident. This is important, because it indicates what must be corrected: the unsafe condition or the unsafe act of an employee. The use of ambiguous terms such as "carelessness," "inattention," "awkward," or "taking an unsafe position" should be avoided because they mean something different to different people, and recommendations for correction are frequently phrased in generalities. Too frequently the unsafe act or unsafe condition is not specifically identified.

The investigator of an accident must find out what the injured person did or did not do, or what was or was not present in the environment, that caused the accident. If the investigator fails to determine the exact cause, it is probable that another accident of the same kind will eventually occur.

Why did the employee commit or perform an unsafe act? Why was an unsafe condition allowed to exist? Answers to these questions are important. They provide the information needed not only to take the required corrective measures, but also to improve operating methods if necessary. These reasons, which are called "contributory causes," are actually the basic deficiencies that produced the accident.

Another way to arrive at the reasons for an accident is for the supervisor to talk to witnesses about it. The supervisor must, first, seek them out and, second, question them thoroughly. There is no better way for a supervisor to strengthen management techniques than by talking to those who were involved in an accident or saw it. Until we get to the point where supervisors are truly involved in the investigation and prevention of accidents, we have no assurances that real progress is being made. Only with the supervisor's cooperation will management be able to establish controls that will prevent accidents.

RESPONSIBILITIES AND ACCOUNTABILITY OF MANAGEMENT

The establishment of responsibility for safety at each level of management forges an unbroken chain of accountability from the line or job supervisor directly to the president of a company or—in the case of a small plant—to the owner (see Figure 1-2). In a large company, the chain will link the line supervisor to his or her supervisor, to the general supervisor, to the superintendent or superintendents, and then to the plant manager. The plant manager is linked to a vice president who in turn is linked to the president of the company. All levels of managers are then account-

FIGURE 1-2
Simplified outline of a typical management setup, showing all levels from the president on down.

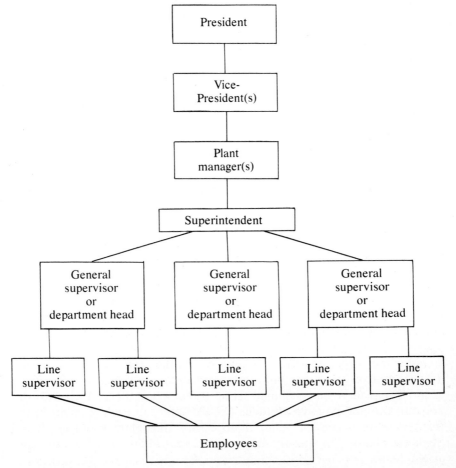

able to someone at the next higher level of command, in the true managerial fashion. In turn, a manager at each level should be able to evaluate the safety efforts of each subordinate manager. A system of responsibility and accountability like this exemplifies true team effort and sound managerial business practice.

The system should operate as follows, below the vice-presidential level.

Supervisor

The supervisor is responsible for the safety of all employees assigned to work for him or her, and for the safe condition of the work area he or she is assigned to operate in or to control.

General Supervisor or Department Head

The general supervisor or department head is responsible for the training and development of each supervisor who is accountable to him or her. There are also developmental duties, including the responsibility for establishing good housekeeping practices in the department.

Superintendent

The superintendent is responsible for all safety functions in his or her area of administration. In this managerial capacity, the superintendent maintains the chain of command from first-line supervisor to plant manager. It is at this level of management that one can see at first hand, and most accurately assess, the safety performance of the plant management. The superintendent's long-term objective is to administer the plant's safety activities in accordance with the corporation's safety program.

Plant Manager

The responsibility for safety performance and objectives of a plant is vested in the plant manager, who is generally accountable to a vice president at the corporate level. The plant manager must establish and administer the corporate safety program within the plant, and must also establish controls to ensure uniform compliance in all departments of the plant. The plant manager should set up training programs which will continue to develop positive attitudes toward safety in each member of the supportive management team. If necessary, the plant manager should spell out clearly the specific duties and responsibilities of each member of the team.

FIGURE 1-3
A well-kept plant and a commitment to good housekeeping will make for fewer accidents, higher production, and greater return on invested capital.

PRINCIPLES OF INDUSTRIAL HOUSEKEEPING

Housekeeping should be ranked at the top of the list of things to be done to prevent accidents and increase production. It can be said that housekeeping is the one activity which, more than any other, is the spearhead of a safety program (Figure 1-3).

Management must have a sincere desire for a clean and orderly plant and must be able to express this desire clearly, by actions, by direct order, or by both. Perhaps it may adopt and publicize a housekeeping policy, which should simply state that the company wants a clean and orderly plant and should be translated into specific orders from executives to department heads and superintendents, to line supervision, to employees.

Steps to Take

The steps which management must take in organizing and maintaining a good housekeeping program can be summarized as follows:

1 Make clear to all members of the organization what is required of them.

2 Enumerate the advantages of good housekeeping for the employees (personal safety) and for the operating managers (profitability).

3 Make managers and supervisors responsible and accountable for the good housekeeping of their departments, shops, and work areas.

4 Tell managers what should be done and how; then enforce the rules.

5 Provide the equipment and personnel necessary to get the job done.

6 Reward when a job is being done well and penalize when a job is being done poorly.

Advantages

The advantages of a well-kept plant are many. The most important can be summarized as follows:

1 A reduction in injuries sustained on the job.

2 Increased efficiency, brought about by an easier flow of materials and improved performance.

3 Improvement in employees' morale. Employees want to work in an orderly environment.

4 Better community relations; winning the esteem of the community.

5 Better labor relations; reduction in complaints by employees.

6 Minimization of losses from fires.

Attitudes to Guard Against

Certain attitudes retard progress toward good housekeeping.

1 Guard against objections by operating personnel that lack of space makes good housekeeping impossible.

2 Poor housekeeping conditions should not be blamed on employees if their supervisors have allowed such conditions to exist.

3 The notion that a dirty plant is a profitable plant, and that busy plants are dirty, is outmoded.

4 Last but not least, the theory that good housekeeping costs money should also be refuted.

The Need for Continued Efforts

After developing a good housekeeping program in the plant, management should not relax its efforts. Like every other element of a safety program, the maintenance of good housekeeping will require constant, repetitive attention. Once the first forward step has been taken, management should never take a step backward or allow any other level of the organization to take a step backward.

QUESTIONS AND EXERCISES

1 Name the three essentials for a successful safety program.
2 What is the purpose of accident prevention?
3 Who should investigate accidents in the workplace? Why?
4 Give some examples of unsafe acts committed by employees in the workplace.
5 Give some examples of unsafe conditions in the workplace.
6 Why is good housekeeping such an important part of safety programming?
7 Name some of the advantages of good housekeeping for both management and employees.
8 What can an organization do to promote good housekeeping among its employees?
9 In your opinion, which of the ''Accident Types'' are responsible for the greatest percentage of all accidents? Explain.
10 What are some means by which a supervisor can maintain workers' interest in safety?
11 Should the board of directors of a corporation be concerned about the safety of its employees? Why?
12 Name three ways in which the supervisor can help to develop a positive attitude toward safety by his or her employees.
13 Why may it be worthwhile to collect data on ''no-injury'' or ''near-miss'' accidents?
14 What are the four basic principles for writing a good accident report?
15 What are the six basic questions that should be answered during the investigation of an accident?
16 What is the Occupational Safety and Health Administration's definition of ''serious physical harm''?
17 Write a case study showing an injury to an employee that was caused by an ''electrical contact.''
18 If a supervisor is to be held accountable for safety, what authority must management allow the supervisor to have?
19 Does the use of incentives in an organization's safety program help to promote safe work practices by employees? Explain.

20 Name three ideas that can be utilized in a safety promotion program. Do not use money as an incentive.
21 When trying to understand human behavior, managers must learn what people's basic needs are. Although different people have different needs, there are some basic needs that we all have. Name them.
22 The great majority of unsafe conditions are created by employees at the workplace. Name the methods that management must use for correcting unsafe conditions.
23 Do you believe that a dirty plant is a profitable plant, and that busy plants are dirty? Explain your reasons for accepting or rejecting this theory.
24 Do you believe that good housekeeping costs money? Explain your reasons for accepting or rejecting this theory.

CASE STUDY

The operator of a pedestal grinder—an operator who is a trained employee—wears all of the required personal protective equipment, and has been taught what the hazards are on this job. In your opinion, what types of hazards should the operator be aware of?

CASE STUDY

A supervisor made an inspection within the department and found the following conditions:

a frayed electrical cord
a pile of oily rags and some scrap wood around an employee's work area
an employee nailing some boards without wearing eye and face protection

What actions must the supervisor take to correct the unsafe acts and unsafe conditions that were discovered during the inspection?

CASE STUDY

A machine operator moving a tub of material noted an electrical cord lying on the floor in her way. Instead of picking it up and placing it out of her way, she kicked it with her foot, causing the plug to explode. What was the cause of the accident? What would be the corrective action required to prevent recurrence?

BIBLIOGRAPHY

Compliance Operations Manual, U.S. Department of Labor, Occupational Safety and Health Administration, OSHA 2006, January 1972.

Guidelines for Setting Up Job Safety and Health Programs, U.S. Department of Labor, Occupational Safety and Health Administration, OSHA 2070, October 1972.

"Organizing an Accident Prevention Program," *National Safety News,* Vol. 112, no. 4, October 1975.

CHAPTER 2

MANAGEMENT LEADERSHIP

During the past decade, industry has seen an unparalleled demand for safety in the workplace. Of the many pieces of legislation passed by the Congress of the United States, the one most significant to industry as a whole was the Occupational Safety and Health Act of 1970. The enforcement of this act has added a legal compulsion to what was formerly motivated either by humanitarian purposes or by profit motives.

The following figures give an indication of the importance of the safety of the employee at the workplace. Every year there are approximately 14,000 work-related fatalities in the United States. The annual number of disabling injuries varies from 2.2 to 2.5 million. The evidence is clear that a well-managed safety program can significantly reduce occupational injuries and illnesses and their attendant operating costs.

Management must be willing to accept its responsibility for the safety of its employees as an integral part of doing business. It must further accept the responsibility of stimulating awareness of safety of its employees and must demonstrate its own interest in the safety of its employees, if those employees are expected to cooperate in making workplace conditions safe.

Representatives of management, both line and staff, must reflect this same interest in the safety of their employees. Each operating manager must assume the responsibility for his or her own plant, division, or department and must be given the necessary authority to fulfill that obligation.

A safety program must always start with top management—management that realizes its accident problems and demands safe operation. Top management's attitude toward accident prevention is almost invariably taken up by supervisors and employees. If the top executives are not genuinely interested in preventing accidents and the resultant injuries, no one else in the organization is likely to be. Any accident control program, therefore, must start with the top management's announced and demonstrated interest, if employee cooperation and participation are expected to be obtained. Employers or managers owe it to themselves and to their employees not only to promote safety and safe working conditions, but also to provide a safety policy implemented by rules that are enforced. Mere lip service—regardless of any motivational principles—will eventually undermine the program, which will lose its effectiveness. The time is right for American industry to do what it should have done long ago, and that is to properly educate and train workers concerning the safety and health aspects of their jobs.

MANAGEMENT SAFETY POLICY

A prime requisite for any safety program is to leave no doubt in the mind of the employee that management is concerned about the prevention of accidents at the workplace. Management's attitude toward the safety of its employees must be demonstrated in the form of a written policy statement and made known to all levels of management and employees alike. This policy should outline the organization's aims and objectives for its safety program and should designate the authority and responsibilities for achieving them. Because of the importance of an organizational safety policy and of making that position clear, many companies regard the safety and health policy as the cornerstone of their safety program. The policy should be given wide publicity and should set the pace for both management and worker responsibilities to the program. Its style of presentation should be as important as the clarity with which it identifies functional authority and responsibilities.

Declaration of a Safety Policy

Why have a safety policy? For one thing, good policy makes it easier to enforce safety rules and procedures. Also, it makes it easier for supervisors to comply with company policy. And it makes it easier for employees to follow safety instructions and safety rules.

Within the language of the organization's safety policy, a statement should declare that management has the ultimate responsibility and authority for the safety performance of the company and that it will hold equally accountable all levels of management and employees in imple-

---

COMPANY XYZ SAFETY POLICY

We, the management of this organization, believe that each employee is entitled to receive recognition as an individual, constructive leadership, equitable compensation, and an environment free of conditions that could cause bodily harm.

The safety and welfare of our employees continue to be the prime consideration of this organization. Working conditions must be maintained in a clean and orderly condition so as to encourage safe and efficient operations for all employees.

FIGURE 2-1
Company XYZ Policy.

THE PRESIDENT'S 3-POINT PLAN FOR A SAFE ABC CORPORATION

The safety of our employees is entrusted to the management of this corporation.

All levels of management shall actively participate in their respective functions related to the safety of the employees of this organization.

We, the management, further require that our employees fulfill their obligations—as regards their own personal safety—in order to accomplish the goals of this mission.

FIGURE 2-2
The President's Three-Point Plan for a Safe ABC Corporation.

menting those objectives. It is not enough to have a policy statement regarding safety. There must be a procedure for keeping it in a state of high visibility. This can be done by placing it in the employee's handbook or safety rule-book, by posting it on bulletin boards, and by giving it periodic publicity in company publications. It should also be posted in all management offices to remind managers of their obligation to this aspect of company operations. The effectiveness of a safety policy varies in direct proportion to the support given to it by management.

Management Controls

Since management controls all aspects of the workplace including hiring, training, production, quality control, and a variety of many other activities common to its operations, it must also control the recognition, evaluation, and control of the hazards. The same standards for achieving production, quality control, and a host of other business-related objectives should also be used for achieving job health and safety objectives.

All levels of management must be involved in the activities required for planning, organizing, and controlling job-related health and safety activities. Each level of management must be held accountable for all specific safety responsibilities which cannot be delegated downward. To be most effective, the focus of management's safety efforts should be on hazard control, rather than on "accidents." The control of hazards at the workplace requires the application of good, sound, basic management skills. Cost-conscious organizations have learned that they must control accidents and their costs if they are to do business in today's highly competitive marketplace.

Assumption of Responsibility

The establishment of responsibility for safety at each level of management forges an unbroken chain of accountability from the chief executive officer of the organization down to the first-line supervisor. This accountability must be extended in direct line through the operating departments to the employees. Management must see to it that this responsibility is fully accepted and then in turn, hold supervisors accountable for the safety performance of their respective areas of responsibility.

A successful safety program must have the backing of top management and the cooperation of its employees. If top management is not interested in accident prevention, it is most likely that others in the management structure will reflect that attitude. And so will the workers—because they can sense and detect what's going on.

Employee safety programs are successful only when there is a genuine commitment by all of the management groups. Such commitment must filter down through the entire organization—from plant manager, to the line supervisor, to the employee. Mere lip service, regardless of any motivational principles, will eventually undermine the program and cause it to lose effectiveness. Safety, perhaps more than any other effort, requires constant and unrelenting vigilance.

Assignment of Responsibility

For many years it has been said that safety is everyone's responsibility. This statement may be generally true, but both common and statute law dictate that the safety of the worker is a management responsibility. Those in ultimate control of the organization must regard the provision of a safe workplace as a fundamental principle in their relation with their employees. Successful programs have one thing in common: there is a deep-seated commitment by top management. Such commitment filters down through the hierarchy to the workers on the line.

It has always been an important gesture for top management to endorse safety programs. It made for good public relations. Today, it is imperative that top management become involved and participate in its safety programs because of the vast scope and potential consequences of new comprehensive state and federal legislation dealing with occupational safety and health. The role of safety in the American workplace is exemplified by the number and variety of regulations, laws, and court decisions.

Therefore, the elements that top management must provide in order to have an effective safety program are:

1 Top management must provide a forceful and continuous leadership role in the safety program.
2 Plant layout, machines, and equipment must be made safe.
3 Supervisors must be competent and must be safety trained.
4 Employee cooperation in accident prevention must be maintained.

CATEGORIES OF SAFETY RESPONSIBILITIES, BY LEVEL

Employer's Safety Responsibilities

A Supreme Court decision in *United States v. Park* pointed out that a top manager can delegate tasks to be accomplished but, in effect, he or she cannot delegate the responsibility for ensuring that those tasks have been done. In this case, the Food and Drug Administration (FDA) charged that Acme Markets and its president, Park, had violated the Federal Food, Drug, and Cosmetic Act by permitting food held in a warehouse to be contaminated by rodents. Acme pleaded guilty but Park did not. He admitted he was responsible for seeing that sanitary conditions were maintained, but argued that the responsibility was one he had assigned to "dependable subordinates." The case was tried in a federal court where Park was found guilty; the decision was reversed in a Court of Appeals. It was reversed again in the Supreme Court, when in 1975, the Court indicated that company officials may delegate responsibilities but continue to bear the responsibility of following up such assignments. The Supreme Court's ruling makes it clear that an executive who has the responsibility and authority to prevent a violation, and who fails to do so, may be held criminally liable for the law's violation.

The employer's safety responsibilities to his or her employees can be summed up as follows:

1 Providing a safe workplace for the employees.

2 Making a policy statement regarding accident prevention.

3 Maintaining an on-going safety program.

4 Providing medical and first-aid systems.

5 Establishing a safety training program on a periodic basis for supervisors and employees alike.

6 Providing adequate budgets for all safety-related objectives.

7 Complying with the Occupational Safety and Health Act.

Employer's Responsibilities under the Occupational Safety and Health Act (OSHA)

1 Providing a workplace free from recognized hazards that are causing or are likely to cause death or serious physical harm to employees, and complying with the standards, rules, and regulations issued under the Act.

2 Becoming familiar with the mandatory OSHA standards and making copies available to employees for review upon request.

3 Informing all employees about OSHA.

4 Examining workplace conditions to ensure they conform to applicable standards.

5 Minimizing or reducing hazards.

6 Ensuring that employees have, and use, safe tools and equipment (including personal protective equipment) and that such equipment is properly maintained.

7 Using color codes, posters, labels, or signs to warn employees of potential hazards.

8 Establishing or updating operating procedures, and communicating them so that employees follow safety and health requirements.

9 Providing medical examinations when required by OSHA standards.

10 Reporting to the nearest OSHA office any fatal accident or one which results in the hospitalization of five or more employees.

11 Keeping OSHA-required records of work-related injuries and illnesses, and posting a copy of the totals from OSHA Form #200 during the entire month of February each year. (This applies to employers with 11 or more employees.)

12 Posting, at a prominent location within the workplace, the OSHA poster Form #2203, informing employees of their rights and responsibilities. (In states operating OSHA-approved job safety and health programs, the state's equivalent poster and/or OSHA 2203 may be required.)

13 Providing employees, former employees, and their representatives access to the Log and Summary of Occupational Injuries and Illnesses (OSHA #200) at a reasonable time in a reasonable manner.

14 Cooperating with the OSHA compliance officer by furnishing names of authorized employee representatives who may be asked to accompany the compliance officer during an inspection.

15 Not discriminating against employees who properly exercise their rights under the Act.

16 Posting OSHA citations at or near the work site involved. Each citation, or copy thereof, must remain posted until the violation has been abated, or for three working days, whichever is longer.

17 Abating cited violations within the prescribed period.

Safety Responsibilities of Operating Managers

The most effective means by which operating managers, superintendents, and department heads can implement the organization's safety program is to do the following:

1 Audit and monitor all of the safety and health activities under their jurisdiction.

2 Ensure that every accident and near-miss and all safety complaints are investigated as promptly as possible.

3 Ensure that the OSHA standards requirements pertaining to their area of responsibility are fulfilled.

4 Provide safety training for supervisors and employees alike on a timely basis.

5 Hold departmental safety meetings with all supervisors at least monthly.

6 Ensure that supervisors are conducting safety meetings and are also making individual daily safety contacts with their employees.

7 Ensure that supervisors are fulfilling management/union contractual obligations relating to safety.

8 Review the plant's accident record and insist on appropriate action when accident trends are unfavorable.

9 Give leadership and direction in the administration of safety activities.

Safety Responsibilities of the First-line Supervisor

The success of an accident prevention program depends on the sincere and constant efforts of the supervisor. While it is agreed that most accidents can be prevented by employees working safely, it the supervisor who is in the position of making employees work safely. Because of this, it can be said that the supervisor is the most important person in the organization's safety program. His or her importance to the total safety effort can be summed up as follows:

1 The supervisor is in such close contact with the workers that he or she knows them on a first-name basis.

2 The supervisor has the necessary job knowledge relating to all of the machines and processes under his or her supervision.

3 The supervisor knows the safety rules and safe job procedures for all of the operations under his or her jurisdiction.

4 Last, it can be said that the supervisor is a direct link between the workers and the company.

Supervisors are responsible and held accountable for the following activities within their jurisdiction:

1 Investigating accidents and safety complaints
2 Enforcing safety rules and procedures
3 Correcting unsafe acts and unsafe conditions
4 Instructing employees in safe job procedures
5 Ensuring that all personal protective equipment is used in accordance with job requirements
6 Making planned daily individual safety contacts with employees
7 Encouraging employees to report unsafe conditions, faulty equipment, or defective safety devices which would adversely affect their safety performance
8 Making daily inspections of the work areas for unsafe practices and conditions, including poor housekeeping

Because of the direct control that supervisors have over employees, and because of their direct action in maintaining a safe workplace, the first-line supervisor must be given the appropriate authority and management support to carry out this responsibility. By not allowing the supervisor to assume responsibility for the safety of the worker, upper management is denying the supervisor one of his (or her) basic managerial rights.

If the supervisor has knowingly allowed an unsafe act to occur and an unsafe condition to exist, then the supervisor must share in the responsibility of its occurrence. In fixing responsibility, the seriousness of the incident is not the factor to be considered; what is important is the conduct of the supervisor.

The prevention of accidents at the workplace is of serious importance. Since employees do not have the responsibility or the authority to function in a supervisory capacity, they cannot be held responsible for controlling hazards in the workplace. The worker, with rare exceptions, has no part in managerial responsibilities, nor does he or she have any obligation for the safety of the workplace unless specifically authorized to do so by the management.

Safety responsibilities should be included in the supervisor's job description and in the performance evaluation. However, evaluating safety performance can be difficult because it is based on the attitude of the supervisor toward employee safety or perhaps on the safety performance of the supervisor's workers.

Today, modern management insists on operating policies based on measured performance tasks, enforced standards, and appropriate corrective measures when performance is poor. Therefore, by holding supervisors responsible for safety, the end result can be good for both the employee and the organization.

Safety Responsibilities of Staff Organizations

Staff organizations generally encompass both supervisory and non-management persons such as engineers, time-study people, labor relations people, and "front-office" types. They generally give some form of service or direction to production and other shop-related departments within the organization. Hence, they, too, are exposed to the same workplace environment as are the employees. Subsequently, the nature of their decisions or recommendations can at times create an impact on the safety performance of the departments they service or partially control. Therefore, it is equally important that they understand the responsibilities and accountability in the various levels of line management and act accordingly.

Employee Safety Responsibilities

Management must make it clear to all employees that they are expected to follow all safety rules, procedures, and instructions as responsibly as they would any other company directive. Employees must regard safety measures as part of their job requirements. A frequent difficulty for supervisors is securing the cooperation of employees regarding their personal safety on the job. It is only when employees are convinced of the supervisor's honest and sincere interest in their personal welfare that the supervisor can establish trust. If the supervisor is to require employees to follow the safety rules and to work safely, then he or she must set a good example of safe performance.

Employee responsibilities include the following safety and health requirements:

1 Follow the employer's safety and health rules, procedures, and regulations.

2 Wear or use the prescribed safety equipment as required by the employer.

3 Report hazardous conditions or equipment to the supervisor.

4 Report job-related injuries or illnesses to the supervisor, and seek medical or first-aid treatment immediately.

WORD POWER

OSHA   Occupational Safety and Health Administration.

risk   An expression of an exposure to a hazard.

occupational injury   An injury resulting from a work-related accident.

occupational illness   An illness caused by environmental factors at the workplace (silicosis, dermatitis, asbestosis, etc.).

employee/worker   A person who performs work for an employer for wages or salary and does not hold supervisory or management status and responsibilities.

employer   One who employs others to work for him or her for wages or salary.

supervisor   A person in charge of an employee, or group of employees, engaged in a form of work.

manager   A person who directs supervisory personnel to attain the goals of a company or organization.

QUESTIONS AND EXERCISES

1 Design a safety policy which you, as the plant manager, would use as the foundation of your plant's newly created safety program.

2 In your opinion, what are the events that led to creation of the Occupational Safety and Health Act?

3 Why is the supervisor considered as the most important person in the organization's safety program?

4 Explain why hazard recognition is a management function.

5 What was the violation that led to *United States v. Park,* and what was the Supreme Court's final ruling?

6 In your opinion, what are the duties and responsibilities of the safety professional in a manufacturing organization?

7 Should the safety professional occupy a staff or line position?

8 Name some manufacturing processes that you would consider as hazardous operations.

9 What are the two most important ingredients of a successful safety program?

10 Name five safety responsibilities of employers under the Occupational Safety and Health Act.

11 List the safety responsibilities of the first-line supervisor.

12 Discuss the safety responsibilities of staff organizations.

13 What obligation should be placed on the employee for helping the employer to keep the workplace a safe place of employment?

14 Would the use of "safety performance" as a criterion in the supervisor's evaluations be of any consequence to the safety program? Explain.

15 What should a union's responsibility be toward the organization's safety program?

16 In your opinion is there such a thing as a "freak" accident or an accident without cause? Explain.

17 What does the concept, "management must be willing to accept its responsi-

bility for the safety of its employees as an integral part of doing business''
mean to you? Elaborate.

18 Should accident prevention be the focus of the safety department only, or
should it be the focus of all of the management in an organization? Explain.

19 Indicate something of the magnitude of the costs of accidents on a national
scale.

20 Has the role of the federal government through the Occupational Safety and
Health Act been of any significance in the reduction of workplace accidents in
industry?

21 Is your state under the federal act or does it have its own state OSHA program?

22 Explain the contribution to an organization's safety program by: (a) the chief
executive office, (b) the plant manager, (c) department heads and superintendents, (d) supervisors, and (e) employees.

23 What are some logical steps that management can take for controlling hazards
in the workplace?

24 Should management place major emphasis for controlling accidents on human
behavior or on the control of physical hazards? Explain.

BIBLIOGRAPHY

All About OSHA, rev. ed., U.S. Department of Labor, Occupational Safety and
Health Administration, OSHA 2056, 1985, p. 40.

Hammer, Willie: *Occupational Safety Management and Engineering,* 3d ed.,
Prentice-Hall, Englewood Cliffs, N. J., 1985, p. 80.

HAZARD RECOGNITION: A MANAGEMENT RESPONSIBILITY

Managers must ensure the proper recognition of hazards by training both supervisors and employees concerning the real and potential health and safety hazards of the working and manufacturing environment. Furthermore, they must ensure active participation by all supervisors in reviewing, auditing, and inspecting the workplace for unsafe conditions and unsafe work practices.

To prevent accidents and to control losses, it is first necessary to identify all of the hazards in order to determine those activities in an operation where losses can occur. Employers must develop the expertise to make such evaluations. Accident prevention represents, above all else, a form of control:

1 Control of the worker's performance.
2 Control of the tools and machines.
3 Control of the working environment.

An accident is something that is unplanned, uncontrollable, and in some way undesirable; it disrupts the normal functions of a person or persons and causes injury, near-injury, or property damage.

CAUSES OF ACCIDENTS

The causes of accidents are either environmental or behavioristic: unsafe conditions or unsafe acts. Many accidents are a combination of both factors. It is accurate to state that there is no single cause of an accident; rather, there are underlying contributing factors leading up to it. Sometimes an accident is caused by a series of errors on the part of a worker, co-workers, or the supervisor, combined with poor design, poor maintenance, the conditions within the plant itself, or the ill-conceived layout of a job or process. When an accident is being investigated or analyzed, all of these factors should be taken into consideration. If they do not appear on the surface, then a more thorough investigation should be made to ensure that they are not overlooked. An employee's mental condition is also important; for example, if employees knowingly or unknowingly "trapped" themselves in an unsafe condition, this may be the result of emotional makeup or problems at home or on the job. Careless attitudes and ignorance are also contributing causes of accidents. Faults on the part of managers and supervisors may also be contributing factors and must be recognized as such by management.

The key to accident prevention is:

1 Determining the cause.
2 Preventing recurrences.

Many workplace environments contribute to accidents because they have poorly maintained equipment, unguarded moving machine parts, defective floors and aisles, toxic fumes, and literally hundreds of unsafe conditions that are constant contributors to accidental injuries at the workplace.

The Unsafe Act

An unsafe act is any act on the part of a person which will increase his or her chances of having an accident. Some unsafe acts are so hazardous that it takes very little repetitive action before the occurrence of an accident.

There are probably hundreds of reasons why workers take unnecessary risks on the job. Many do so because they firmly believe they will not have an accident. "It can't happen to me."

To borrow a phrase from the behaviorists, there appears to be a problem of negative reinforcement. For example, thousands of automobile drivers have exceeded the speed limit, run red lights, passed on hills, and darted in and out of lanes of traffic. They have been "rewarded" for this unsafe driving behavior in two ways: they have achieved whatever their goal was, and they have had no accidents and are thus still alive. Clearly, their "unsafe acts" or behavior have paid off for them; the more they have succeeded in careless methods, the more reinforcement they seem to have received. The mass media communications would appear to indicate that "speed *always* kills," "passing *always* leads to accidents," and "running red lights is *always* dangerous," but driving experience reinforces that these statements are not true. Hence the statement, "It can't happen to me." This form of negative reinforcement has always been a major problem in the war against accidents at home, on the streets and highways, and in industry.

When trying to determine if an unsafe act was the cause of an accident, investigators may ask of themselves: what did someone do or fail to do? Or, was the injured employee following procedure? What specific tasks need to be eliminated or added to the existing process in order to prevent a recurrence of the accident?

Some of the most commonly found unsafe acts observed at the workplace are:

1 Using broken or defective hand tools.
2 Not wearing the prescribed personal protective safety equipment.
3 Not following safety procedures or obeying the safety rules.
4 Poor housekeeping practices on the part of the worker around the work area.

It has been consistently found that many unsafe acts occur because workers were not properly trained or motivated by their supervisors. Some workers have not learned that certain actions in some situations will increase their chances of accidental injury. This is especially true of employees who are unfamiliar with a particular task or process. Sometimes experienced workers do not recognize that a specific action on their part is unsafe. Therefore, it should never be assumed by management that only new employees have all the accidents and it is they who need safety training.

The Unsafe Condition

An unsafe condition is a condition within the working environment which increases the worker's chances of having an accident. Among the greatest contributors of unsafe conditions at the workplace are the actions of the employees themselves.

Management's investigation of an accident at the workplace may often uncover that an unsafe condition was the cause or the contributing factor of an accident. When determining if an unsafe condition existed, the investigator must consider such possibilities as:

1 The mechanical and physical condition of the equipment.
2 The condition of the walking and working surfaces (floors and working platforms).
3 Illumination, ventilation, sound and vibration, etc.

What is behind the unsafe condition? There are many reasons for unsafe conditions, but generally they fall under one of the following categories:

1 The actions or inaction of the workers at the workplace.

2 The deterioration of tools, machines, and equipment due to normal wear and tear.

3 Poor or inadequately designed tools, machines, and operating equipment.

4 The omission of safety features during the engineering or maintenance of the equipment.

CONTRIBUTING CAUSES OF ACCIDENTS

The contributing causes of accidents are those acts or conditions which led up to the cause of the accident, but in themselves were not the determining cause of the accident. Some typical examples of contributing causes of accidents are:

1 Inadequate codes or standards.
2 Failure by management to enforce safety rules.
3 Faulty design or lack of maintenance.
4 Inadequate personal protective equipment.

The control of the contributing factors and the need to eliminate the unsafe acts and unsafe conditions at the workplace can only be accomplished by a strong management team and a cooperative safety-minded work force.

THE ANATOMY OF AN ACCIDENT

There are four distinct parts in the anatomy of an accident. They are:

1 Contributing causes
2 Immediate causes
3 The accident
4 The results of the accident

These four parts of the accident sequence are outlined as follows:

1 Contributing causes
  a Supervisory safety performance
    (1) Safety instructions inadequate.
    (2) Safety rules not enforced.
    (3) Safety not planned into the job.
    (4) Infrequent employee safety contacts.
    (5) Hazards left uncorrected.

     (6) Safety devices and equipment not provided.

  b Mental condition of worker

     (1) Lack of safety awareness and training.

     (2) Lack of coordination.

     (3) Improper attitude.

     (4) Slow mental reaction.

     (5) Inattention.

     (6) Lack of emotional stability.

     (7) Nervousness.

     (8) Temperament.

  c Physical condition of worker

     (1) Extreme fatigue.

     (2) Deafness.

     (3) Poor eyesight.

     (4) Heart condition.

     (5) High blood pressure.

     (6) Lack of physical qualifications for the job.

2 Immediate causes of accidents

  a Unsafe acts

     (1) Protective equipment or safety equipment provided but not used.

     (2) Hazardous method of handling materials (failure to allow for sharp or slippery objects and pinch points, wrong lifting methods, loose grip, etc.).

     (3) Improper use of tools or equipment although proper tools were available.

     (4) Hazardous movement (running, jumping, stepping or climbing over, throwing, etc.).

     (5) Horseplay.

  b Unsafe conditions

     (1) Ineffective safety device.

     (2) Safety device required but not provided.

     (3) Poor housekeeping (material on floor, poor piling, stacking, and storage, congestion of aisles).

     (4) Defective equipment, tools, machines, and electrical systems.

     (5) Improper dress or apparel for job.

     (6) Improper or inadequate illumination, ventilation, etc.

3 The accident

  a Struck by.

  b Struck against.

  c Caught in, on, or between.

  d Fall from above.

  e Fall at ground level.

f Strain or overexertion.
g Electrical contact.
h Burn.
4 Results of the accident
   a Annoyance.
   b Production delays.
   c Reduced quality.
   d Spoilage.
   e Minor injury.
   f Disabling injury.
   g Fatality.

ACCIDENT CONTROL STEPS

An accident is usually considered by many as an indicator that some part of the management system (in this case, the accident prevention program) is not functioning as well as it should be. It is also considered a reflection on the efforts and capability of the management and indicates that some control steps must be taken. These steps are as follows:

1 Supervisor safety performance
   a Job-hazard analysis.
   b Enforcement of safety rules.
   c Adequate safety knowledge.
   d Promotion of employee participation in the safety program.
   e Proper job placement.
   f Development of safe working conditions.
2 Mental condition of worker
   a Daily safety contacts by the supervisor.
   b Adequate safety indoctrination and on-the-job safety training.
   c Safety promotion and publicity.
   d Employee participation in the safety program.
   e Regularly scheduled safety meetings.
   f Adequate supervisor-employee communication on all matters concerning safety on the job.
3 Physical condition of the worker
   a Preplacement physical examinations.
   b Periodic reexaminations.
   c Proper job placements.
   d Adequate medical systems.
   e Recognition of physical limitations of workers newly placed on a job.

CLASSIFICATION OF ACCIDENT TYPES

During the course of an accident a worker may strike his or her body against an object, or the worker might be struck by an object or substance. Or, the worker may incur a burn, an electrical shock, or perhaps a strain or overexertion. Whatever the case, these occurrences are known as basic accident types. These accident types are generally classified as follows for purposes of clarity and uniformity when preparing the accident report and when describing accidental injuries.

Struck by: A "struck by" accident is one where the worker has been unexpectedly struck by or contacted by a moving object or substance. Being struck by or contacted by a vehicle, a hammer blow, or a foreign piece of material in the eye—all are common examples of "struck by" accidents.

Struck against: A "struck against" accident is always the case of a moving worker contacting or striking against some object or substance. Some typical examples of "struck against" accidents are striking against a corner or a sharp edge, striking against a hot pipe, running or walking into a moving vehicle, or running or walking against or into another person.

Caught in, on, or between: Although three accident types are involved here, it is common practice to consider them as one type. A "caught in" accident type would occur if a worker's foot was caught between some broken boards in the floor. A "caught on" accident would involve a worker's shirt sleeve being caught on a wire fence. An example of a "caught between" accident would be a worker's leg or arm caught between revolving gears or moving machine parts.

Fall from above: Many falls occur from a higher level to a lower level. For example, falls from a work platform or a ladder or down a flight of stairs fit into this category.

Fall at ground level: Accidents involving "fall at ground level" are those related to slipping, sliding, tripping, or falling to the floor or ground from the same level.

Strain or overexertion: Workers strain or overexert themselves while carrying, pushing, or pulling objects or materials beyond their physical limitations.

Electrical contact: An injury of this type results from a contact by the body with an electrical current or with any electrically charged equipment or fixtures.

Burn: This condition is caused by a part of the body coming into contact with a spark, open flame, or a hot surface or substance.

INJURY DISTRIBUTION

The factors that determine who will be injured through accidents, when, and why, have been the subject of almost endless discussion. Many research studies have been made in efforts to discover and appraise these factors, and although much has been learned and many findings announced, there is still far from general agreement about the validity of many of them.

Three theories of injury distribution have been widely discussed. Each has its adherents. Stated very simply these three theories are:

1 *Theory of chance distribution:* According to this theory, each hazard or unsafe act, however slight the degree of hazard involved, will, if a sufficient number of exposures occur, yield an injury. Which exposure will do so is, in each instance, purely a matter of chance.

Examples:

(*A*) A repairman left a large bolt on the flange of a crane grinder in a steel plant. Later, the motion of the crane dislodged it and it struck a worker on the head, killing him. That it fell at just the right moment to hit anyone was pure chance. That it hit the worker in the right way to kill him was also a matter of chance.

(*B*) A worker nailing up a heavy crate hit the nail a glancing blow. The nail flew across the aisle striking another worker in the cheek, narrowly missing his eye. That the nail flew instead of bending over was largely chance. That it missed the worker's eye was chance again.

2 *Theory of biased distribution:* This theory assumes that a person once hurt is thereby likely to be either more or less apt to become an accident victim again. His or her susceptibility to accidents will either have been increased by nervousness or fear, or decreased by greater caution and improved judgment.

Examples:

(*A*) A window cleaner in a plant, the windows of which could be cleaned from a step ladder, failed one day to set a ladder properly. It fell, and the window cleaner received a brain concussion. The cleaner recovered, but thereafter became so nervous on a ladder that he had to be placed on other work.

(*B*) The time-honored practice of requiring a student pilot who makes a bad landing to go up again at once if uninjured indicates the applicability of this theory to high hazard situations.

3 *Theory of unequal liability:* This theory assumes that some persons are much more liable to be involved in accidents than others, that is, they are "accident prone."

Examples:

This can be physical, psychological, or both, as the following three examples exemplify:

(*A*) A laborer was, because of faithful work, promoted to crane hooker-on. The laborer suffered a crushed finger on the second day on the job. Two more minor injuries and several close calls quickly followed a return to work. An eye test showed that this worker lacked binocular vision, a non-correctable defect. The laborer was changed to another job that did not require such vision and received no more injuries.

(*B*) A young man who had been an armature winder was hired as a motor repairman. He learned the more diverse work quickly and, since the repair work did not take his full time, he was soon being used in the maintenance of the electrical equipment generally, some of which was work done "hot." All went well until he had a violent quarrel with his teammate. This was patched up, but he remained sullen and uncommunicative. He met with two accidents (electrical shock) on succeeding days—the second of which necessitated artificial resuscitation. Investigation showed that he had a record of being hot-tempered, and that a period of moodiness and inattention followed each outburst. He was kept on the motor repair work, but his spare time was changed to work he could do safely alone. The whole matter was discussed with him privately and helpfully. He accepted the change. No further trouble developed.

(*C*) A man who, because of his industry and willingness, was given all types of work in a machine shop, seemed always to be getting hurt, though not seriously. He pinched his fingers handling the steel parts, he bumped his shins, and dropped things on his toes. He fell down stairs, and he ran into a column. Finally, a thorough examination, physical and psychological, showed that his distance perception was poor, his muscular coordination very poor, his mental reactions were slow, his reasoning power was of a very low order, and he had a serious inferiority complex. He was fitted with glasses that corrected his vision and was assigned to the stockroom keeper. Under this stockroom keeper's sympathetic supervision, the worker eventually showed a change for the good in his safety performance.

Analyses of accidents throughout industry show a wide range of safety performances and also verify that all of these three theories apply. The extent to which they apply varies widely among industries, establishments, and types of activities. Obviously, an analysis of accident reports reveals that chance plays a major role in determining whether or not in each accident someone is hurt, and if so, how seriously.

WORD POWER

hazard   A condition with the potential of causing injury or property damage.
control   to command, dominate, or to exercise direction or restraint.
analysis   The separation of a whole, whether a material substance or any matter of thought, into its constituent elements.
liability   An obligation to rectify or recompense for any injury or damage for which the liable person has been held responsible.

QUESTIONS AND EXERCISES

1 In your own words (and not the language used in this chapter), define "accident."
2 What are the causes of accidents?
3 Define the unsafe act and the unsafe condition.
4 What is considered as among the greatest contributors to accidents in the workplace?
5 Give some examples of unsafe conditions most commonly found in a manufacturing plant.
6 In your opinion, what are the major reasons for accidents at the workplace? Explain.
7 What is the key to accident prevention?
8 What are the categories under which unsafe conditions are created?
9 What are the four parts of the anatomy of an accident?
10 Much publicity has been given to the theory of unequal liability (accident proneness). What are your feelings about it? Discuss in class.
11 What four controls are required for the prevention of accidents?
12 Give some examples of unsafe acts caused by the existence of unsafe conditions in the workplace.
13 Give four examples of contributing causes of accidents.
14 Give some examples of contributing causes of accidents by the supervisor.
15 The mental condition of the worker can also become a contributing cause of accidents. Name five conditions that apply.
16 Name five physical conditions of a worker that can become contributing causes of accidents.
17 Give an example of the theory of biased distribution.
18 In assessing your present workplace, what injury sources would be of most concern to you? If you had the authority, what preventive measures would you instigate?
19 Debate the following statement in class: Unsafe acts are harder to correct than unsafe conditions in the workplace.
20 Debate in class the topic of unsafe acts by employees—from the side of management and the side of the employees.

21 If you were required to prepare a list of workplace hazards, where would you list electrical hazards?

22 In your opinion, how important is safety training to employees? Elaborate.

23 Write a job description, including safety responsibilities, for a supervisor in a small light-manufacturing plant.

24 Design a performance evaluation for supervisors and show the safety responsibilities you would use as criteria for performance evaluation.

25 Develop a safety program for a small light-manufacturing plant. Present the program to the class for appraisal and discussion.

BIBLIOGRAPHY

Blake, Roland P.; *Industrial Safety,* 3d ed., Prentice-Hall, Inc., Englewood Cliffs, N.J., 1963, p. 68.

De Reamer, Russell: *Modern Safety Practices,* John Wiley and Sons, Inc., New York, N.Y., 1985, p. 20.

Mendelsohn, H. A.: "A Critical Review of the Literature and a Proposed Theory," in *The Denver Symposium of Mass Communications,* Murray Blumenthal, ed., National Safety Council, Chicago, Ill., 1964, p. 94.

SAFETY PROGRAM MANAGEMENT

In the past fifty years, industrial technology has developed more new equipment and processes than were developed in the preceding thousand years. Technology has created expanded requirements and need for accident prevention at the workplace. Many tasks have become increasingly complex and demanding, and their potential for serious injury has become heightened. As a result, management has broadened its functions to include the application of safety program management into its systems. It further means that safety programming has become a function of the management group, from the line supervisor up to the chief executive officer of the organization.

The severe effects upon an organization's efficiency brought about by increased injury losses, the escalating size of bodily injury awards, and the replacement costs for people and equipment are among the factors that endanger the profit potential and increase the probability of business failures today.

In most organizations, the safety program has evolved from an obvious need for its services. Serious hazards, bad safety performance, the Occupational Safety and Health Act, and the continuous prodding by the labor movement have caused the promulgation of most safety programs.

Safety program management is the control of the working environment, the equipment and processes, and the workers for the purpose of reducing accidental injury and losses in the workplace.

The design of a safety program follows a variety of styles and concepts. Some of them are simple in structure; others rely on engineering concepts; and yet others generally call on management strategies to achieve their objectives.

Experienced management knows that safety management programming is good business and is good for business. It also knows that it must establish policies, procedures, and practices under which every member of management will clearly understand that the safety of the worker is a prime responsibility.

CORPORATE POLICY ON SAFETY

Management's acceptance of its responsibilities for safety can best be expressed by setting a policy to provide a safe and healthful workplace for its employees. Usually such a policy comes from the president of the company or corporation but it may come from someone else, even a plant manager. In all probability the policy will delegate some responsibilities to the subordinates of the writer and will lay out a plan for promulgating the program from each level of management down to the next level. This is a commitment, and it should be designed to lay down step by step the company's plan for making a safe workplace. It should also establish the safety program as a whole: that is, it should require the keeping of a thorough and effective system of accident investigation and recording; and it may also direct that a training program be made available for employees and supervisors. Specific goals should be established and maintained to ensure that all personnel responsible for safety matters are kept abreast of new standards or procedures as they are published by the corporation and by the Department of Labor.

A safety program must always start at the top management level. Each management level must reflect an interest in the company's safety objectives and set a good example of compliance with safety rules. Employees will believe in safety only to the degree that they think their supervisors believe in it. Management's interest must be audible, visible, and continuous.

For years the business community has championed the slogan "Safety pays." True; it always has paid. Personnel managers and safety professionals are quick to point out that a good safety program improves morale, saves money by reducing the need for replacing injured employees, reduces the cost of workers' compensation insurance, and in general is a sound way to do business. It has always been important for management to endorse and participate in safety programs. Today it is imperative for top management to become involved because of the vast scope and po-

FIGURE 4-1
Industry must not only provide a safe workplace for employées but also maintain a clean environment for the surrounding community. When new plants are built in suburban areas, managers are responsible for obeying all laws which set environmental standards for those areas.

tential consequences of comprehensive new state and federal legislation dealing with occupational health and safety.

THE SAFETY PROFESSIONAL

While top management has the ultimate responsibility for the safety of its organization, it delegates authority for safe operations down through all levels of management. The line supervisor is the key element in the organization's safety program because he or she is in direct contact with the employees at all times. Members of management who are delegated to act in the capacity of safety managers or safety engineers (whatever the title may be) should function in a staff capacity. Their role is to administer policy, to provide technical expertise, to train supervisors in safety techniques, to conduct safety promotion, and to keep the management informed of the progress of the program. When investigating accidents, they should be able to give management an accurate account of the causes, and to recommend the proper course of action to prevent any further recurrences.

The safety professional in an organization with diversified activities, as many of them are today, must be a generalist knowledgeable in a wide range of technical, legal, and management activities. In essence, safety professionals must constantly keep themselves informed of the latest developments that relate to the manufacturing processes within the industry and organizations they are employed in. The safety professional is a vital member of the management team. This person is in the organization to aid supervisors in maintaining one of their job-related responsibilities—the prevention of accidents and property damage, while responding to environmental concerns.

The safety manager is responsible for the functions of the safety department and its activities. He or she may have several people under immediate supervision, people who are also knowledgeable in safety. Although the safety manager has been given the responsibility for the safety activities within the organization, the ultimate responsibility is with the manager in control of the organization as a whole.

Specific Functions of Safety Personnel

There are specific safety functions that safety managers must carry out either as their direct responsibility to the safety program or on behalf of the top management. These responsibilities are as follows:

1 Establishing programs for detecting, correcting, or controlling hazardous conditions, toxic environments, and health hazards. Ensuring that proper safeguards and personal protective equipment are available, properly maintained, and properly used.

2 Establishing safety procedures for: employees, plant design, plant layout, vendors, outside contractors, and visitors.

3 Establishing or approving the establishment of safety procedures for purchase and installation of new equipment and for the purchase and safe storage of hazardous materials.

4 Maintaining an accident recording system to measure the organization's safety performance.

5 Staying abreast of, and advising management on, the current federal, state, and local laws, codes, and standards related to safety and health in the workplace.

6 Carrying out the company's safety obligations as required by law and/or by union contract.

7 Conducting investigations of accidents, near-misses, and property damage and preparing reports with recommended corrective action.

8 Conducting safety training for all levels of management and newly hired and current employees. Emphasizing the importance of continuous safety training for all management and employees.

9 Assisting in the formation of both management and union/management safety committees, as well as conducting monthly meetings of the executive safety committee (department heads and superintendents) and attending monthly departmental safety committee meetings.

10 Keeping informed on the latest developments in the field of safety such as personal protective equipment, new safety standards, workers' compensation legislation, new literature pertaining to safety, as well as attending safety seminars and conventions.

11 Maintaining liaison with national, state, and local safety organizations and taking an active role in the activities of such groups.

12 Accompanying OSHA Compliance Officers during plant inspections and insurance safety engineers on audits and plant surveys. The safety engineer further reviews reports related to these activities and—with management—initiates action for necessary corrections.

13 Distributing the organization's statement of policy as outlined in its organizational manual.

In conducting the safety program the safety professional performs a distinct service to management. He or she serves as a consultant to the operating managers and to the line organization, but it is ultimately their safety program, because it is designed to help them fulfill their responsibility for safety.

Administration of the Safety Program

The individual responsible for the safety program should have a job title consistent with the other job titles within the organization. This title may be Safety Manager, Safety Engineer, Safety Supervisor, or Risk Manager. This person should definitely be a supervisor with the authority to carry out the functions as outlined in this position's job description. This person should be able to communicate with all levels of management and should have some basic expertise in the fundamentals of the type of industry he or she is working in. These individuals must establish objectives for improving the safety of the work environment—a task which should be divided into short-range and long-range programs. They must also establish standards for safety performance and then must measure that performance. They must also measure how well supervisors are fulfilling their own safety responsibilities to the organization.

The safety professional must also communicate with top management for the purpose of promoting his or her safety programs and related activities. The professional has the responsibility of maintaining attention to safety. At this level of management, the prime commitment of the corporation is to production and costs. This implies that the safety professional must also be able to talk the language of costs. He or she must show what accidents and injuries are costing, as well as what the medical costs and workers' compensation costs are. Only then will management listen to the professionals.

Another important function of the safety professional is that he or she must constantly reevaluate the safety program, because the work environment is constantly in a state of change. For example, there are new employees hired into the organization, new equipment to contend with, new work processes, and the constant stream of new or revised OSHA standards that must be interpreted.

Relationships

Inside the Company   The safety engineer will establish and maintain the following relationships within the company.

With the Plant Manager   The safety engineer is accountable to this executive for the proper interpretation and fulfillment of the duties and responsibilities of this position and related authority, in concurrence with the dictates of the corporate manager of safety and hygiene (if such a position exists within the corporation).

With Department Heads and Supervisors   The safety engineer is responsible for providing advice and guidance about safety and industrial hygiene appropriate to the processes, installations, and procedures of the plant.

With Employees   The safety engineer is responsible for providing advice and guidance about any employee's specific job or work area in the interest of preventing accidents and controlling property damage.

With Unions   The safety engineer is responsible for fulfilling his contractual obligations regarding matters of safety and health.

Outside the Company   Here, the safety engineer must establish appropriate relationships with professional and organizational groups.

Accountability

The safety engineer is accountable to the plant manager for his or her actions and their consequences. Performance will be judged on the following criteria:

1 Reduction of the frequency and severity of accidents. The same criteria for measurement must be consistently used throughout the company.

2 Reduction of costs stemming from accidents. "Weightings" must be used to correct for dissimilarities between operations in different areas of the company.

3 The efficiency and smoothness of a department's operations within operations of the plant as a whole.

Need for Safety Professionals

It should be noted that the number of people employed in a plant should not be the only factor determining whether the safety program should be in the hands of a full-time safety professional. The nature of the operation should indicate what the need should be. The trend is to employ full-time safety professionals for any or all of the following reasons:

1 The passage of the OSH Act.

2 The high degree now developing of union involvement in safety and environmental health.

3 Changes in machine design and plant layout; product safety and the great need for fire prevention and security; and the way people think about the profit motives of a company.

THE SMALL COMPANY

The active management and control of a safety program in a small plant may usually be delegated to the personnel manager, the general manager, the industrial engineer, or possibly a supervisor in the organization who may have authority and good rapport with both top management and the employees. Then again, the plant's safety program may be in the hands of both management and employees through the formation of a joint employee-management safety committee. The owner of a small company or plant may have special problems in safety or in medical services, and is not likely to be in a position to hire trained experts in these fields such as a full-time safety professional, a full-time doctor, or even a full-time nurse. Still, technical information about safety is available to the owner of a small plant from many sources. First, the insurance carrier can provide the expertise of its own safety engineers. Second, help is available from the National Safety Council, whose services cover a wide range of safety activities in all areas of industry, transportation, agriculture, and many more areas too numerous to mention here. Third, the state division of labor can be of help. Fourth, there are part-time as well as full-time safety consultants. Finally, colleges with safety curricula can be consulted. By using outside agencies, the owner may hold the cost of technical services below what it would be if a full-time safety professional were hired.

A full-time or part-time nurse is more significant to the safety program in a small organization with few employees than in a larger plant with a medical department. Nurses carry much responsibility and are in a position to be of great help to the safety program because the nature of their job brings them the people who should have been more concerned with safety in the first place. In other words, whether an injury is slight or serious, the employee will be telling the nurse how the injury occurred.

Surprisingly, OSHA does not require formal company safety programs. Therefore, even the small company must take the lead and provide a safety program of its own. In situations where no policies have been established, a company or plant must create them. Where policies do exist, they should be reviewed and, if necessary, improved. If a safety professional is hired to implement a new program, he or she must take the responsibility of obtaining the full support of management and must continue to hold its interest in the day-to-day problems and possibilities of the program. The professional may have to remind managers that OSHA establishes only minimum standards. A plant must better the standards if accidents are to be eliminated. Since OSHA was initiated, companies and plants have had to be altogether different types of operations. They must be multifaceted and action-oriented where safety is concerned. Someone in the organization—even the small organization—must become a student of OSHA standards. By now, many a plant has found this out for itself.

ATTITUDES TOWARD SAFETY

The background against which attitudes favorable to safety must be formed is very much different today from what it was in the past. The difference becomes apparent when we compare the attitudes and values of the old and the young. Such a comparison is very important because young people make up the work force of today and the next decade.

Their values and attitudes have been determined by the institutions that influenced them as they were growing up and by the experiences they have undergone.

The changing effects of institutions can be summed up as follows: (1) The family has less influence (perhaps because in more households both men and women are working). (2) The church has less influence. (3) Schools have changed their teaching methods, and they too appear to be exerting less influence; particularly, memorization as a method of learning has given way to thinking out problems for oneself.

Experiences which have had enormous influence include the war in Viet Nam, the nuclear threat, the civil rights movement, the growing emphasis on ecology, and the communications explosion. These have all contributed to the change in values and attitudes of young people today. One important characteristic of many young people is a distaste for social and institutional rigidity. They also fear "losing themselves" to technology. They are intolerant of hypocrisy, and they have different life styles and different work ethics. They don't like doing things the hard—even when that is the safe—way if there is a simpler way. Furthermore, they resent the supervisor saying, "Do it my way." They believe in individuality.

Where does all this fit into a safety program? Anyone responsible for a safety program must recognize that young people's values and attitudes call for different kinds of safety programs.These people and their values are not going to be changed. Companies and plants must learn to accept them for what they are, and to live with them. A safety program must be structured so that it will appeal to modern young people. A company must indicate specifically to managers what employees are to do and how it can be done safely, and should build safety into appraisal systems. Both positive and negative rewards must be used, and there must be a connection between safety and rewards. Employees want to be involved in the decision-making process. If a safety program is structured around the things that satisfy people, workers will cooperate with it. In all of their relationships, including work, today's workers view themselves as equals, not subordinates.

SAFETY PROMOTION

Although safety can to some extent be engineered into equipment and processes, it is still necessary to motivate employees to perform their work safely. Safety promotion is persuasion through motivation. Persuasion, by definition, is the attempt to form or change a person's opinion, attitude, or point of view. An effective program of safety promotion must be persuasive; it must provide a stimulus—or stimuli—to which employees will respond positively. The goal of any such program is participation by employees.

Two points should be made at the outset about safety promotion: one, scare tactics are to be avoided; two, safety promotion should always take the form of mass communication, since communication directed at individuals can produce embarrassment and, as a consequence, a rejection of personal involvement.

Unfortunately, there is no simple formula for effective mass communication, because communication involves complex interactions of physical, psychological, and social factors. It must always be remembered that different people will react differently to the same communication; and that what appears reasonable to the communicator may appear unreasonable to the recipient. But in general it can be said that employees are able to tell the difference between a good safety program and a poor one, and that they will react favorably to management's interest in their safety and to a sound, well-diversified safety program. Management must provide a safety program that will arouse and maintain the interest and awareness of all concerned.

Making employees interested in their own safety and well-being is the responsibility of the safety professional who plans the program and the supervisors who carry it out. The safety professional must realize that a safety-promotion program can succeed only through the efforts of management and employees alike, and must therefore use every available means to maintain their enthusiasm.

Keeping employees interested in the safety program is an exacting job; it requires ingenuity and imagination, since new motivational programs must be frequently produced. The safety professional must know what types of people he or she is dealing with, and what kinds of appeals will be most effective. In all cases, of course, information provided to employees must be precise and correct; and all programs must be backed up by facts. Those responsible for the programs must be realistic in their expectations: it is not realistic, for example, to expect that every idea will be adopted immediately by everyone. New ideas will be adopted gradually over time, as the result of rather complex processes, although adoption may be speeded up by various devices, such as dramatization of the need for a program.

Interest in the safety program can be created in various ways. In many companies, safety contests have been very effective; usually, such contests are held for the purpose of reducing the frequency of accidents. Any good idea about safety promotion can be used as the basis for competition, if it is properly organized. Employees can compete on a departmental or plant-wide basis, or—in a large corporation—between plants. Such competition can also improve communication and thus improve team spirit.

However interest is created, the safety professional must bear in mind that employees—like everyone else in our society—are almost constantly bombarded with messages: buy brand X, avoid aerosols, jog daily, vote for Y, save energy, don't eat sugar. The safety program has to compete with all these messages, and many more, for space in the employee's consciousness and memory bank. Therefore, it may be necessary for the safety professional to use some of the tactics of other professional communicators, including "Madison Avenue" techniques.

Regardless of what techniques are used, however, the ultimate goal is the same: the safety of the employees. Safety promotion, when carried out intelligently, represents a uniquely effective tool for preventing accidents.

Safety Consultants and Supervisors

While top management has the ultimate responsibility for the safety of its organization, it delegates authority for safe operations down through all management levels. As was noted in Chapter 1, the supervisors on the line are actually the key persons in a safety program because they are in direct contact with the employees at all times. Persons who are delegated to act as safety engineers or safety consultants should function in a staff capacity. Their role is to administer policy, to provide technical expertise or information, to train supervisors in safety, to set up promotional materials, and to inform top management of the results of the program. When investigating accidents, they should be able to give management an accurate account of causes, and to recommend the proper course of action to prevent similar accidents.

In general, the safety program is administered by supervisors or other persons holding line positions in a small company and staff positions in a large company. In a large corporation, the safety professional and his or her organization should have staff status and authority. The exact organizational status of the safety staff must be determined by each firm in terms of its own operating policies.

Objectives of the safety program as a staff function should include the following:

1 Keep direct responsibility for safety at the supervisory level.
2 Always maintain the advisory status of the safety professional.
3 Enable the safety professional to spread efforts over as great an area as possible.

4 Ensure more ready compliance with the safety program by working through supervisors, since employees are accustomed to going to their immediate supervisors for direction.

RECORD KEEPING

One of the most important parts of an accident program is record keeping. Accident records are one of the primary means any company has for measuring the effectiveness of its accident prevention program. There are essentially two different kinds of accident records: (1) injury records, of which the OSH Act requirement (discussed below) is only a part; (2) accident investigation records. All companies should keep both types.

Injury Records

The OSH Act requires an employer to keep at least three forms of records: a log of occupational injuries and illnesses; a supplementary record of each occupational illness and injury; an annual summary of occupational injuries and illnesses.These, as was noted above, cover only a part of the accident records that should be kept by a company.

Injury records are kept as a measurement of performance. By comparing current injury records with past injury records a company can see if it has made any progress with its safety programs. This type of record can also be used to determine problem areas or trends, and it will pinpoint areas that warrant immediate or long-range attention.

Accident Investigation Records

Accident investigation records are also a tool for measurement. This form of record should serve two purposes: (1) to determine the surface causes and the underlying causes of an accident; (2) to evaluate the supervisor's ability to prevent accidents. Accident investigation records should be concerned not with fixing the blame for an accident, but with determining the causes of the accident. It is the elimination of the causes and "subcauses" of an accident that will prevent mishaps in the future. The supervisor's ability to discover these causes acts as a measurement of the ability to prevent accidents.

OSHA Requirements

Log of Occupational Injuries and Illnesses Each employer must maintain a log of all recordable occupational injuries and illnesses. The employer must enter each recordable occupational injury and illness on the log as early as practicable but no later than six working days after receiving information that a recordable case has occurred. For this purpose, OSHA Form 200 or any private equivalent may be used. If an equivalent to OSHA Form 200 is used, such as a printout from data-processing equipment, the information must be as comprehensible to a person not familiar with the data-processing equipment as OSHA Form 200 itself.

Period Covered   Logs are to be established on a calendar-year basis. The initial log must include recordable occupational injuries and illnesses occurring on or after July 1, 1971. A log must be maintained in each establishment for five years following the end of the year to which it relates.

Supplementary Record   In addition to the log of occupational injuries and illnesses each employer must have available for inspection a supplementary record for each occupational injury or illness for that establishment. The record must be completed in detail as shown in the instructions accompanying OSHA Form 101. Workers' compensation reports, insurance reports, or other reports are acceptable alternatives if they contain the information required by OSHA Form 101.

Annual Summary   Each employer must compile an annual summary of occupational injuries and illnesses for each establishment. Each annual summary is to be based on the information contained in the log of occupational injuries and illnesses for the particular establishment. OSHA Form 200 is to be used for this purpose.

The summary must be completed no later than one month after the close of each calendar year, and must be posted in each establishment. The summary covering the previous calendar year is to be posted no later than February 1 and must remain in place until March 1. A failure to post a copy of the establishment's annual summary may result in the issuance of citations and assessment of penalties pursuant to the OSH Act.

Access to Records   Log of Occupational Injuries, OSHA Form 200; Supplementary Record, OSHA Form 101; and Annual Summary, OSHA Form 200, must be available for inspection and copying by Compliance Safety and Health Officers of the Occupational Safety and Health Administration, U.S. Department of Labor, during any occupational safety and health inspection.

Reporting of a Fatality or Multiple Hospitalization   Within 48 hours after the occurrence of an accident which is fatal to one or more employees or which results in the hospitalization of five or more employees, the accident must be reported either orally or in writing to the nearest office of the Area Director of the Occupational Safety and Health Administration, U.S. Department of Labor. The report must relate the circumstances of the accident, the number of fatalities, and the extent of injuries. The Area Director may require additional reports, in writing or otherwise, concerning the accident.

Falsification of or Failure to Make Records or Reports   The OSH Act provides that "whoever knowingly makes any false statement, representation or certification in any application, record, plan or other document filed or required to be maintained pursuant to this Act shall, upon conviction, be punished by a fine of not more than $10,000, or by imprisonment for not more than six months, or both."

Definitions   The following definitions should be noted:

1 *Establishment.* An establishment is any separate physical location where economic activity is carried on and where employees report directly to work.

2 *Lost workdays.* Lost workdays are those days, exclusive of the day of injury, which the employee normally works. All days (consecutive or not) which meet the definition of lost workdays (as outlined below) are counted as full lost workdays, if: *(a)* the employee would have worked but could not; *(b)* the employee was temporarily transferred to another job; *(c)* the employee could not perform the full duties of his or her normally assigned job; *(d)* the employee could not work full time.

3 *Medical treatment.* Medical treatment includes treatment by a physician or registered professional under the standing orders of a physician, but does not include first-aid treatment even when administered by a physician. Injuries requiring first aid are not recordable.

4 *Nonfatal cases without lost workdays.* A nonfatal injury or illness which does not result in lost workdays is to be recorded on the log only if it meets *one* of the following criteria: *(a)* it results in the permanent transfer of an employee or the termination of employment; *(b)* it requires one or more medical treatments other than first aid; *(c)* it results in a diagnosis of occupational illness; *(d)* it causes loss of consciousness; *(e)* it results in restriction of work or motion.

The unsafe physical condition is often a product of the environment—the general conditions surrounding the workplace, the equipment, and materials used, or the process employed. Improvement in the environment can help eliminate the unsafe physical condition; it can also improve working conditions, leading to better production.

Improvement in physical conditions should be carefully planned and tailored to operating requirements. Minimum standards of safety for all physical conditions—either in terms of specification or performance—should be determined in advance, and every effort made to meet these standards.

Survey of the Plant

Before improvements in the physical conditions are attempted, a complete survey of the plant itself should be undertaken and an analysis should be made of past injuries to determine their relationship to possible physical hazards. In making the survey, all factors which may influence the operation should be considered. The number of such factors will vary, depending upon the size of the plant and the product manufactured.

Familiarity with the plant may mean that hazards can go unrecognized. It may be advisable to get assistance in making the survey from someone relatively unfamiliar with the plant, such as the insurance carrier's engineer or the state safety inspector. It may even be helpful for the supervisor of one department or shop to inspect another supervisor's department for hazards which are overlooked by those who have been associated with them.

Following the initial survey, a complete list should be made of all physical changes necessary. If these changes are extensive, it may be impossible to make them all at once. The most important ones should be made as soon as possible; the remainder should be planned over a period of time. Periodic follow-up surveys of the plant should be made to check on action taken, and to make certain that physical conditions are maintained safely.

Plant and Facilities   Many hazards are directly connected with the physical condition of the plant and its facilities; that is, they stem from the condition of the buildings themselves. For example, hazards may exist in storage areas, the location of machines, the flow of materials, and other such factors. Some factors may be difficult to change or control because of limitations imposed by poorly designed buildings, lack of space, or obsolete equipment. But with study and planning, improvements can often be made which will eliminate dangerous conditions. Improvements in the work environment will usually aid production because it allows greater efficiency.

Buildings The most logical place to start a survey is with the buildings and the grounds surrounding the plant. Many old buildings, and some modern ones, are poorly designed from the standpoint of safety. Also, a change in occupancy or manufacturing methods may mean that a building which was once adequate in terms of safety is no longer free of hazards.

The first consideration should be whether the buildings are structurally safe for present or future use. Buildings designed for light manufacturing, for example, may not be strong enough for heavy manufacturing or heavy storage.

The condition of floors and stairs should be carefully noted. Rough, uneven walking or working surfaces which could cause slips or falls should be corrected. Stairs should have a slope of from 30 to 35 degrees. Treads should be about 9½ inches wide, with a 1-inch nosing; and should not exceed 8 inches in height. The height and width of all treads in a flight of stairs should be uniform. Stairs should be provided with handrails and amply illuminated.

Adequate exits should be available. In general, at least two exits should be provided from each floor, and there should be no dead-end halls or corridors where persons might be trapped in a fire. The Building Exits Code of the National Fire Protection Association should be consulted for acceptable standards for exits in buildings of various sizes and occupancies.

Fire protection also should be considered. The need for standpipes and hoses, sprinkler systems, and fire extinguishers—usually considered in terms of protecting property—are also essential to protect the life of employees. Local regulations applying to fire protection should be complied with. Where there are no local regulations covering fire protection, applicable codes of the National Fire Protection Association should be followed.

Layout of the Workplace As part of the survey of the plant and its facilities, a detailed study and analysis of each workplace should be included. The purpose of such an analysis is to determine if machines and related equipment are located so that they do not, in themselves, create hazardous conditions. Among the factors to be considered are:

1 Relation of one machine to another.
2 Space, access, and means of exit for the operator.
3 Space for materials on the job
4 Space for storage of produced material.
5 Width of aisleways.

In theory, a straight-line flow of materials from the receiving platform through the shop to the shipping dock is the safest and most efficient.

Practically, this is seldom possible. But an efficient layout of the workplace should eliminate backtracking, crisscrossing, and unnecessary delays which cause excessive handling. Where a study of the layout shows marked deviation from a straight line, long delays, or excessive handling, further investigation should be made to see if a more efficient route could be established.

Access to the Workplace   In providing for an efficient workplace, consideration should be given to aisles and passageways. Aisles are needed for the safe and efficient movement of people, equipment, and materials from one workplace to another and between departments.

Aisles should be as straight as possible, with corners rounded or diagonal and with no obstruction of visibility at the corner. Where vehicular aisle traffic is heavy, aisles should be at least twice the width of the vehicle or load, plus 3 feet, if the aisles are also used by pedestrians. Where fork lifts or other stacking trucks are used, room must be provided for maneuvering. Traffic signs for vehicles should be provided where there is a significant amount of traffic. Aisles should be plainly marked.

Aisles should not be looked upon as waste space, but rather as an important part of the layout. The same care and planning should be used in determining their location as for any other facility. Aisles determine the traffic pattern of the plant and should be looked upon as an important highway system involving main traffic roads and the necessary feeder road to provide a smooth flow of materials from one job station to another, and from one department to another. They should be kept free of obstruction at all times.

Illumination   Illumination of the workplace is an important factor in accident prevention. It has been estimated that inadequate lighting is the cause of 15 to 25 percent of all industrial accidents. Proper lighting can prevent accidents; it can also increase efficiency and improve employees' morale.

Lighting standards will be determined by the manufacturing method and the product. For most manufacturing a combination of general and localized lighting is advisable. Criteria for good lighting include proper intensity of light and absence of glare, shadows, and too-sharp contrast. The installation of an adequate lighting system for various conditions in the workshop may be a complicated matter; thus, expert technical assistance from a competent lighting engineer is advisable. The recommendations of the Illuminating Engineering Society, as contained in the American Standard Practice for Industrial Lighting, should be used as a guide.

The installation of an adequate lighting system is only the first step in providing proper illumination. Lamps and fixtures need periodic cleaning and replacement to keep them in top condition, and a schedule for such cleaning and replacement will be necessary.

Equipment The equipment used, such as machines, tools, mechanical material-handling equipment, and related accessories, is an important part of the workplace and influences the general environment in which work is performed. In establishing a safe physical environment, control over equipment should take first priority.

Guarding of Machines After a satisfactory layout for machines has been determined, all machines should be properly guarded. Sometimes the proper guardians for machines cannot be determined without having the employees stand in their operation positions so that the surveyors can have a second look. All applicable state and federal regulations governing the guarding of mechanical equipment should be complied with. The most up-to-date information on guarding mechanical power-transmission apparatus and point-of operation guarding is available in the Occupational Safety and Health Standards of the OSH Act. Here is one example of the thousands of occupational safety and health standards: "Aisles and passageways shall be kept clear and in good repair, with no obstruction across or in aisles that could create a hazard." This form of wording is typical of such standards.

Tools and Equipment Injuries from tools and equipment stem from several causes: using the wrong tool or equipment for the job to be done; using the right tool or equipment but using it unsafely; using unsafe tools or equipment.

The causes of injury suggest the measures necessary to control accidents. Stated briefly, they are:

1 Select the right tool or equipment for the job.
2 Train employees in the safe use of tools and equipment.
3 Maintain tools and equipment in a safe condition.

Where tools are issued to employees by the employer, the safety of tools can be ensured by a system of inspection and maintenance by the toolroom or tool crib. Where tools are owned by the employees, a periodic inspection by management will be helpful in eliminating unsafe tools. Mushroomed heads, split handles, sprung jaws, poorly sharpened or poorly set cutting heads, broken insulation, and lack of proper grounding are examples of unsafe conditions requiring control measures. The safe use of tools can be accomplished only through training in safe work practices and through proper supervision by management.

Maintenance To keep machines, tools, and equipment safe, a definite schedule of preventative maintenance should be set up. Greasing, oiling,

FIGURE 4-2
The safe storage of these conveyor rolls is an example of insistence by management on good housekeeping practices.

and other routine maintenance should be done on a regular schedule, based on time or some other measure of operation. Some equipment may need preventative maintenance daily, while other equipment (such as lifts, tow trucks, and other in-plant vehicles) may need maintenance after so many hours of use or so many miles of travel. Schedules for maintenance should be established by management.

Wherever the failure of equipment may cause an accident or loss of valuable time, the equipment should be inspected and maintained on a very comprehensive basis. For equipment whose failure would not be so serious, maintenance can be less comprehensive. Parts subject to failure (such as the clutch and brake on a press, the limit switch on an overhead crane, and the hoisting mechanism of a forklift truck) should be adjusted at regular intervals and regularly inspected to determine the amount of wear, and a definite schedule for replacement of parts should be adhered to. In this way, worn parts will be replaced before they wear out and obsolete equipment can be renewed before it can cause damage or inconvenience. Preventive maintenance offers a rich field for efficient management and a safe working environment.

Materials   Materials used in the manufacturing process may also affect the general environment of the workplace. The same care and planning should be used in appraising the hazards of the materials as are used in determining the hazards of the workplace or the machines and tools.

Material hazards may originate from the way materials are handled or from their inherently hazardous nature.

Flow of Materials   Management should study the effect of materials and their movement on the environment of the shop or plant by tracing the raw materials used through the various manufacturing steps to their completion as finished products. This study should include every operation in the manufacturing process in each place where work is done, and also each place where the material comes to rest while awaiting the next step. Where several different departments (such as a shear department and a press department) are involved in the same product, the flow of materials of each should be tracked through the various manufacturing steps, the subassemblies and final assembly of the finished product.

A simple method of tracking the flow of materials is by means of a flow chart (Figure 4-3 is an example). Such a chart should show the location of each operation, the machines or equipment used, the means of transporting materials from one operation to the next, disposal of scrap (if any), and other pertinent data. An analysis of the hazards of each operation or movement of materials should be made and recorded on the chart showing the flow of materials through the plant. A careful study of the chart by managers will often reveal to them where improvements can be made to control or eliminate hazard-creating operations or movements of materials used in the manufacturing process.

Another method sometimes used to pinpoint the hazards of an opera-

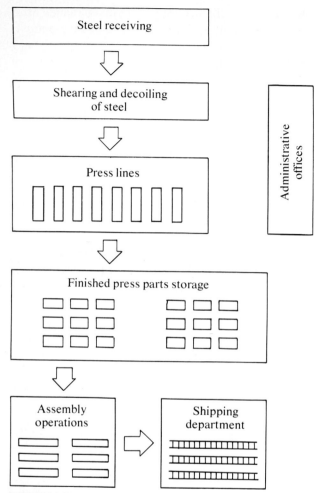

FIGURE 4-3
Product flow in an automobile plant.

tion or process is recording pertinent data on a process chart. A process chart differs from a flow chart in that no sequence of operations is followed; instead, similar machines are grouped together because the hazards are similar.

Flow charts and process charts can be very simple or they can be very elaborate. In either case, they afford a satisfactory method of analyzing and recording information helpful in preventing injuries.

Handling Methods Tracking the flow of materials offers an excellent opportunity to study methods of handling and storing materials. For example, a study might show that mechanical handling may be not only

FIGURE 4-4
There should always be a guard under overhead material that is to be conveyed or handled so that it cannot fall off the conveyor.

cheaper but also safer than manual handling. Also, if delays and stop pages can be eliminated, some handling operations might be discontinued entirely. In the matter of storage of materials, standards must be established, and then adhered to, for width of aisles, height of material, and methods of stacking or piling. Standards should also be established for storing materials at machines, with full consideration being given to safety, ease in handling, and exits for workers.

Stacked or piled materials (Figure 4-5) should be kept as low as possible. If crates, bins, or baskets are used, a limitation should be set on how high these materials should be stacked.

Handling of materials can be controlled by:

1 Substitution of mechanical handling devices for manual handling.

2 Careful placement and training of personnel, with adequate supervision.

3 Strict adherence to established standards for height of stacked or piled material, width of aisles, operation of trucks and battery lifts, and other methods of handling materials.

Hazardous Materials No appraisal of the environment is complete without an analysis of the hazards created by the nature of materials used in the manufacturing process.

Most plants use chemicals in one form or another. Some are harmless; others may be dangerous. Unfortunately, trade-name labels do not always give the chemical composition, and the user may be unaware of the hazardous nature of the materials.

As a matter of policy, for every chemical or chemical compound used, the composition should be ascertained, the hazards should be determined, and measures of control should be established to protect employees against the hazards.

Hazards to health are usually more difficult to recognize and control than factors which might produce accidents. Therefore, it is desirable to secure expert opinion on possible health hazards from the department of the state government having jurisdiction over occupational health problems, from OSHA, or from the National Institute of Safety and Health (NIOSH).

Detailed information on specific chemical materials is beyond the scope of this book, but the following are broad principles of control for three major classifications of materials frequently encountered in industry.

Corrosive materials: Caustics and acids destroy tissue by chemical action upon contact with the skin or upon inhalation of their fumes or vapors. Preventative measures include:

FIGURE 4-5
Good materials-handling engineering includes the safe storage of stacked material.
Material stacked properly as it is here, is a part of good housekeeping, efficient materials
flow, and the clean, modern appearance of a plant.

1 Prevention of spillage.
2 Safe methods of handling.
3 Personal protective equipment and clothing.
4 Emergency provisions such as deluge showers, eye baths, and respiratory
protective equipment.

Toxic materials: Many chemicals and chemical compounds used in industry are highly toxic. These injurious substances reach the body and
can cause damage by (1) inhalation, (2) absorption through the skin, and
(3) ingestion. They may be in the form of liquids, solids, or gases. Liquids and solids may be present as airborne substances in vapors, mists,
dusts, fumes, and smoke.

Control measures will depend on conditions. Recognized methods include:

1 Substitution of less toxic materials.
2 Enclosure of harmful processes.
3 Isolation of harmful processes.
4 Local exhaust ventilation.
5 Use of wet methods.
6 Use of personal protective devices.
7 Decreased daily exposure of workers.

Flammable materials: Flammable materials and strong oxidizing agents present a further problem, that of fire and explosion. Vapors of some solvents, in addition to being toxic, are a serious fire hazard. Preventive methods against fire are largely based on the idea of keeping the concentration of vapors well under the lowest limits at which they will burn. In many cases even these lowest limits are too high with respect to the worker's health and should be used only to assess the danger of fire or explosion, not as a measure of health hazards.

The Process   The process, like the materials used, may contribute to the environmental hazards of the workplace. Hazards may be traumatic in nature or they may involve long-range deterioration of health. Possible sources of injury closely related to the process would include the following:

1 Methods of feeding or operating machines.
2 Methods of handling materials.
3 Extremes of temperature or humidity.
4 Repeated motion, shock, vibration, or noise.
5 Electromagnetic radiation.

The survey for the elimination of environmental hazards should include a study of the process to assess the hazards it creates and to evolve control measures.

In some cases, the process can be changed to eliminate the hazards: for example, wet drilling instead of dry in quarry work, or complete enclosure of a cleaning process where toxic solvents or caustics are used. The automatic feeding of machines (such as punch presses) offers many possibilities for eliminating hazards. Mechanical handling may be substituted for manual handling, as it is both more efficient and less dangerous. The fact that a certain process has always been done in one way does not necessarily mean that a better and safer way cannot be devised. Imagination and ingenuity will often suggest ways in which the process can be changed to eliminate some of its hazards.

Plant Safety Committees

Top Management   Management safety committees at various levels depend primarily on the size of the plant.

In general, there should be an executive safety committee with members from top management, superintendents, and department heads or general supervisors. This committee should meet at least once a month and should review and discuss the safety record of each department, with special emphasis on the cause and correction of any lost-time accidents that have occurred.

The executive safety committee should include in the minutes of its meetings, and as part of its agenda, such matters as safety projects involving the entire plant, safety promotion aimed at employees, and any other safety-related matters that could be carried out only by the level of management represented in this group. Normally, the members of this committee would be appointed by the plant manager.

Departmental Department heads should meet with their supervisors, at all levels, at least once a month to review the safety record of their departments. The departmental safety committee is a management committee, and if necessary, it should act on any recommendations or policy delegated to it by the executive safety committee. Meetings held by this group should take into consideration such items as the instruction of new employees, the correction of unsafe acts and conditions, and safety instructions for all operations.

The departmental safety committee meeting should have a chairperson, and regular minutes should be written up by a member of the committee designated as the secretary. The department head should designate a rotating chairman and appoint a secretary for the coming month at the end of each month's meeting. The minutes should be completed, typed, and distributed as soon as possible after a meeting so that members can carry out any assignments delegated to them before the next meeting of the committee.

Joint Management-Labor Is a joint labor and management safety committee necessary? Since the establishment of OSHA (and in many cases since the sheer necessity has arisen), unions have negotiated joint local safety committees composed of members of management and members of the local union. The provisions of agreement negotiated by management and labor vary considerably from local to local and from contract to contract. Since safety is a matter of great concern to both employees and management, it should be the intent of both parties to provide (within the limits of their agreements at negotiations) a tool for preventing accidents and improving the workplace for all employees. This is a tool from which both parties will benefit.

Both managers and employees chosen to work on such a committee should have a sincere interest in the prevention of accidents and should leave political aspirations behind them. The success or failure of such a committee depends on the good faith of all parties involved.

A joint committee functions in an advisory capacity unless negotiated agreements dictate stronger functions.

FIGURE 4-6
New jobs should be carefully processed, and new equipment well guarded, to eliminate hazards to employees.

The joint safety committee should review the plant's preceding month's safety record, discuss lost-time accidents and their causes, and consider any pertinent statistical data. Finally, it should submit to the plant manager recommendations for the improvement of plant safety.

Investigating accidents to discover their causes is usually not a routine part of the joint safety committee's job, but there is hardly a committee that has not investigated the circumstances of an accident to find out what caused it. Again, the main purpose of an accident investigation is to find out how to prevent similar accidents in the future—not to place blame but to correct the situation.

Safety Committee Functions

A basic format for all safety committees should include the following minimal functions:

1 Review and discuss all lost-time accidents and make recommendations to upper management to prevent recurrence.

2 Review and discuss the plant's safety record and make necessary corrections or recommendations.

3 Make safety inspections for the purpose of discovering unsafe conditions and practices and make recommendations for their abatement.

4 Make good housekeeping inspections and follow-up to ensure that ˙ recommended corrective procedures were acted upon.

5 Participate in safety promotional activities for the purpose of arousing and maintaining the interest of employees and management alike.

These functions can be used as a format for any of the types of safety committees shown here, including the Management-Union Safety Committee. For any of these committees, the purpose is the same—promoting safety in the workplace. It should be recognized, though, that even under favorable circumstances, safety committees have certain limitations.

It is important that the people selected to serve on safety committees have a sincere interest in safety and a desire to serve. To maintain that interest requires constant effort and commitment by the chairperson. Committee membership should rotate at intervals long enough so that each member has had time to make a positive contribution and to gain personal experience and interest. Committee size should be small enough to be effective but large enough to provide the required expertise to serve the mission.

Safety Rules

Safety rules are codes of conduct designed for the purpose of avoiding injury and property damage. They should be prepared in realistic and easily understood language. Employees cannot be expected to respect and follow illogical, unfair, or unrealistic safety rules. Safety rule books and safety manuals should contain specific instructions as to the employee's safety responsibilities to the organization. Whenever safety rules are ignored, other plant rules are usually ignored, too. Experienced supervisors have found that when employees conduct themselves according to the safety rules of the organization, the work will be done more safely and more efficiently. Hence, there are many strong arguments for establishing rules and seeing that they are enforced.

Management should follow the following criteria when preparing to write safety rules to govern the conduct of employees in the workplace:

1 Require that safety rules be clear and easily understood.

2 Keep the number of general safety rules to a minimum.

3 Stipulate only those rules that are currently required.

4 Stipulate only those rules that can be strictly enforced.

Supervisors have the responsibility for enforcing the safety rules. Here are four common-sense methods for gaining compliance by employees:

1 Set a good example by always following the safety rules.

2 Never fail to correct an employee who is not complying with the safety rules—that is, don't play favorites.

3 Show the same concern for unsafe conditions as you would for unsafe acts.

4 Follow company policy regarding disciplinary actions. (If corrective action is not taken, the rule will eventually cease to exist.)

General Safety Rules

1 In case of sickness or injury, no matter how slight, report immediately to your supervisor and seek first aid.

2 Good housekeeping is your responsibility. Keep your work area and tools cleaned up.

3 Safety glasses and hearing protection are required in designated areas. Obey warning signs when entering these areas.

4 Running on company premises is forbidden.

5 Fighting, horseplay, throwing objects, and distracting the attention of fellow workers is prohibited.

6 Safety devices and guards are for your protection. Never operate machines or other equipment unless they are in place.

7 The possession or use of a controlled substance or intoxicating liquors will be dealt with accordingly, up to and including discharge.

8 Never use makeshift or defective ladders, scaffolding, platforms, or rigging.

9 Report unsafe equipment and unsafe conditions to your supervisor.

10 Do not wear gloves while operating machinery unless authorized by your supervisor.

11 Do not wear rings, bracelets, wrist watches, or any other jewelry while operating machines.

12 Do not wear long sleeves, ties, and loose, ragged, or torn clothing around moving machinery.

13 Anyone with long hair should wear a cap or other hair covering while working around moving machinery

14 Never use defective chisels, hammers, punches, wrenches, or other defective tools. Exchange or see that defective tools are repaired.

15 Pile material, trucks, skids, racks, crates, boxes, ladders, and other equipment so as not to block aisles, exits, fire fighting equipment, alarm boxes, and electric lighting or power panels.

CATEGORIES OF INSPECTIONS

Depending upon the company's safety organization and the interest of the safety manager, various methods of carrying out inspections have been devised. Listed here are the three most generally used methods for carrying out such inspections:

1 Informal inspections
2 General (planned) inspections
3 Critical parts inspections

Informal inspections or spot inspections are made on a daily basis by the supervisor within his or her department to check and see that tools, machinery, and equipment are in safe operating condition; aisles and passageways are clear and unobstructed; good housekeeping is maintained; and last, employees are complying with established safety rules. From time to time, a member of the safety department will make such an inspection within the various departments of the plant.

General inspections are planned wall-to-wall inspections of the entire plant by a management team or safety committee members at regular intervals. A written report of this group's findings should be made during the inspection and processed for action through established organizational channels.

Critical parts inspection is the daily inspecting and checking of operating equipment by operating personnel as a part of their working procedure. The items that should be inspected are any piece of equipment or machinery which could cause bodily harm to employees. Examples of equipment that should be inspected on a daily basis are overhead crane cables and hooks, punch presses, fire extinguishers, elevators, personal protective equipment, etc.

PLANNED INSPECTIONS

Loss exposures are created by the day-to-day activities in any type of organization. Equipment and facilities do wear out. At some point, wear and tear make the risk of accidents too high. Inspections are needed to detect such exposures in a timely manner. They also provide feedback on whether equipment purchasing and employee training are adequate. Also, conditions change. People, equipment, materials, and the environment are constantly changing. Some changes remove previous hazards, others create new ones. A prominent management philosophy is that "all problems result from changes." Inspections focus on these changes and help identify and solve problems.

Nothing is risk-free. A good inspection program can identify:

1 Potential safety problems.
2 Potential equipment deficiencies.
3 Potentially unsafe acts by employees.

Potential safety problems can include unsafe equipment, poor house-keeping practices, and fire and explosion hazards.

Potential equipment deficiencies includes unsafe design, inadequate guards on machines and equipment, and defective work tools and equipment.

Potentially unsafe acts by employees could include improper lifting, using equipment unsafely, making safety devices inoperable, and failing to use personal protective equipment when required. Again, these are but a few of the many examples that could be included here.

EMPLOYEE SAFETY TRAINING

Employee safety training should begin on the first day at the workplace and should continue periodically for the length of the worker's affiliation with the company. It should begin in the form of a formal safety indoctrination program on the worker's first day at work. Training is one of the best methods that can be used to influence human behavior for the purpose of developing sound and safe work habits.

A safety training program is needed:

1 for newly hired employees
2 for employees reassigned to other jobs
3 for employees returning to work after an extended lay-off period or medical leave
4 when new equipment and processes are introduced or installed in the workplace
5 whenever the need arises to improve and update safe work practices and procedures

There is no set of objectives that will serve all organizations and all situations. Each set must be designed and developed to meet its own circumstances and special needs. When employees are properly trained to do their job or task, they will do it safely. Many organizations with good safety records believe that continuous safety training, supported by a sincere positive attitude on the part of management, is the key to a successful safety program.

New Employee Indoctrination

Indoctrinating new employees begins at the time of employment—before the employee starts working. On this first day in a new and strange environment, the new employee needs to begin acquiring knowledge in the following areas:

1 Company safety rules.
2 Employee's responsibilities under OSHA.
3 Types of personal protective equipment available to them.
4 Location of medical and first-aid facilities.
5 Procedures for reporting: (*a*) job-related injuries, (*b*) on-the-job hazards, (*c*) defective or unsafe equipment and conditions.

Departmental Indoctrination

When new employees reach their own department, the supervisor should give them additional safety training before they begin work. Some of this training may cover some of the materials given to them earlier during their indoctrination period, but is now applied specifically to the work they will be doing.

The supervisor should discuss the following subject materials with his or her new employees:

1 Hazards inherent on the job—or within the department.
2 The safeguards and precautionary measures for those hazards.
3 Personal protective equipment required on the job.
4 Instructions for the care and proper use of the personal protective equipment.
5 Location of emergency exits and telephone.
6 Location of fire equipment and other emergency equipment.

Supervisors should make provisions for training in safe work procedures and should follow up to ensure that these procedures are being used.

Many organizations have very elaborate indoctrination programs. These may include a safety department sponsored safety talk and viewing of a safety film, a plant tour, discussion of the company's products, and listening to talks by the department heads of various departments.

EVALUATING INJURIES AND ILLNESSES

We all know the value of a good safety record. Work-related injuries and illnesses are expensive in terms of both decreased productivity and increased premiums for worker's compensation insurance and related costs. They become especially costly in human terms when they lead to the permanent loss of a skilled worker or when they destroy employees' morale. Clearly, a good occupational safety and health program is good business, and an accurate system of evaluation is a part of such a program. An accurate evaluation is best done by the use of incidence and

severity rates which can compare overall performance of a plant's existing safety record to the previous year's record. Using these rates, comparisons can also be made between different plants in a particular industry or departments within a particular plant. Thus, when the injury data
for a particular unit are higher than for other units, the safety professional must decide what action to take for applying corrective measures.

Computing the Incidence and Severity Rates

An increasing tendency of modern management is to measure all industrial results and all forms of effort. In safety performance, injury statistics are used to measure that effort. The method used to measure
disabling-injury experiences needs to be understood fully because of the
wide use of such statistics throughout the United States.

Disabling-injury *incidence rate* is defined as the number of lost-time
~~days away from work~~ for every 200,000 work hours. For example, a plant
employing 850 employees worked 1,750,000 work hours last year. It incurred 7 lost-time accidents during that period of time. The *incidence
rate* for the plant last year is computed as follows:

$$\text{Incidence rate} = \frac{\text{Number of lost-time accidents} \times 200{,}000}{\text{Employee hours worked}}$$

Thus, the computation for the above plant is:

$$\text{Incidence rate} = \frac{7 \times 200{,}000}{1{,}750{,}000} = 8$$

The 8 in the computation represents 8 lost-time accidents per every
200,000 work hours.

The 200,000 hours in the formula was established by the Bureau of Labor Statistics. It is assumed that a base of 100 full-time employees would
work 200,000 hours per year (40 hours per week per worker, 50 weeks
per year).

Computing the Severity Rate

Severity rate is the number of lost-time days charged for disabling injuries per every 200,000 work hours. Just as in the incidence rate, work
hours also means work hours of exposure to injury. For example, a plant
employing 50 employees clocked 600,000 work hours. It experienced 75
lost-time workdays during this year. The *severity rate* for that period of
time is computed as follows:

$$\text{Severity rate} = \frac{\text{Number of lost-time days} \times 200{,}000}{\text{Employee hours worked}}$$

Thus, the computation is as follows:

$$\text{Severity rate} = \frac{75 \times 200,000}{600,000} = 25$$

The 25 in the formula represents 25 lost-time days for every 200,000 hours worked.

A plant may have a high incidence rate, and the injuries may be minor, or, a plant may have a low incidence rate, but the injuries are severe. Thus, a measurement of the severity of injuries is an important part of the overall measurement of disabling-injury experience.

CONTROLLING THE PHYSICAL ENVIRONMENT

A safe and healthy place to work is the foundation on which every successful safety program is built. To provide such a workplace requires control of the physical environment—the surroundings and external conditions which influence the day-by-day operation of the establishment, including the possibility of injury to employees. Control of the physical environment is management's responsibility, and the organization and administration of every safety program must be predicated on the fact that management has made every effort to provide a safe workplace.

The importance of a safe physical environment cannot be overemphasized. Most industrial injuries, as has been noted, are caused by the combination of an unsafe physical condition and an unsafe act. Eliminate the unsafe physical condition, and one contributing cause is eliminated.

QUESTIONS AND EXERCISES

1 ABC Construction Company worked 500,000 hours and had 25 lost-time accidents last year. What was the incidence rate?
2 ABC Construction Company worked 500,000 hours and had 60 lost-time days last year. What was the severity rate?
3 Company X worked 800,000 hours last year and had 20 lost-time accidents. Company Y worked 400,000 hours and had 10 lost-time accidents last year. Which company had the better safety record?
4 What are the general functions of a plant safety committee? What are the functions of a departmental safety committee?
5 Other than what you have read in this chapter, can you name some other functions of the safety professional?
6 What type of safety training should be given to supervisors?
7 What is the purpose of the wall-to-wall inspection?
8 Do you feel that the incidence and severity rates are good measuring devices? Explain.
9 Do the manufacturers of machinery and processing equipment have any responsibility to the employees of their customers? Explain.

TRUE OR FALSE?

1 One of the most effective means by which experienced employees can benefit from additional job safety training is through periodic safety meetings.

2 Designs for pressure vessels and air systems must meet American Society of Mechanical Engineering standards.

3 Corrosives are defined as chemicals that effect the central nervous system.

4 Members of the safety department should act as enforcers and apply disciplinary action when necessary.

5 It is not necessary for management to design safety rules; it can best be done by the members of the union safety committee.

6 The critical-parts inspection is the inspection of operating equipment such as hoists, grinders, tools, electrical equipment, etc.

7 One of the best methods for determining accident trends is by accumulating and studying accident and injury data.

8 To compute the incidence rate, multiply lost-time days times 200,000 and divide by the total number of work hours.

9 A material or substance may be considered as toxic when only a small amount of it will cause injurious effects to the average adult person.

10 Two general methods of appraising plant safety are plant inspections and statistical comparisons.

BIBLIOGRAPHY

Accident Prevention Manual for Industrial Operations, 7th ed., National Safety Council, Chicago, Ill., 1974.

Bird, Frank E., Jr. and George L. Germain: *Practical Loss Control Leadership,* Institute Publishing, Loganville, Ga., 1986, p. 122.

Consultative Approach to Safety, U.S. Department of Labor, Office of Technical Services Division of Safety, Bureau of Labor Standards, Bulletin 223, 1960, p. 52.

Control of the Physical Environment, U.S. Department of Labor, Office of Technical Standards, Bulletin 211, 1960.

OSHA Guide to Evaluating Your Firm's Injury and Illness Experience: 1974, Manufacturing Industries, U.S. Department of Labor, Bureau of Labor Statistics, Report 475, 1976.

"Safety Management in a Small Business," *National Safety Council Transactions, Industrial Subject Sessions,* Vol. 12, National Safety Council, Chicago, Ill., 1975.

5

ACCIDENT INVESTIGATION: A MANAGEMENT FUNCTION

An accident investigation collects information and interprets facts relating to an accident. It is also a process to explore ways and means to prevent or minimize a recurrence of the accident and resultant injuries or property damage. The investigation must be for fact-finding, not fault-finding; otherwise, it may do more harm than good. This is not to say that responsibility may not be fixed on persons whose personal failure has caused injury, or that these persons should be excused from the consequences of their actions. But the investigation itself should be concerned only with facts.

Accidents should be investigated by management as soon as possible. Depending on the nature of the accident or other conditions, the investigation may be made by the supervisor, the safety representative, or the safety committee. If the accident involves special features, the presence of a member of the plant engineering department may be warranted. Photographs of the accident scene are a necessary part of the investigation. Reports or recommendations can often be profitably supplemented by photographs because they can show clearly the need for corrective action.

The most important person in the investigation of an accident is the supervisor. This person is usually in constant daily contact with the workers and should be thoroughly familiar with all of the possible hazards and working conditions that may exist in his or her area of responsibility.

Gathering all of the facts in the investigation of an accident is no small responsibility. Collecting information and related data from witnesses to an accident at the workplace is clearly important to the investigation. Again, it is the first-line supervisor who knows the employees personally and how best to approach them for those facts. In addition, the supervisor is the person upon whom management must rely to interpret and enforce such corrective measures as are devised to prevent recurrence of the accident.

CASES TO BE INVESTIGATED

It is obvious that accidents causing death or serious physical injury should be investigated. In addition, the "near-miss" accident that might have caused death or serious physical injury should be considered equally important and should be investigated. Also, any epidemic of minor injuries demands study. A particle of steel dust in the eye or a scratch from handling sheet metal may be a very simple case; the immediate cause may be obvious, and the loss of time may not exceed a few minutes. However, if cases of this or any other type occur frequently in the plant, or in any one department, an investigation should be made to determine the underlying causes.

The chief value of such an investigation lies in uncovering contributing causes. The wise supervisor is constantly alive to the advantage of this kind of accident investigation, which may prove more valuable, though less spectacular, than one following a fatal injury.

THE ACCIDENT REPORT

The first step after the accident investigation is the preparation and submission of the accident report. The investigating supervisor should be able to assemble and record the facts in such a way that anyone can understand what had occurred during the course of the accident and how it had occurred. A poorly written report not only reflects a poor attitude by the investigating supervisor but also a lack of interest by upper management. After writing the accident report, the supervisor never knows on whose desk it may land, within or outside the company. Therefore, the report should be accurate and must include all of the facts and information that can be gathered by the investigating supervisor.

The three most important parts of the accident report are:

1 The description of the accident.
2 The recommended action to prevent recurrence.
3 The corrective action taken.

The most commonly recommended methods of preventing recurrence include:

1 *Repairing* the unsafe condition.
2 *Replacing* the unsafe condition.
3 *Designing* or engineering a new process to prevent recurrence.
4 *Redesigning* the cause of the unsafe condition.
5 *Retraining* the injured worker if necessary.
6 *Transferring* the injured worker to a type of work commensurate with his ability to perform the work.
7 *Undertaking disciplinary action,* if required.
8 *Writing up a new safety procedure,* if necessary.
9 *Investigating a new, pertinent safety rule* to reduce or eliminate the cause of the accident.

GUIDELINES FOR CONDUCTING INVESTIGATIVE INTERVIEWS

The fundamental purpose of interviewing is to obtain from the person being interviewed an accurate and comprehensive account of all pertinent facts, interpretations, and opinions relating to the accident under investigation. The best interview meets the criteria of being complete, accurate, and pertinent. The goal is to hear and record all of the given information. The person being interviewed must be free to describe the accident and provide information without being influenced either by the interviewer's personality or the setting in which the interview is taking place.

In many accident investigations, interviews may be the primary source of investigation. The following guidelines are recommended for conducting the interview:

1 Have a plan and know where the interview is going to lead. If possible, prepare questions ahead of time.
2 Make sure you understand the technology of the equipment or process involved in the accident.
3 Hold the interview in private, in order to avoid distraction. Though a quiet room is best, the interview may have to conducted at the scene of the accident in order to have technical matters explained.
4 Put the person being interviewed at ease. Avoid being overbearing in speech, voice, or manner. Be careful not to talk down to the person or use language that is above his or her understanding.
5 Avoid asking any questions that suggest an expected answer and avoid questions that can produce only "yes" or "no" answers. Always keep control of the interview.

6 Close the interview in a courteous, but firm, manner. Encourage the witness to contact you if any other pertinent information comes to mind.

Key Questions in the Accident Investigation

Six basic questions provide for opening the investigation and lay the foundation upon which it can be developed. These key questions are:

1 *WHO* was injured? Who saw the accident? Who installed the faulty equipment? Who was responsible for installation?

2 *WHAT* happened? What was the cause of the accident? What did the injured worker do or fail to do? What equipment or facilities were involved?

3 *WHERE* did the accident happen? Where was the injured worker's supervisor at the time of the accident? "Where" questions help to determine what caused the accident and the events leading up to it.

4 *WHEN* did the accident occur? The answer to "when" questions should contain more information than just a clock reading. Though the time of the accident occurrence is important to know, relationships are often more important. "When" questions ask such questions as: when did the injured worker take the guard off his or her machine? When was the injured person last given safety instructions? "When" questions also elicit information or relationships between activities or events.

5 *HOW*? Questions beginning with the word "how" also provide information on the interaction and relationship between activities and events. How did the accident happen? How well was the worker instructed to perform the task? How did the guard fall off the machine?

6 *WHY* didn't the worker perform the job as he or she was instructed? Why didn't the worker perform the job in a safe manner? Answers to "why" questions should give some indication as to the corrective measures that should be taken, since the answers here will generally focus on unsafe conditions or unsafe acts.

Steps in the Accident Investigation

Much of the information concerning the accident investigation will be obtained from witnesses and, if possible, the injured worker. Some of the people interviewed may have witnessed the accident. Other witnesses may be able to provide only one or two facts relating to the accident. Last, there will be witnesses who will provide only hearsay information.

All of this information will have to be sorted out in some form of logical sequence in order to put the facts of the investigation in proper perspective. In order to do this, the investigator must proceed as follows:

1 If possible, discuss the accident with the injured employee.

2 Determine what the injured worker was doing prior to and at the time of the occurrence of the accident.

3 Determine if this action was in pursuit of his or her regular duties.

4 Determine if the injured worker was instructed how to properly perform the job.

5 Determine if the task was performed in accordance with those instructions.

6 Determine if the actions or inactions of co-workers contributed to the cause of the accident.

After an Accident

In order to assure the rehabilitation of injured workers as soon as possible and reduce medical and workers' compensation costs, employers should:

1 Have an injury-reporting procedure and ensure that supervisors and employees understand it. An example of such a procedure is: "All employees should report all injuries to their supervisor and report for medical treatment as soon as possible."

2 Accept the employee's injury report as submitted without implying doubt about its validity until a full investigation has been completed.

3 Require that supervisors, if possible, together with the employee, investigate the accident as promptly as possible.

4 Maintain supervisory contact with the employee and counsel him or her through rehabilitation. Ensure that the insurance carrier has also initiated contact with the injured worker. If there is a labor contract, adhere to it.

5 Contact the physician involved and advise him or her of any pertinent information regarding the case. Assess the credibility and rapport between the employee and the physician.

6 Initiate injury rehabilitation with the worker as soon as possible. If so advised by the doctor, offer "light" duty and gradually increase work requirements as rehabilitation progresses until the worker is back to full duty.

7 Follow up with the employee to ensure that he or she is complying with the requirements of the rehabilitation program.

8 Follow up with the insurance carrier for an equitable and timely claim settlement.

Analysis of the Accident Investigation

Causes of accidents and the appropriate action to prevent recurrence can only be ascertained through proper investigation of all of the factors that contributed directly or indirectly to the accident. The investigator's findings will reflect on the thoroughness and effectiveness with which information is collected and analyzed. Throughout the investigation there are two goals that the investigator must keep in mind:

1 To find out what happened and why.
2 To prevent it from happening again.

Deductive reasoning, which begins after disclosure of the basic facts and continues through the process of analysis, should be the basis for all investigative findings. It may be necessary to resort to a process of elimination to arrive at conclusions as to what occurred in the accident sequence.

An often overlooked by-product of accident investigations is the identification of potential causes of unforeseen future accidents. Such factors may have little to do with the investigation; however, the investigator should be aware that such factors do exist and often precipitate a possible future accident of greater magnitude.

OSHA WORKPLACE ACCIDENT INVESTIGATIONS

The Occupational Safety and Health Administration has the authority to investigate workplace accidents for the purpose of determining:

1 If there has been a violation of federal safety and health standards.
2 Whether any such standards require revision.
3 Whether any new standards need to be developed.
4 If corrective actions will prevent recurrence in the future.

To warrant investigation, accidents should meet one of the following criteria:

1 One or more fatalities, five or more employees hospitalized in the course of one accident, or any combination thereof.
2 Frequently occurring incidents of a like nature.
3 Accidents or events of national importance that involved extensive property damage and could have caused death or multiple injuries.
4 Accidents occurring in industries that are the subject of any special federal program.
5 Accidents involving significant publicity.

ACCIDENT ANALYSIS—TRENDS

It has been proven that an effective method to reduce accidents is to analyze accident reports for trends. The analysis of trends can unmask potential problem areas in the workplace. By grouping similar data from accident reports or even first-aid cases, a pattern of potential accident causes can be tabulated. For example, some of the most common or most useful categories are grouped as follows:

1 *Accident trends caused by unsafe acts:* How many accidents can be traced to the unsafe acts of the injured workers? What was the corrective action taken?

2 *Accident trends caused by unsafe conditions:* How many accidents were caused by unsafe conditions? Are they found in any particular area or department or are they widespread? What was the corrective action taken?

3 *Accident trends caused by occupation:* Are there certain occupations in the plant that incur more injuries than do others? Are the causes of accidents to this group due to the lack of a specific type of knowledge, safety rule, or safety procedure? Does this mean there is a deficiency in the training program?

Of course, merely obtaining and categorizing all of this information will not prevent recurrence of the accidents. The conditions that were the causative factors in the accidents must be corrected.

There are additional uses for the statistical information derived from making the accident analysis. It furnishes data for some of the following:

1 New safety rules or safety procedures.
2 Changes in work practices.
3 Upgrades in the level of safety training.
4 Upgrades in the level of supervisor's safety contact with employees.
5 Purchase of a particular type of personal protective equipment.

The accident analysis reveals the flaws in the environment and also in the management's policies and programs. It tells management what can happen. The accident analysis is a good method of giving interested managers solid data for preventing accident recurrence. It is a good tool.

BASIC ACCIDENT INVESTIGATION EQUIPMENT

Every organization should have in its possession an accident investigation kit. If supervisors or safety department personnel don't gather the evidence or make good observations at the time of the initial accident investigation, it is almost impossible to accurately reconstruct the facts later. This kit should include, at a minimum, the following items which could easily fit into a medium size briefcase:

1 A tape recorder to record conversations with witnesses or for recording personal impressions.

2 Pens, pencils, white chalk, note pads, and plain drawing paper.

3 A 35-mm. camera or Polaroid camera, extra film, extra camera batteries, and plenty of flash bulbs (if applicable).

4 A tape measure, rule, rope and fluorescent tape to isolate or rope off the accident scene, and warning and caution signs.

5 Large and small manila envelopes and plastic bags for collecting evidence and any other needed samples.

6 Personal protective equipment such as a hard hat, safety glasses, ear plugs, safety shoes, and coveralls.

Photographic Documentation

Photography can serve as a valuable method of recording conditions that may change during the investigation. It can also aid in preparing the accident report and in analyzing conditions at the site of the accident.

Photographs of the overall scene, wreckage area, and pertinent hardware should be made prior to any adjustments to the scene of the accident. They are helpful in determining what happened as well as in providing illustrations for reports. In instances where unusual wreckage patterns exist, color photographs are of value to the investigation. An example would be in differentiating between smoke and oil discolorations or among variously colored paint smears.

Do not use a Polaroid camera for other than reference pictures. Polaroid pictures do not blow up well and eventually fade. Always date pictures, identify locations, and indicate the time the photograph or photographs were made. At times it may be necessary to describe what the photograph is trying to identify. Photographs obtained from other sources also should have the above information, as well as the name of the source.

Accident Scene Sampling

There are many reasons for taking samples of materials during an accident investigation. Samples may reveal the cause of death or injury.

In cases where it is suspected that some mechanical or structural defect contributed to an accident, it may be necessary to take samples of soil, concrete, or parts of machinery suspected of failure. For example, samples of material taken from an accident may reveal that insufficiently cured concrete contributed to the collapse of a building or that unstable soil conditions led to a trench or excavation cave-in.

LEGAL RAMIFICATIONS OF ACCIDENTS

People want to believe that someone else is at fault for every accident that occurs. And chances are today that a serious accident will end up in court.

Since the company's lawyer usually isn't called until after a lawsuit is filed, the lawyer must rely on the investigation and subsequent accident report that was prepared by a supervisor or the safety department. The information gathered by the supervisor at the accident scene is critical to the lawyer's ability to defend the company in a court of law. The supervisor's attitude and the manner in which he or she investigates an accident can create a case for litigation or prevent it.

If a work-related case ends up in court, chances are the supervisor may be called to testify as a witness. Therefore, when investigating an accident, supervisors should be careful about what they say. Statements of company employees are usually admissible into evidence.

Another important factor is careful attention to the preservation of accident investigation records. The safety department may have done an outstanding job of investigating a serious accident, but if no one can find the records several years later, the investigation becomes worthless.

In matters such as this, the supervisor is the key point person. If he or she has investigated the accident thoroughly, collected all the evidence, and taken statements of all pertinent witnesses, the supervisor has not only enhanced the company's safety program but also increased its chances of winning a lawsuit and saving money.

CASE STUDIES

Studying actual cases is one of the best methods for becoming acquainted with accident investigation techniques. Almost 90 percent of all work-related accidents are due to employee-related activities or simply to pure negligence. Accident investigations should, in addition to producing facts about the accident, lead to improvements in accident investigation techniques. In almost every case, the preventive action ultimately recommended or taken indicates the direction management must take to prevent recurrence of the accident. Therefore, expertise in accident investigation techniques is an important tool in the prevention of accidental injury in the workplace.

Case Study 1

Extent of injury: Amputation of middle finger of right hand.

Description of accident: While wearing gloves, a drill-press operator was drilling holes in metal fasteners to be used in aircraft wing gas tank assemblies. She then attempted to make a tool change while the machine was operating at a slow speed. While she was doing so, the glove on her right hand caught on the revolving drill and caused an amputation of the middle finger on her right hand.

Cause of accident: (1) Not shutting down the drill press before attempting to change the tooling. (2) Wearing gloves while performing work around revolving machine parts.

Corrective action required: (1) Instruct employees to shut off and lock-out equipment before making tool changes. (2) Instruct employees not to wear gloves around revolving machinery. (3) Post a sign on or near the affected equipment instructing employees not to wear gloves around revolving machinery.

Follow-up action: Employees should be further instructed not to wear rings, watches, bracelets, or long sleeves around rotating machinery; shirts and blouses should be tucked into pants or trousers; and employees with long hair should wear protective head covering.

Case Study 2

Extent of injury: Fracture of skull.

Description of accident: An employee in a foundry was using an overhead wall-mounted electrically controlled crane to move a heavy casting from one position to another at his workstation. The casting weighed approximately 3,000 pounds. While he was moving the casting, it fell, causing the hoist cables to snap and strike the employee a glancing blow to his head. Fortunately, he was wearing protective head gear, or the blow could have been fatal when the hoist eyebolt assembly failed.

Cause of accident: The nut holding the eyebolt on the crane hook had loosened, causing the eyebolt threads to become stripped and thereby causing the crane hook to fall.

Corrective action required: (1) Make monthly or quarterly inspections of all hoisting equipment in the plant. (2) Design an inspection checklist for this program, making sure that eyebolts and lifting lugs are included. (3) Make it a plant procedure that employees visually inspect all lifting components before using the equipment. (4) Publicize this accident throughout the plant to ensure that all employees are aware of the causes of the accident and action taken by management.

Case Study 3

Extent of injury: Amputation of left thumb.

Description of accident: A shipping department packager in a small manufacturing plant placed a gear on a layout table and sprayed it with rust preventative before packing it for shipment. After the employee had sprayed several gears, the spray gun became clogged and failed to operate. The employee tried to clean the clogged spray gun tip. At that instant, the gun discharged, and the employee's left thumb was severely lacerated and subsequently had to be amputated.

Cause of accident: Cleaning clogged spray gun while under 40 pounds pressure.

Corrective actions required: Remove pressure from spray gun before attempting repair work.

Follow-up action required: (1) Caution employees to remove pressure from spray gun before attempting repair work. (2) Warn employees of the possibility of dangerous substances penetrating the skin. (3) Publicize this accident throughout the department and other departments that use spraying equipment.

Case Study 4

Extent of injury: Radiation burn on outer surface of right leg about 2 inches in diameter. No other ill effects found.

Description of accident: A laboratory employee used a half-gram radium isotope to test a casting in a foundry. After finishing the test, she laid the radium isotope capsule on a nearby work table, forgot about it, and went back to her department. Later on during the day, realizing her error, she returned to the area where she had been testing castings and found the capsule missing. All of the employees in the department were promptly notified of the danger. The plant safety engineer with the use of a radiation detector located the capsule in an employee's clothes locker. The locker belonged to a maintenance man who later explained that he had taken the capsule because it looked like a plumb bob. He said that he had it in his front pants pocket for about an hour. He was immediately examined by the plant physician. Two days later, he reported back to the plant doctor with what appeared to be a sunburn over a small area on the outer surface of his right leg.

Cause of accident: (1) Negligent use of radioactive isotope by laboratory worker. (2) Lack of knowledge by employee regarding radioactive materials in the plant. (3) Lack of proper procedures governing the use, handling, and storage of radioactive materials.

Corrective action required: (1) Establish a radiation safety education program for all employees. (2) Establish such rules and procedures as are necessary to prevent the exposure of personnel to greater-than-permissible amounts of radioactive material. (3) Establish a schedule of periodic medical examinations for employees working in the areas, preempted by the use of radiation film badges. (4) Establish a radiation detection monitoring program for the plant on a scheduled basis.

Case Study 5

This case study should be used for classroom participation and discussion.

Description of accident: John Brown, a newly hired employee of Smith Steel Corporation, was assigned to work as a laborer in an iron ore furnace while it was shut down for repairs. This was his first day on the job. His job consisted of the following tasks: (1) Knocking brick loose from the furnace wall with a pick. (2) Loading the bricks into a wheelbarrow. (3) Wheeling them outside of the furnace and dumping them on a pile on the floor. This was his job assignment for the entire shift. The next morning his wife called the personnel department and stated that her husband was in great pain during the night and that she took him to the local hospital emergency room, and that he was subsequently admitted as a patient. She concluded by stating that tests at the hospital revealed that her husband had a herniated disc in the fifth lumbar region of his back.

Problem: You have been assigned as the company representative to investigate the alleged accident and to make the following determinations: (1) Is this a work-related injury? (2) Did the employee sustain the injury while in the employ of a previous employer? (3) Could the employee's injury have occurred at home? (4) Is this a workers' compensation case?

WORD POWER

investigate To search; to inquire into; to examine in detail for the purpose of ascertaining facts.

facts Something known to have happened; something said to be true or supposed to have happened.

fact-finding Engaged in determining or searching for facts.

interview The conversation of an investigator with a person or persons from whom facts are being sought.

disciplinary action Punishment inflicted for the purpose of correction and training.

workers' compensation A form of no-fault insurance designed to recompense injured workers while recuperating from a work-related injury or illness.

QUESTIONS AND EXERCISES

1 What is the purpose of the accident investigation?
2 What are the three most important parts of the accident report?
3 What are the six key questions that must get answered in order to make a complete and thorough accident report?
4 Define "accident analysis."
5 Why can we not totally rely on the accident analysis to predict future accidents?
6 List at least three reasons why the line supervisor should make accident investigations.
7 What are some good guidelines for interviewing witnesses on an accident?
8 List several benefits of personal involvement by supervisors in accident investigations.
9 Are accident reports a reliable indicator of the degree of risk in a particular operation?
10 Because showing is usually better than telling, should investigators use reenactment whenever possible?
11 Under what circumstances should middle and upper management participate in accident investigations?

TRUE OR FALSE?

1 To be statistically solid, accident data must be collected either over long periods of time or from a large number of similar activities.
2 When safety department members participate in accident or property damage investigations, they should serve as advisers to the line management investigators.
3 The immediate actions of a well-trained supervisor at the site of an accident can gain evidence that might take others many more hours to get through accident reconstruction.
4 It can be assumed, with reasonable accuracy, that the loss potential is sometimes more important than the actual loss which occurred in an accident.
5 A great percentage of all activities in the workplace are caused by the unsafe acts of the persons involved.
6 Witnesses to an accident should be interviewed as a group rather than as individuals, in order to save time and get all of the information as quickly as possible.
7 Under no conditions should middle and upper management be participants in an accident investigation.
8 It is not important for an organization to have an accident reporting procedure as long as employees report to the medical department after an injury.
9 The Occupational Safety and Health Administration has no authority whatsoever, under any circumstance, to investigate workplace accidents.
10 The accident analysis is not a good method of searching for trends as to the causes of accidents.

BIBLIOGRAPHY

Accident Prevention Manual for Industrial Operations, 7th ed., National Safety Council, Chicago, Ill., 1974, p. 157.

Investigating Accidents in the Workplace: A Manual for Compliance Safety and Health Officers, U.S. Department of Labor, OSHA 228, 1977, pp. 1, 7–8, and 10.

Tiedt, Thao and Roger Kindley: "A Lawyer's Perspective on Accident Investigations," *Professional Safety,* August 1987, p. 11.

CHAPTER 6

MEDICAL AND HEALTH SURVEILLANCE SYSTEMS

Medical and health surveillance systems may range from the elaborate to the bare minimum required by OSHA. One establishment may have a full-time staff of physicians, nurses, and technicians; another may have only the required first-aid kit with a trained first-aid person, while others may fall within intermediate stages.

Regardless of the size of the program, the prime function of such a system is to:

1 Maintain the health of the work force.

2 Prevent or control occupational and nonoccupational diseases.

3 Prevent and reduce job-related disability and the resulting lost time incurred.

Since 1970, an increasingly large net of federal and state laws has imposed worker health responsibilities on industry. It requires industry to detect and eliminate health-damaging processes and products. Concern over the influence of corporations on worker health and safety has reached high proportions. Not too surprisingly, this area has become a prime target of federal and state regulations.

The union movement in the United States has basically concerned itself with wages and working conditions. But union philosophies are slowly changing. Today, unions also include in their negotiations with management such things as union-management safety committees, employee health and safety programs, and "rights to information." Bargaining between unions and management over safety and health issues has grown considerably. The labor movement was also one of the prime supporters of the passage of the Occupational Safety and Health Act of 1970.

OSHA has established requirements for medical monitoring in the workplace. There is a need for medical surveillance of the health of workers whose jobs may require both chemical and physical exposures to potentially hazardous concentrations of toxic materials. Medical surveillance, as required by OSHA, should be consistent with the risks of the exposure involved and designed to help monitor and protect the health of the workers.

HEALTH SURVEILLANCE SYSTEMS

Health surveillance systems are designed by monitoring information created from data gained from employee exposure, employee exposure records, and from company processes. Health surveillance systems within the work force can detect higher than normal incidences of occupational disease arising from a pattern of exposure or job function. Medical surveillance examinations consist of general tests to assess the general health of the worker and specialized tests to detect the early effects of specific exposures which an employee is known to have had. The creation of such a system is designed to include the following set of medical information.

1 Preemployment physical examination
2 Employee medical records
3 Employee exposure records
4 Periodic health examinations
5 First aid systems

Preemployment Physical Examinations

Physical examinations are designed to determine the general health of an individual. The preemployment physical examination is made to determine the physical condition of prospective employees in order to determine their ability to perform a job in accordance with their mental ability and physical capabilities. It will also help to determine if their disabilities, if any, will affect their personal efficiency, safety, and health, or the safety of others. The prospective worker should be able to meet the physical demands of the job. The physical examination should also detect any disease in its early stages. Job applicants with nondisabling and noncommunicable disorders can often be employed while being treated by their personal physicians. Prospective workers who will be exposed to noise levels above OSHA requirements should be examined for hearing acuity to determine prior hearing loss, if any, and periodically thereafter to detect any possible early hearing loss due to noise in the workplace. Finally, if the preemployment physical examination reveals that the prospective employee cannot meet the physical demands of the job, then at this point the company should notify the applicant that his or her services will not be required. A reasonable explanation for turning down the job applicant should also be given.

Employee Medical Records

Employee medical records are records containing the health status of an employee which are made or maintained by a company physician, nurse, or other health care personnel. These records would include the following information:

1 Medical and employment questionnaires or histories.

2 The results of medical examinations, including the preemployment physical examination, periodic health examinations, laboratory tests, X-ray examinations and all biological monitoring.

3 Medical opinions, diagnoses, progress notes, and physician's recommendations.

4 Descriptions of treatments.

5 Prescriptions.

6 Employee medical complaints.

Employee medical records do not include any of the following:

1 Blood specimens and urine samples which are routinely discarded as a part of normal medical practice and need not be required to be maintained by other medical or legal requirements.

2 Records concerning health insurance claims and workers' compensation claims if maintained separately from the employer's medical program.

3 Records containing voluntary employee assistance programs such as alcohol, drug abuse, or personal counseling programs, if maintained separately from the employer's medical program and its records.

Employee Exposure Records

Employee exposure records are records documenting an employee's exposure to a toxic substance or a harmful physical agent in or through any route of entry into the body such as inhalation, ingestion, skin contact, or absorption. This includes both past exposure and potential exposure (meaning accidental or possible), but does not include situations where the employer can demonstrate that the toxic agent is not used, handled, stored, or generated in the workplace.

Employee exposure records consist of the following:

1 Harmful physical agents (any chemical substance or biological agent such as bacteria, virus, and fungi).

2 Physical stresses (noise, heat, cold, vibration, repetitive motion, and ionizing and nonionizing radiation).

Preservation of Medical and Exposure Records by the Employer
Unless a specific occupational safety and health standard provides a different period of time, employers shall assure the preservation and retention of records as follows:

Employee medical records: Employee medical records shall be preserved and maintained by the employer for the duration of the employee's employment plus 30 years, except for health insurance claims records that are maintained separately from the employer's medical program.

Employee exposure records: Employee exposure records shall be preserved and maintained by the employer for at least 30 years, except for background data for workplace monitoring or measuring, such as laboratory reports and worksheets. These need only be retained for one year, provided that the sampling results and other background data relevant to the interpretation of the results obtained be retained for at least 30 years.

Access to Employee Medical and Exposure Records   Under OSHA standards, employers are required to allow access to an employee's medical and exposure records to employees, their designated representatives, and to OSHA. These records must pertain to the employee's working conditions and the workplace only.

Whenever an employee requests access to his or her medical records and the physician representing the employer believes that the information contained in the records shows a diagnosis of a terminal illness or a psychiatric illness that could be detrimental to the employee's health, the employer may inform the employee that access to the records will only be provided to a designated representative of the employee with specific written consent. "Specific written consent" means a written authorization containing the name and signature of the employee authorizing the release of the medical information, the name of the designated representative (individual or organization) that is authorized to receive the released information, and a general description of the medical information that is authorized to be released.

Transfer of Records   Whenever an employer will cease to do business, he or she shall:

1 Transfer all medical and exposure records to the successor employer. The successor employer shall receive and maintain these records.

2 Whenever an employer ceases to do business and there is no successor employer, the employer shall notify the affected employees of their rights of access to the records at least three months prior to the closing date of the business.

3 Or, the employer shall transfer the records to the Director of the National Institute for Occupational Safety and Health (NIOSH), if so required by a specific occupational safety and health standard.

Periodic Health Examinations

Periodic health examinations of all employees may be on a required or voluntary basis.

A required program should be applied to workers who are exposed to health-hazardous processes or materials, or whose work involves responsibility for the safety of others, such as vehicle operators. Substances like lead or carbon tetrachloride that are capable of causing occupational disease are usually subjected to process controls that will keep the workers safe from poisoning; however, caution dictates the advisability of periodic health examinations of such workers, to be certain that the engineering and hygiene controls are effective and continue to be so. This procedure also enables early detection of the hypersusceptible individual and the worker whose personal unsafe practices defeat the control measures.

Frequency of the examination must vary in accordance with the quality of the engineering control, the nature of the exposure, and the findings on each examination. Thus, exposures to some substances might justify examinations or laboratory tests on a weekly basis; others, on a monthly or quarterly basis, whereas annually or biannually may be adequate in some dust exposures.

In many cases, laboratory tests of blood or urine will suffice as the major portion of a periodic examination program, with complete examinations being made less frequently. The type of special examination (laboratory, X-ray, etc.) necessary for any exposure and the interpretation of the results are decisions requiring the most expert medical personnel.

Special Examinations

Many organizations make "return to work" physical examinations of employees who have been absent from work more than a specified number of days. This is done for two purposes: the control of communicable diseases and to determine the employee's suitability for return to work after a nonoccupational illness or injury. There is a wide difference in the effects of the same disease on different persons. One person will go to bed at the slightest sign of discomfort, whereas another worker must be totally overwhelmed before he or she will become absent from work.

First-Aid Systems

Definitions   The American Red Cross defines first aid as the immediate attention given to a person suffering from injury or illness before the services of a doctor can be obtained. OSHA defines first aid as the attention given to such injuries as cuts, minor scratches, and bruises that are so minor that a person would not ordinarily seek medical attention.

OSHA Requirements   OSHA standards require that employers be responsible for the availability of medical personnel in the absence of a physician, medical clinic, or a hospital within the near proximity of the workplace.

First-Aid Training   OSHA standards further require that the employer shall have a person trained in first aid in order to fulfill the above standards requirement. This person may be a supervisor, a current employee, or a person trained and hired specifically for that purpose. The permissible sources of training are the American Red Cross, United States Bureau of Mines, State or Federal OSHA Training Divisions, and the employer's insurance carrier.

First-aid kits shall be inspected and approved by a physician. The physician must provide a letter of approval to the employer.

Retention of Records   A record must be made of all first-aid cases. These records shall be maintained by the employer for 30 years beyond the termination date of the employee. It is important that the recording forms and filing systems be simplified to the extent that they can be interpreted by a physician or nurse.

Role of the Occupational Physician

The expectations of an occupational physician in an organization are based primarily upon the role given him or her by the employer. But in order to do the job effectively, the physician must have a thorough knowledge of what is manufactured in the plant, how it is made, what raw materials are used, the potential and actual health hazards associated with the manufacturing, and the physical requirements of the various types of jobs. The occupational physician must have this information so that he or she can carry out the preplacement health appraisals, periodic health examinations, and health education programs. Application of occupational health principles also helps to assure that workers are placed in jobs according to their physical capacities, mental abilities, and emotional makeup. Occupational safety and health programs are concerned with all aspects of a worker's health and safety and interface with the work environment. These programs cannot succeed in their role without the assistance of the occupational physician and without the full support of management. Furthermore, the physician or the medical director must be given enough authority so that workers will respect his or her judgment and follow instructions on personal health and safety.

The Occupational Health Nurse

Occupational health nursing today is a specialized branch of the nursing profession that requires expertise beyond that of the nursing profession itself.

In order to fulfill their duties as partners in the management health and safety team, nurses need to have a knowledge of such subjects as workers' compensation laws, health and safety laws, occupational disease, and a fair knowledge of the types of hazards inherent in the organization that they are working in.

Working with the plant physician, the occupational health nurse can provide a variety of nursing services, such as initial care for work related injuries and illnesses, counseling, health education, consultation about sanitary standards, and referrals to community health agencies. The nurse may also be required to participate in programs for evaluating employee health, such as health examinations, or programs for the prevention of disease, such as immunizations.

Planning for Medical Emergencies

The management safety and health safety team—the industrial physician, nurse, and safety professional—should, with the aid of top management, plan for the emergency handling of injured employees in the event of an emergency or a disaster, such as fire, explosion, power failure, or other catastrophe. This is called preplanning for emergencies. OSHA standards require that employers shall have a plan of action, should such an occurrence befall the plant or organization. Such a plan involves the selection by upper management of a coordinator or director who can provide a procedure for:

1 The selection and training of first-aid workers.

2 The transportation and care of injured workers within the plant.

3 Providing for the further evacuation to hospitals of the more seriously injured workers.

A company's chance of survival and recovery is greater when the knowledge and expertise of a management health and safety team is coordinated in the company's preplanning for emergencies.

SUBSTANCE ABUSE

One of the most serious problems facing industry today is the growing abuse of drugs and liquor. Very little work has been done to find the statistical evidence of the effect of drug abuse on occupational injuries. One reason is that users have been fearful of job dismissal. Another reason may be that accident investigators do not have the expertise to discover drug abuse at the workplace, thereby not associating it with an accident when one occurs.

Therefore, many companies have dealt with the drug problem through preemployment screening so as not to hire persons with drug problems or else through discharging them when drug addiction is discovered after employment. It appears that industry will have the burden of rehabilitating the drug and alcohol abuser. The prevalence of drug and alcohol abuse seems to be increasing at a high rate among young people—and it is these same young people who are the reservoir of this country's work force.

Drug abusers in industry are usually young adults ranging in age from age 18 to the early 30s. An interesting, but unproven concept is a phenomenon called "maturing out" which claims that most addicts are around 30 years of age or younger, and for reasons unknown, there appears to be a tapering off in drug usage after age 30.

A company's attitude toward workers impaired by drugs depends on the presence and type of health care coverage, and on top management's attitudes toward the subject of chemical dependency.

DRUGS AND DEPENDENCE

Drugs cause physical and emotional dependence. Users may develop an overwhelming craving for specific drugs, and their bodies may respond to the presence of drugs in ways that lead to increased drug use.

Regular users of drugs develop *tolerance,* a need to take larger doses to get the same initial effect. They may respond by combining drugs—

frequently with devastating results. Many drug users calling a national cocaine hotline report that they take other drugs just to counteract the unpleasant effects of cocaine.

Certain drugs, such as opiates and barbiturates, create *physical dependence*. With prolonged use, these drugs become a part of the body chemistry. When a regular user stops taking the drug, the body experiences the physiological trauma known as *withdrawal*.

Psychological dependence occurs when drug taking becomes the center of the user's life. Among drug users, psychological dependence erodes work performance and can destroy family ties, friendships, outside interests, values, and goals. In children, the child goes from taking drugs to feel good to taking them to keep from feeling bad. Over time, drug use itself heightens the bad feelings and can leave the user suicidal. More than half of all adolescent suicides are drug related.

Drugs and their harmful side effects can remain in the body long after use has stopped. The extent to which a drug is retained in the body depends on the drug's chemical composition, that is, whether or not it is fat-soluble. Fat-soluble drugs such as marijuana, phencyclide (PCP), and lysergic acid (LSD) seek out and settle in the fat tissues. As a result, they build up in the fatty part of the body, such as the brain. Such accumulations of drugs and their slow release over time may cause delayed effects (flashbacks) weeks, and even months, after drug use has stopped.

SPECIFIC DRUGS AND THEIR EFFECTS
Cannabis

All forms of cannabis have negative physical and mental effects. Several regularly observed physical effects of cannabis are a substantial increase in heart rate, bloodshot eyes, a dry mouth and throat, and increased appetite.

Use of cannabis may impair or reduce short-term memory and comprehension, alter sense of time, and reduce ability to perform tasks requiring concentration and coordination, such as driving a car. Research has shown that students do not retain knowledge when they are "high." Motivation and cognition may be altered, making the acquisition of new information difficult. Marijuana can also produce paranoia and psychosis.

Because users often inhale the unfiltered smoke deeply and then hold it in their lungs as long as possible, marijuana is damaging to the lungs and pulmonary systems. Marijuana smoke contains more cancer-causing agents than tobacco.

Long-term users of cannabis may develop psychological dependence and require more of the drug to get the same effect. The drug can then become the center of their lives.

Drug Type: *Marijuana*   Before 1960, the use of marijuana was restricted almost to certain subcultures, such as jazz musicians and artists in the big cities and towns. By 1960, however, college students had discovered marijuana, and since that time the number of marijuana users in the general public has increased to the point that it has become an over billion dollar business, annually, in the United States.

Called: Pot, Grass, Weed, Reefer, Dope, Mary Jane, Sinsemilla, Acapulco Gold, Thai-Sticks.

Looks Like: Dry parsley mixed with stems that may include seeds. Sinsemilla is a so-called top-grade marijuana plant which contains no seeds.

How Used: Can be eaten or smoked.

Drug Type: *Hashish*

Looks Like: Concentrated syrupy liquid varying in color from clear to black.

How Used: Can be smoked or mixed with tobacco.

Inhalants

The immediate negative effects of inhalants include nausea, sneezing, coughing, nosebleeds, fatigue, lack of coordination, and loss of appetite. Solvents and aerosol sprays also decrease the heart and respiratory rates, and also impair judgment. Amyl and butyl nitrate cause rapid pulse, headaches, and involuntary passing of urine and feces. Long-term use may result in hepatitis or brain hemorrhage.

Deeply inhaling the vapors, or using large amounts over a short period of time, may result in disorientation, violent behavior, unconsciousness, or death. High concentrations of inhalants can cause suffocation by displacing the oxygen in the lungs or by depressing the central nervous system to the point that breathing stops.

Long-term use of inhalants can also cause weight loss, fatigue, electrolyte imbalance, and muscle fatigue. Repeated sniffing of concentrated vapors over time can permanently damage the nervous system.

Drug Type: *Nitrous Oxide*

Called: Laughing gas, Whippets.

Looks Like: Propellant for whipped cream in aerosol spray can, or also a small 8-gram cylinder sold with a balloon or pipe (buzz bomb).

How Used: By inhaling vapors.

Drug Type: *Amyl Nitrite*
Called: Poppers, Snappers.
Looks Like: Clear yellowish liquid in ampules.
How Used: By inhaling vapors.

Drug Type: *Hydrocarbons*
Called: Solvents.
Looks Like: Cans of aerosol propellants, gasoline, paint thinner, or airplane glue.
How Used: By inhaling vapors.

Stimulants: Cocaine

The stimulants produce excitement; specifically, they can decrease fatigue, increase talkativeness and physical activity, enhance physical performance, produce a state of alertness, diminish appetite, and induce a longtime elevated mood, often leading to euphoria. Use of cocaine can cause death by disrupting the brain's control of the heart and respiration.

Cocaine stimulates the central nervous system. Its immediate effects include dilated pupils and elevation of blood pressure, heart rate, respiratory rate, and body temperature. Occasional use can cause a stuffy or runny nose, while chronic use can ulcerate the mucous membranes of the nose. Injecting cocaine with unsterile equipment can cause AIDS, hepatitis, and other diseases. Preparation of freebase, which involves the use of volatile substances, can result in death from fire or explosion. Cocaine can produce psychological and physical dependency, a feeling that the user cannot function without the drug. In addition, tolerance develops rapidly.

Crack or freebase rock is extremely addictive, and its effects are felt within 10 seconds. The physical effects include dilated pupils, increased pulse rate, elevated blood pressure, insomnia, loss of appetite, hallucination, paranoia, seizures, and eventually death.

Drug Type: *Cocaine*
Called: Coke, Snow, Flake, White, Blow, Nose Candy, Big C, Snowbirds, Lady.
Looks Like: White crystalline powder, often diluted with other ingredients.
How Used: Inhaled through the nasal passages, injected, or smoked.

Drug Type: *Crack*
Called: Crack, Rock, Freebase Rocks.
Looks Like: Light brown or beige pellets, or crystalline rocks that resemble coagulated soap—often packaged in small vials.
How Used: Smoked.

Other Stimulants

The abuse of amphetamines such as "speed," "meth," "uppers," and all of the other stimulants in this group is a very serious problem. A typical example is the respectable but overweight individual who begins taking "diet pills" to lose weight, but often continues to use them because the pills give him or her extra energy. Another example is the college student who relies on stimulants to stay awake all night to write a paper or "cram" for an exam. Although both of these people are misusing drugs, society usually ignores the problems of stimulant abusers it thinks respectable, but condemns as criminal someone on society's fringe, the "mainliner."

Stimulants can cause increased heart and respiratory rates, elevated blood pressure, dilated pupils, and decreased appetite. In addition, users may experience sweating, headache, blurred vision, dizziness, sleeplessness, and anxiety. Extremely high dosages can cause a rapid or irregular heartbeat, tremors, loss of coordination, and even physical collapse. An amphetamine injection creates a sudden increase in blood pressure that can result in stroke, very high fever, or heart failure. In addition to the physical effects, users report feeling restless, anxious, and moody. Higher doses intensify the effects. Persons who use large amounts of amphetamines over a long period of time can develop an amphetamine psychosis that includes hallucinations, delusions, and paranoia. These symptoms usually disappear when drug use ceases.

Drug Type: *Amphetamines*
Called: Speed, Uppers, Ups, Pep Pills, Copilots, Bumblebees, Hearts, Benzedrine, Dexedrine, Footballs, Bephetamines, Black Beauties.
Looks Like: Capsules, pills, or tablets.
How Used: Taken orally, injected, or inhaled through the nasal passages.

Depressants

The effects of depressants are in many ways similar to the effects of alcohol. Small amounts can produce calmness and relaxed muscles, but larger doses can cause slurred speech, staggering gait, and altered perception. Very large doses can cause respiratory depression, coma, and death. The combination of depressants and alcohol can multiply the effects of the drugs, thereby multiplying the risks.

The use of depressants can cause both physical and psychological dependence. Regular use over time may result in a tolerance to the drug, leading the user to increase the quantity consumed. When regular users stop taking large doses, they may develop withdrawal symptoms ranging from restlessness, insomnia, and anxiety to convulsions and death.

Babies born to mothers who abuse depressants during pregnancy may be physically dependent on the drugs and show withdrawal symptoms shortly after they are born. Birth defects and behavioral problems may also result.

Drug Type: *Barbiturates*
Called: Downers, Barbs, Amytal, Blue Devils, Yellows, Yellow Jackets, Red Devils, Tuinals, Seconal, Membutal.
Looks Like: Red, yellow, blue, or red and blue capsules.
How Used: Taken orally.

Drug Type: *Tranquilizers*
Called: Valium, Librium, Equanil, Miltown, Serax, Tranxene.
Looks Like: Tablets, or capsules.
How Used: Taken orally.

Hallucinogens

Hallucinogenic drugs are so called because one of their main effects is the production of hallucinations. The best known of these is LSD, a common street drug which is illegally manufactured, sold, and ingested. A panic reaction, in which users feel trapped in the experience and unable to get in touch with reality, is a common side effect and, if long lasting, can cause the person undergoing the "bad trip" to have a serious accident.

Phencyclidine (PCP) interrupts the functions of the neocortex, the section of the brain that controls the intellect and keeps instincts in check. Because the drug blocks pain receptors, violent PCP episodes may result in self-inflicted injuries.

The effects of PCP vary, but users frequently report a sense of distance and estrangement. Tine and body movements are slowed down. Muscular coordination worsens and senses are dulled. Speech is blocked and incoherent.

Chronic users of PCP report persistent memory problems and speech difficulties. Some of these effects may last 6 months to a year following prolonged daily use. Mood disorders—depression, anxiety, and violent behavior—also occur. In later stages of chronic use, users often exhibit paranoid and violent behavior and experience hallucinations. Large doses may produce convulsions and coma, heart and lung failure, or ruptured blood vessels in the brain.

Lysergic acid (LSD), mescaline, and psilocybin cause illusions and hallucinations. The physical effects may include dilated pupils, elevated body temperature, increased heart rate and blood pressure, loss of appetite, sleeplessness, and tremors.

Sensations and feelings may change rapidly. It is common to have a bad psychological reaction to LSD, mescaline, and psilocybin. The user may experience panic, confusion, suspicion, anxiety, and loss of control. Delayed effects, or flashbacks, can occur even after use has ceased.

Drug Type: *Phencyclidine*
Called: PCP, Angel Dust, Loveboat, Lovely, Hog, Killer Weed.
Looks Like: Liquid, capsules, white crystalline powder, or pills.
How Used: Taken orally, injected, and smoked. Can be sprayed on cigarettes, parsley, and marijuana.

Drug Type: *Lysergic Acid*
Called: LSD, Acid, Green Dragon, Red Dragon, White Lightning, Blue Heaven, Sugar Cubes, Microdot.
Looks Like: Brightly colored tablets, impregnated blotter paper, thin squares of gelatin or clear liquid.
How Used: Taken orally.

Narcotics

Narcotics initially produce a feeling of euphoria that often is followed by drowsiness, nausea, and vomiting. Users also may experience constricted pupils, watery eyes, and itching. An overdose may produce slow and shallow breathing, clammy skin, convulsions, coma, and possibly death.

Tolerance to narcotics develops rapidly and dependence is likely. The use of contaminated syringes may result in diseases such as AIDS, endocarditis, and hepatitis. Addiction in pregnant women can lead to premature, stillborn, or addicted infants who experience severe withdrawal symptoms.

Drug Type: *Heroin*
Called: Smack, Horse, Black Tar, Brown Sugar, Junk, Mud, Big H.
Looks Like: Powder, white to dark brown; or a tar-like substance.
How Used: Injected, inhaled through nasal passages, or smoked.

Drug Type: *Methadone*
Called: Dolophine, Methadose, Amidone.
Looks Like: A solution.
How Used: Taken orally or injected.

Drug Type: *Codeine*
Called: Codeine, Empirin compound with codeine, Tylenol with codeine, Codeine in cough medicines.
Looks Like: A dark liquid varying in thickness, or capsules, or tablets.
How Used: Taken orally or injected.

Drug Type: *Morphine*
Called: Pectoral syrup.
Looks Like: White crystals, hypodermic tablets, or injectable solutions.
How Used: Injected, taken orally, or smoked.

Drug Type: *Opium*
Called: Paregoric, Dover's Powder, Parepectolin.
Looks Like: Dark brown chunks, or powder.
How Used: Can be smoked or eaten.

Other Drugs

A number of drugs help fight fatigue, add strength, and create relaxation. Beta blockers, steroids, narcotics, and diuretics—used by various athletes—are now finding their way into the workplace. A drug which works for a laborer may not work for a person assembling fine, intricate computer parts. For example:

Beta Blockers Beta blockers reduce the heart rate and act like tranquilizers. They help reduce tension in high-pressure jobs.

Steroids Steroids provide power and strength. They are generally abused by some athletes because they can give the needed edge for weight lifters, wrestlers, those involved in basketball, volleyball, soccer, and track and field competition.

Diuretics Diuretics provide quick weight loss and could mean the difference for a worker taking a preemployment physical examination or for those whose work requires an annual physical examination.

ALCOHOL

Alcoholism is the most serious drug problem in the United States. Alcohol is a chemical depressant of the parts of the brain that suppress, control, and inhibit thoughts, feelings, and actions. It erodes the capacity to think and act responsibly. In large enough quantities, it is a general anesthetic, capable of producing coma and death. Problem drinking ranks with heart disease, cancer, and mental illness as one of the nation's four most serious health problems. As such, it is a matter of national concern.

The statistics and characteristics of alcoholism are known and, for the most part, accepted. These facts include:

1 One out of every 6 to 10 persons is an alcoholic.

2 At a conservative estimate, the disease costs industry $15 billion a year.

3 Alcoholism can hit anyone—from the fork lift operator on the plant floor to the plant manager.

4 With a company-sponsored alcohol abuse program to start the recovery and an aftercare program that includes a self-help program such as Alcoholics Anonymous, a job rehabilitation rate of 80 percent is achievable.

IMPLICATIONS FOR SAFETY MANAGEMENT

Injury prevention, operating efficiency, and general social responsibility are the major reasons for recovery programs for alcohol and drug abusers. Obviously, the larger companies are in a better position than small ones to develop substantial recovery programs. Even the smaller concerns, however, should attempt to identify workers in this category and arrange to refer them to agencies within their local communities. The prevalence of alcohol and drug abuse is increasing at a high rate among the young people who are the reservoir of the future work force. Statistical evidence of the influence of alcohol and drug abuse on occupational injuries is slow in materializing because users and their fellow workers have been fearful of legal retribution by civil authorities as well as job dismissal by employers.

The involvement of safety personnel in the organization's alcohol and drug abuse programs is a desirable reflection of management's concern for the health and welfare of its employees. By their training and expertise in studying the work habits of employees, safety personnel can generally recognize undesirable behavioral characteristics in people long before others can. This does not mean they are to supersede the line supervisor's right of control over his or her employees. As members of the organization's alcohol and drug abuse program committee, safety personnel can be of invaluable assistance in promoting its activities.

PREEMPLOYMENT DRUG TESTING

Surprise drug testing has been routine for the military since 1982. More than one-fourth of the nation's largest corporations, including IBM,

DuPont, Exxon, and Federal Express test job applicants for drug abuse, and more companies are considering doing so. Companies that employ people in jobs where drug-impaired judgment could result in lost lives—such as airline pilots and train operators—conduct periodic unannounced drug testing. Mandatory drug testing is required by the Department of Transportation for operators of public vehicles. All companies should have some form of an employee assistance program, called EAP. If the company is too small to afford an in-house EAP, it can contract with outside agencies such as mental health centers or hospitals.

A PLAN OF ACTION

In order to help combat employee drug use most effectively, companies must become involved with the schools, social service agencies, and the union. All must transmit the message that drug abuse is wrong, dangerous, and will not be tolerated. Companies should have a written policy on drug abuse and ensure that managers and employees alike understand its message. Job performance of all workers and managers should be monitored, so that warning signs of drug abuse can be detected early and treated.

Employers can go a step further by providing support to the local schools' prevention programs. They can carry the message to the schools about the effects of drug use on employment. They can further help schools obtain curriculum materials for drug prevention programs and can even provide incentives in drug prevention programs. Employers must demonstrate to students that they will not hire prospective employees who are not drug free, and that they will not tolerate drug use after employment.

Industry must take a strong stand with schools. Schools are the source of future labor supply. In public and private schools, drugs and alcohol run rampant. School principals should not expel students for drug abuse. Every school should have a rehabilitation program in order to combat drug abuse before it can get to the workplace.

A Substance Abuse Policy

Written policy on substance abuse: This policy should outline the company's position on both drug and alcohol abuse and should stress the policy's importance to safety and health in the workplace. It should be communicated to all employees and members of management.

Company's position on drugs and alcohol in the workplace: The use, possession, sale, and distribution of drugs or other controlled substances for nonmedical reasons are prohibited in the workplace. The use, possession, sale, or distribution of alcohol without management authorization is also prohibited in the workplace.

Condition of employment: The policy should give employees and managers alike the choice of seeking help or else being subject to disciplinary action up to and including discharge.

Mandatory random drug testing: The policy should outline in detail the company's position on periodic mandatory drug testing. It should also inform employees that they may be subject to substance testing upon the request of a supervisor.

Defining the problem through consultation: It is the policy of the company to help the employee who has a substance abuse problem by providing consultation and opportunities for rehabilitation. Consultation can help define the problem, answer questions about company policy, and determine what options are available for treatment. Without professional treatment, the condition may get worse and result in disciplinary action up to and including discharge.

The guidelines presented in this book should be modified to fit the needs of a particular organization.

Role of the Safety Department

The role of the safety department in the alcohol and drug abuse programs of an organization is twofold. First, it is important that the members of the safety department familiarize themselves with the symptoms of drug and alcohol abuse. The safety department should be involved in the organization's policy for dealing with employees suspected of alcohol or drug abuse.

Second, the safety department, because of its extensive knowledge of job procedure and plant operations as a whole, can in its plant surveys and inspections be another watchful eye for management. Those supervising the suspected employee(s) can be alerted.

The handling of questionable employee behavior requires the close cooperation of the line supervisor and safety, medical, and personnel departments.

Employee Assistance Programs

Employee assistance programs, a concept growing in popularity in both business and industry, are designed to help an employee whose problems—alcoholism, substance abuse, marital problems, or financial difficulty—are interfering with safe job performance. Without the proper attention, these problems usually become worse and the consequences are often unpleasant and expensive for employer and employee alike.

EAPs are designed to provide confidential, professional assistance to help employees resolve personal problems having an undesirable impact on an individual's life. The programs are voluntary—allowing the employee to seek help on his or her own before a problem becomes a crisis situation. EAPs provide benefits to the employer because they generally reduce insurance costs, absenteeism, and job-related accidents and increase job performance.

Many large companies with 3,000 or more employees can usually afford an in-house EAP program. Smaller businesses, however, generally contract with private EAP operators for the service. Today, because of the great concern about substance and alcohol abuse in the workplace and its consequences, more and more companies are considering the EAP concept. Many employers feel that providing such a company-paid benefit for employees and their families is the best thing to do for all parties involved.

Perhaps the soundest justification for Employee Assistance Programs, other than their positive impact on people with personal problems, has been their ability to weld seemingly diverse interests toward a common purpose. This ability to foster positive management/labor relationships promotes problem resolution, maintains the employee's dignity and confidentiality, and provides a return of investment as an enviable accomplishment.

WORD POWER

emergency   An unexpected, serious occurrence that demands immediate action.
communicable disease   A disease which may be transmitted directly or indirectly from one individual to another.
drug abuse   The use, usually by self-administration, of any drug in a manner that deviates from the approved patterns within a given culture.
drug addiction   A condition caused by excessive or continual use of habit-forming drugs.
EAP   Employee Assistance Program.
alcoholism   A personality disorder in which a person is unable to control drinking of alcoholic beverages.
anesthetic   A substance that produces insensibility or lack of sensation.
delirium   A state of temporary mental disorder characterized by confusion, delusions, hallucinations, illusions, and restlessness.
depressant   A substance that causes a lowering of the body's vital activities.
illusion   A distorted or deceiving perception of a physically present object; that which produces a false impression of belief.

QUESTIONS AND EXERCISES

1 What is the difference between employee medical records and employee exposure records?
2 Are periodic health examinations required under the OSHA Act?

3 Must the employer maintain doctors and nurses for employees on the organization's premises?

4 How long must the employer preserve and maintain employee medical records and employee exposure records?

5 May a member of an employee's family seek access to the employee's medical records at the workplace?

6 What must the employer do with employees' medical and exposure records when the employer ceases to do business?

7 What is a "special examination?"

8 What are the OSHA requirements for first-aid training?

9 What are the OSHA requirements for first-aid kits?

10 For what length of time must the employer maintain and retain first-aid records?

11 For class discussion: What can industry do to eliminate drug abuse in the workplace?

12 Do you feel that preemployment drug screening can help eliminate drug abuse in the workplace? Discuss your reasons with the class.

13 What is an employee assistance program?

14 Design a Substance Abuse Policy for a small organization employing 250 employees.

15 Does drug abuse cause accidents? How?

16 In response to a growing public outcry for action, can you recognize the value of a drug-free society—including the workplace? Elaborate.

17 The general feeling toward drugs (including marijuana) is that current laws against the sale and use of them are not strict enough. What is your feeling about this attitude? Elaborate.

18 What strategies can you recommend for making the workplace free of drugs?

19 What are your thoughts on drug testing at the workplace?

20 Do you believe that drug abuse education should begin at the elementary school level? Why?

21 Would a drug and alcohol abuse program at work have any value if it did not have the backing of top management?

22 What guidelines would you establish for the control of alcohol and drugs in the workplace?

BIBLIOGRAPHY

Accident Prevention Manual for Industrial Operations, 7th ed., National Safety Council, Chicago, Ill., 1974, p. 553.

"Alcohol Abuse in Industry," *National Safety News,* April 1981, p. 42.

Karr, Bruce W.: "Proposed Health Standards—Medical Monitoring," *National Safety News,* August 1976, p. 68.

McElroy, Frank E.: *Accident Prevention Manual for Industrial Operations—Administration and Programs,* 8th ed., National Safety Council, Chicago, Il., 1974, p. 569.

"Occupational Safety and Health—A Matter of Teamwork," *National Safety News,* 1978, p. 51.

"What Works: Schools Without Drugs," U.S. Department of Education, 1986, p. 6.

PSYCHOLOGICAL ASPECTS OF SAFE PERFORMANCE

ATTITUDES

One of the crucial factors known to affect safe performance is mental attitude. It is the product of education, life experiences, social contacts, and the working environment. No wonder that industry spends much time and money studying and determining the attitudes of its workers.

Some theories regard attitude as one's disposition to evaluate objects, persons, or situations either favorably or unfavorably. Sometimes those responses may be dependent, in part, upon the individual's previous experiences. For example, seeing someone in uniform—such as a police officer or a person in the armed forces—may cause certain attitudinal reactions, depending upon the individual's previous experiences with police officers or with military personnel. Certain words also create different responses in people. For example, the words "father," "supervisor," and "safety" may touch off different responses in people—again, depending upon the kinds of previous experiences associated with these words.

Family experiences, childhood, schools, church, and the experience of growing up, all provide a variety of stimuli that influence a person's attitudes. Now it can be seen why attitude plays such an important role in everyday relationships.

Many people have had experiences that were associated with fear, sorrow, pain, or happiness. All of these will tend to make them react the same way to anything similar to the original emotion-provoking situation.

Although there is only one positive emotion listed among those just mentioned, one cannot conclude that most attitudes are negative. There are many positive ones.

Many years ago, the writer John Donne was correct in saying that no man is an island unto himself. As old as those words are, their message is as meaningful today as when they were written. We live together and work together. No one lives or works entirely alone. We are involved with people around us, affected by their accomplishments, marked by their failures. If a worker fails, the one beside him or her fails, too. Both share the burden of that loss. One of the most disturbing penalties paid when an accident occurs occurs when a person realizes that because of his or her own failure, someone else was injured. It isn't easy to live with that feeling. By the same token, it isn't easy for us to live with the feeling that someone else has failed us.

Safety is something that management shouldn't have to sell. But management has to keep selling it because employees keep getting hurt. Promotion by management of increased motivation to use safe practices will lessen the number of failures caused by the untrained, the unskilled, and the chance-takers who accept risk unnecessarily. Nobody wants to be a failure, and few of us are without pride. There is embarrassment in failure, but that usually is the least serious aspect of the matter involved.

It is believed that accidents are conceived in improper attitudes and born in moments of actions without thought. Accidents will cease only when proper attitude is strong enough to prevent the act...when the right attitude creates the awareness that controls the act. One of the solutions to accident prevention is personal responsibility. The philosophy of personal responsibility is nothing new. It is an essential part of the American way of life, as is the work ethic. Personal responsibility and the work ethic go hand in hand.

THE WORK ETHIC

The work ethic of the past has fallen short as a source of our industrial strength. Employers complain bitterly about absenteeism, accidents, poor workmanship, and employee theft. We seem to want more while we give less of ourselves to get what we want. Coffee breaks and lunch hours stretch out. We call in sick when we want a day off. When the supervisor gives us a task to complete we tell him, "that's not my job."

People today are not learning sound work habits and appreciating the work ethic. Employers complain that job applicants appear in sloppy clothes and with little regard for the importance of punctuality and courtesy. Many applicants lack a sense of responsibility and a willingness to throw themselves into a job. The Scriptures tell us: "Whatever your hand finds to do, do it with all your might." How can people today fulfill this Biblical injunction? Many of them don't know how to work. No one taught them how to work or instilled sound work habits into them. In times past, youngsters learned to work at a very young age. They learned the work habit and it stayed with them for years. They did many things to help the family make ends meet. They peddled newspapers, sold broken glass and cardboard to the junk man, cut the neighbors' grass, and shoveled snow off the sidewalks. They did anything they could to earn a nickel or a dime. I could go on and on with a long litany describing how youngsters learned to work. Banking the coal furnace at night, cleaning out the clinkers in the morning, and carrying out the ashes were among the most hated jobs that youngsters talked about in earlier days.

What does this have to do with safety? Good work habits instill pride of workmanship in the individual. They also foster self-respect and respect for others, including the employer. Good work habits go hand-in-hand with safe work habits, a characteristic that all employers wish to see in their employees. People with sound, safe work habits are the foundation of a good safety program. Revival of the work ethic could do much to renew the safety and productivity of the workplace.

HUMAN BEHAVIOR

There are relatively few tasks a person can do without some type of tool, machine, or other equipment at the workplace. The person usually manipulates the tool or machine in such a way that together they accomplish a task. This combination is called a "man-machine system." The term "man" refers to the worker and the "machine" in its broadest sense refers to any tool or device used to accomplish a task.

The reliability of the man-machine system depends on the reliability of both the person and the machine or tool. Both are prone to breakdown or failure. The physical abilities of people have been studied and can be predicted fairly also.

But, prediction of behavior is not as easy. Certain human characteristics are fairly consistent, but there are so many variables that affect human behavior that prediction is difficult. Psychologists have developed several theories of human behavior that have been utilized by management. Two of these theories are particularly appropriate to human behavior in the workplace.

Douglas McGregor, one of the earliest and most prominent advocates of the modern approaches to management, proposed two concepts of worker motivation. His "Theory X" assumes that workers in industry are not motivated by any satisfaction they might derive from doing a good job, but rather only by the pay or other rewards and by the fear of disciplinary action. If management wants the worker to perform in a safe manner, it must provide some tangible reward for doing so, and some punishment for failing to do so.

McGregor's "Theory Y" assumes that the worker is, or can be, motivated by job satisfaction. In this case, management's task is to find ways to make the job satisfactory to the worker. Studies have indicated that most workers are motivated to some degree by job satisfaction, but that they also need some external incentive to achieve what is expected of them. This raises the question "Has management done a good enough job in designing the task and working conditions to motivate the worker as much as it can?" This should be continually evaluated.

Abraham Maslow, a clinical psychologist, has proposed another widely accepted theory. He contends that there is a hierarchy of human needs. Starting at the lowest level, a person concentrates on satisfying those needs, and only when this has been done will he or she move on to the next level. Figure 7-1 shows Maslow's need hierarchy. This diagram might suggest two approaches. First, management might try to place a worker in the type of task best suited to his or her level of need. Sec-

FIGURE 7-1
Maslow's need hierarchy.

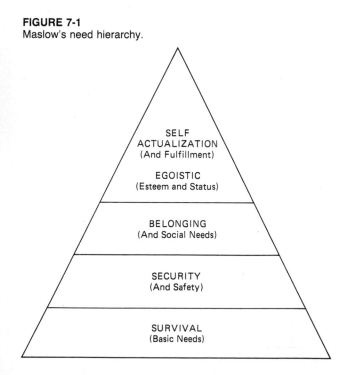

ond, management might try to provide conditions and opportunities that would encourage the worker to ascend to the next level of need.

These theories and concepts may not always explain a worker's response to a given situation, but they help. They become pieces of the jigsaw puzzle of human behavior. The more pieces we can fit into the puzzle, the better we are able to design tools, machines, and a safer environment to work in.

Another approach to behavior focuses on the development of attitudes in the mind of the worker. Each of us is the product of all of the experiences we have had in our lives. The attitudes we developed in childhood remain with us for a long time, probably forever. But attitudes can be changed if a sufficiently strong incentive is provided. Attitudes vary with situations, as well; we do not always hold the same attitudes toward all people, all jobs, or all conditions. A worker can have good feelings about the world in general. Those things that affect needs will have the greatest influence on attitude.

In the work environment, the factors that most affect the worker's attitudes—such as working conditions, tool and equipment design, and plant layout—can generally be controlled or influenced by management. Management's attitude toward the worker's welfare can also play a major role in shaping attitudes. Nothing is more important than attitude in determining behavior and safe performance in the workplace.

Human behavior and attitudes toward the work ethic are different today than they were thirty years ago. The world has become a more complicated place for young people to live in. The traditional form of family life is disappearing, drugs have become more prevalent, and moral values are shifting. Schools today are expected to do more than they have traditionally done in the past. Young people's problems can usually result in inefficient school work and inefficient ways to deal with the real world of working to make a living. Problems of competition and poor self-esteem are accelerating as the family structure changes.

HUMAN BEHAVIOR AND SAFETY

Studies of attitude-formation and attitude changes reveal that there is no 100 percent successful method for changing people's attitudes. One of the least effective ways is to try to convince somebody or a group that he or they should change their attitude. What causes a person to commit an unsafe act and subsequently become injured? Many people would answer that the cause is "poor attitude." This is only partially true. There are many underlying factors behind a poor attitude that lead to committing the unsafe act. Among the outstanding ones are:

Anger
Anxiety
Lack of skill
Physical limitations
Mental limitations
Poor work habits
Unawareness

These factors have been brought to the workplace by the worker because of his or her psychological, physiological, and genetic makeup. There are also other conditions considered factors leading to poor attitude. These factors are found in the workplace. They are a part of, or products of, the manufacturing process itself. They are:

Poorly designed machines and equipment
Unsafe machines and equipment
Poor ventilation and illumination
Poor housekeeping
Poor working conditions
Management apathy
Management-union labor problems

MODIFYING UNSAFE BEHAVIOR

It is well known that all behavior cannot be changed. But behavioral characteristics can be modified under the right conditions and proper environment. If there is a change in the workplace environment, whether good or bad, the worker will most likely psychologically or physiologically respond to that change in a like manner. If there is a change in management style, whether good or bad, this too could provoke a change. The response will most likely appear in the form of attitudes. Management must set the stage for creating attitudinal changes when trying to modify unsafe behavior.

Training employees to work safely is the first step in supervising behavior modification and the formation of acceptable attitudes. While safety training teaches specific skills to workers—often based on a step-by-step job safety analysis of particular duties—it also deals with making employees aware of a wide range of safety information conducive to their own welfare. Considerable research by both government agencies and academic researchers concerning the benefits of safety training has consistently shown that the key to modifying unsafe behavior and changing attitudes is continuous safety training.

An employer can take every precaution, obey every safety regulation, and correct all hazards; yet, many employees will take safety for granted.

The employer can modify this type of behavior through constant and effective safety training. Safety training programs must be designed and executed with the same importance as production or quality control activities.

Much about safety training is changing. A field that used to rely on lectures backed up by safety film has now turned to modern management techniques, featuring employee involvement, in-house produced videos, and computer-generated graphics for illustrating training workbooks to change the entire look of safety training. After employees are taught why they must work safely, supervisors must then motivate the right conduct by reinforcing acceptable behavior and correcting unsafe behavior.

TROUBLED WORKERS: WHO ARE THEY?

A troubled worker is anyone whose personal problems interfere with job performance. While anyone may experience personal difficulties at one time or another, it is at the point where personal problems interfere significantly with job performance that a worker becomes a "troubled employee." Although we face many stresses from day to day in our lives, there are some events which become unbearable and often unmanageable. Safety experts agree that emotional upset plays a significant role in accident causation. As is often the case, emotional upset can have a disturbing influence upon a person's behavior. Unfortunately, this disturbing factor in a person's behavior can be spread to others working in the same situation and create an atmosphere not conducive to safe operations. Troubled workers may do this by unpredictable behavior, by "acting out" inner problems, or by taking out real or imaginary feelings on other employees. These stressors, as they are called, may arise out of personal events or from job-related conditions. Often these stressors may be a combination of both personal and job factors.

Examples of personal life stressors can include any of the following:

Death of a spouse
Divorce
Marital separation
Family member illness
Mortgage on home
Pregnancy
Jail sentence
Alcohol or drug addiction
Financial problems

Examples of job-related stressors can include any of the following:

Changes in work responsibility

Problems with supervisor
Problems with fellow employees
Job transfer
Possibilities of a lay-off
Noise
Temperature, humidity, or air quality
Swing shifts
Job task
Job fatigue

MANAGEMENT CONCERNS

Why should management be concerned about troubled employees? A large segment of our population is experiencing family problems, financial problems, and health problems that create significant levels of stress, but is not seeking any form of relief or treatment. Losses of millions of dollars a year are attributed to work time lost because of these problems. These losses create the following spectrum of problems for management:

Safety problems
Workers' Compensation costs
Absenteeism
Labor turnover
Loss of productivity
Down-time
Low morale

An automobile worker forgets to check an unguarded gear on his production line and is subsequently injured because he failed to heed a co-worker's warning. A warehouse forklift driver damages $50,000 worth of packaged products because he failed to check his brakes before the start of his shift. What were these employees thinking about?

It appears that the problems of troubled employees are becoming an economic timebomb to employers. These problems are threatening the quality of America's workforce.

The front-line supervisor is the point of management control for managing troubled employees. No one else is as familiar with the employees' work habits and most probably their problems as well. Generally, the five most significant keys for detecting problem workers are:

- Their attitudes
- Their behavior
- Their attendance record
- Their safety record

MAINTAINING DISCIPLINE

Many people think of discipline in terms of scolding, reprimanding, or punishing. True discipline involves assisting, guiding, and training people. It should encourage self-development and self-control.

Why don't employees do what they are supposed to do? Why do they commit unsafe acts? There are many reasons. The majority of reasons do not call for punishment; they call for better management and supervisory leadership. The leadership approach relies on coaching, training, and mutual problem-solving between the supervisor and the employee. It is a continuing process.

THE CORRECTIVE INTERVIEW

If and when all else has failed, then it is time for the supervisor to hold a corrective interview with the employee. The purpose of the corrective interview is to determine what really happened and how it might have been prevented, and to analyze the situation to determine if disciplinary action is necessary. The supervisor should hold the corrective interview in the following manner:

1 Hold the interview in privacy.
2 Maintain control of the interview.
3 Don't put the employee on the defensive.
4 Allow the employee to explain the reason for his or her actions.
5 Take the time to get all of the facts.
6 Listen carefully to the employee's explanation.
7 If disciplinary action is necessary, deal promptly and fairly with all violations.
8 Make the corrective action fit the violation and the circumstances.
9 Follow company policy relating to disciplinary action.

Two important questions that supervisors should ask themselves after the corrective interview are, "What could I have done to prevent this offense?" and "What can I learn from this case to help me prevent similar violations in the future?"

ACCEPTANCE OF RISK

Risk is an expression of exposure to a hazard. The acceptance of a risk depends on the benefits that can be derived by the risk-taker. For example, a thing or condition is safe if its risks are judged to be acceptable, and judging the acceptability of that risk is a matter of personal and social value. Perhaps it can be said that safety is not measurable, but that risks are measured. It can be further said that measuring risk is a measure of probability of severity of harm.

Risks are categorized as being voluntary or involuntary. A voluntary risk is one that is freely accepted by an individual on the basis of the individual's own values and experiences. Driving an automobile while not wearing a seat belt is an example of a voluntary risk.

An involuntary risk is one to which an individual subjects himself or herself as the result of relying on another person's judgment. For example, a passenger on a bus, train, or airliner accepts risks over which he or she has no control (other than the decision to take the bus, train, or airliner) and relies on the judgment of the operator. In analyzing the acceptability of risk by the public, it appears that the public is willing to accept voluntary risks more readily than involuntary risk. The same analysis applies in the workplace.

What is it that influences society's acceptance of personal risk?

1 The benefits derived from taking the risk
2 Personal experience
3 Public opinion
4 News media
5 Advertising, usefulness, and the number of participants
6 The feeling that many people have: "it can't happen to me."

ACCIDENT PRONENESS: IS THERE SUCH A THING?

The factors that determine why a person might be "accident prone" have been the subject of almost endless discussion, and aroused a great deal of controversy, much of which persists. Many research studies have been made in an effort to appraise these factors, and although much has been learned, there is far from general agreement about the validity of some of these studies. Attempts to determine just what patterns of personality factors create accident-proneness have not been revelatory. For a while, it was thought to be a personality trait, but studies have not supported that belief.

The term "accident-prone" is generally used to describe or identify a worker who has recurring injuries in the workplace. This generalization is itself subject to question. Furthermore, to label a person as accident-prone would, in all probability, discourage any effort to seek the causative factors of accidents relating to that person.

A term now scientifically acceptable to identify the person whose record shows recurring injuries, first aid cases included, would be "injury repeater." This term describes the person's injury record and avoids the connotation that a person is constantly prone to having accidents. Although there are workers who have more accidents than other workers, these persons are involved in a relatively small percentage of all accidents.

People who might be considered as accident repeaters (based on their safety records) are generally those who have a deficiency, either permanent or short-lived. Such deficiencies might be:

1 A lack of aptitude for the work
2 A lack of certain skills and coordination
3 A possible literacy problem
4 An attitude or personality problem
5 Alcohol or drug problems
6 Personal life stresses (various stresses in a person's life can lead to accidents, heart problems, and other illnesses—see Table 7-1)

TABLE 7-1
PERSONAL LIFE STRESSES

Death of spouse or a child
Divorce
Serious marital separation
Loss of job by breadwinner
Serious personal injury or illness
Long-term jail sentence
Marital reconciliation
Marriage
Retirement
Pregnancy
Sex difficulties
Extreme change in financial status
High home mortgage payments
Foreclosure of home mortgage
Added or change in work responsibilities
Wife stops working
Son or daughter leaves home
Change in work hours or conditions
Minor violations of the law
Change in social or recreational activities
Change in residence
Change in schools
Failing grades in school
Change in eating or sleeping habits
Expenses for vacation or Christmas

Note: These stressors do not necessarily rank themselves in the order shown here for all persons. While one stressor can be a very threatening situation for one person, it may not be as serious for another. Obviously, there are many more stressors than are shown on this list.

ILLITERACY IN THE WORKPLACE

The erosion of worker literacy comes at a time when America's workforce is engaged in competing, in a global economy, for its share of the hi-tech business of the coming century. A substantial fraction of American workers lack the reading and writing skills they need just to find a job. This fact coincides with the demand for workers who can not only read and write, but also use a computer, understand technological concepts, and think independently. Today, a majority of all new jobs will require postsecondary education.

Illiterate employees hurt the employer, as well as themselves, by endangering workplace safety. Today, employees are required to read caution signs, safety instructions, safety procedures, and job instructions. They are also required to read warning labels and hazardous materials information in order to survive on the job. In one case, a feedlot employee killed a pen full of cattle by feeding them poison instead of grain. In another case, employees misordered millions of dollars in equipment parts. There are probably thousands of documented and undocumented cases similar to these.

Personnel managers in many organizations are now testing the education level of all prospective employees. They're not assuming that someone who has a high school diploma and is job hunting can walk in and do the job. The tests not only determine if a prospective employee is able to read and write, but are also designed to foretell if an applicant is promotable to a job requiring greater skills.

Many large companies have organized classes to upgrade worker education by providing basic instruction in reading and writing. Others have contributed money to community efforts for the same purpose. The Literacy Volunteers of America are also doing a great job.

WORD POWER

attitude   An enduring predisposition to respond consistently to given objects, events, issues, and the like.

behavior   The actions, activities, responses, or movements of the individual before or toward others.

behavior pattern   A recurrent way of acting by an individual or group toward a given object or in a given situation.

competence   The ability to perform an action.

habit   An acquired and consistently manifested behavior; a learned act that has become automatic through constant repetition.

incentive   A reward that stimulates or maintains goal-directed behavior.

learning   An enduring change in knowledge or behavior resulting from training, experience, or study; the process that brings about such a change.

motive   A stable personality trait that consists of a tendency to perform certain actions or to seek certain goals.

industrial psychology   The branch of psychology concerned with work, personnel problems, individual behavior, etc., in business and industry.

QUESTIONS AND EXERCISES

1 Name five serious stresses in a person's life that could cause him or her to be less alert on the job and subsequently become injured as the result of an accident.

2 Name five conditions in the work environment that could lead to emotional stress and eventually to a job-related accident.

3 What do you think of the use of the expression "accident-prone?" Is there such a condition? Elaborate.

4 Discuss "emotional states" and injuries.

5 What is your philosophy of the work ethic?

6 Compare some of the capabilities of human beings and machines.

7 Other than what was described in this chapter, can you think of any methods for modifying unsafe behavior? What about the use of disciplinary action when necessary?

8 Name some attitudes on the part of the supervisor that would seriously deter the effects of the organization's safety program.

9 An eager well-motivated person who is undistracted by personal problems can outperform a distracted or poorly motivated person, other things being equal. Why? Elaborate.

10 The traditional family appears to be disappearing, moral values are shifting, and drugs have become prevalent. Why? Elaborate.

CASE STUDY

You, the safety manager, have just received a report that John Smith, a machinist, has lost a finger while operating a milling machine in the machine shop.

His supervisor states that Smith was careless, that he is accident-prone, and that he should be discharged.

Smith's record shows that he has been on his present job for two weeks. He has been with the company for two and one-half years. Three months ago, he was off work for four days with a badly lacerated wrist as a result of an accident while operating a pedestal grinder. A year earlier, he suffered a back injury while he was helping to install a die in a power press. "He constantly goes for first aid for minor cuts, bruises, and scratches," said his supervisor.

Do you agree with the supervisor? How would you investigate John Smith's case relative to possible causes?

Discuss some of the possibilities of this case with the class. Would retraining be a factor in this case? Would disciplinary action be a factor?

BIBLIOGRAPHY

Hammer, Willie: *Occupational Safety Management and Engineering,* 3d ed., Prentice-Hall, Englewood Cliffs, N.J., 1985, p. 189.

Marshall, Gilbert A.: *Safety Engineering,* PWS-Kent Publishing Co., Boston, Mass., 1982, p. 57.

McElroy, Frank E.: *Accident Prevention Manual for Industrial Operations, Administration and Programs,* 8th ed., National Safety Council, Chicago, Ill., 1981, p. 315.

"Talk Topics," *A Supervisor's Guide for Safety Discussions,* National Safety Council, Chicago, Ill., p. 19.

CHAPTER

FUNDAMENTALS OF INDUSTRIAL HYGIENE

Correcting occupational health hazards is the responsibility of management because such hazards can cause legal compensable illnesses, and can also impair employees' health to the extent that time is lost from the job or employees work at less than full capacity. Industrial hygiene concerns itself with the control of occupational diseases that arise out of and during the course of employment. According to the American Hygiene Association, industrial hygiene is "that science and art devoted to recognition, evaluation and control of those environmental factors or stresses arising in or from the workplace, which may cause sickness, impaired health and well-being, or significant discomfort or inefficiency among the workers of the community."

Although safety engineers are likely to be primarily interested in preventing physical damage resulting from accidents—broken bones and bloody injuries—they should also become involved in preventing work related diseases. Diseases can be considered as injuries affecting the lungs, skin, kidneys, liver, and brain, and other organs of the body. Safety engineers do not have the background and training to perform the functions of industrial hygienists, but they should learn to recognize problems in this area, and should know when and where to get help. When engaging the services of industrial hygienists, safety engineers should be able to see that the recommendations made by the hygienist are carried out.

Much has been said and published on this subject, but it has often been presented in specialized, technical, or scientific language that either is meaningless to most persons or creates a good deal of misunderstanding. Also, considerable confusion has arisen from the conflicting opinions that have been widely expressed. This chapter is intended to help clarify the subject of industrial hygiene.

There are three key concepts which must be applied in an effective program of industrial hygiene:

1 *Recognition:* Knowledge of stresses arising out of industrial operations and processes.

2 *Evaluation:* A judgment or decision involving measurement of stress and based on past experience.

3 *Control:* Isolation, substitution, change of process, wet methods, local exhaust ventilation, general or dilution ventilation, personal protective equipment, housekeeping, and training and education.

RECOGNITION

Types of Stresses

Stresses may take the following forms.

Chemical Chemical stresses are created by liquids, gases, dusts, fumes, mists, and vapors in the form of air contaminants and skin irritants. The following definitions are those of the American Standards Institute, and should be noted well (they are very common in industrial hygiene):

Fume: A substance composed of solid particles formed by condensation from a gaseous state; these particles are microscopically small (odorous gases and vapors are not fumes).

Gas: A substance that will diffuse to evenly occupy the space in which it is enclosed. A gas does not appear in the solid state or liquid state at standard temperature and pressure.

Vapor: A gaseous form of a substance that is normally a liquid or solid.

Mist: A suspension in air of very small drops, usually formed by mechanical means (atomization) or by condensation from the gaseous state.

Dust: A substance consisting of solid particles that have been reduced to a small size by some mechanical process.

Physical Physical stresses are created by electromagnetic and ionizing radiation, noise, vibration, and extremes of temperature and pressure. They are defined professionally as follows:

1 *Noise:* unwanted sound (this is covered in detail in Chapter 13)
2 *Temperature:* either high or low extremes
3 *Illumination:* level of intensity
4 *Vibration:* motion condition
5 *Radiation (ionizing):* cell-damaging
6 *Radiation (electromagnetic):* heat-producing
7 *Pressure:* atmospheric, either high or low

Biological Insects, molds, fungi, and bacteria create biological stresses.

Ergonomic Monotony (repetitive motion) and work pressure (such as fatigue) create ergonomic stresses. Monotony or repeated motion means the performing of the same task over and over again. Work pressure can be, in addition to fatigue, problems such as worry and the inability to live up to a standard of performance.

Exposure

"Exposure" is defined as the entering into the body of a health hazard. Such a hazard can affect the nervous system by entering through the eyes, ears, and breathing area of the mouth and nose, or it can be absorbed into the system through the skin on contact and possibly find its way into the digestive tract or other organs of the body.

Except for skin diseases, most occupational diseases are acquired by inhalation. Certain chemical agents that reach the lungs can pass into the blood stream and over a long period of time can be absorbed into various other parts of the body. Other agents may stay in the lungs and cause damage in this organ only. Lung tissue is by far the most efficient medium the body possesses for absorbing materials; and the surface of lung tissue in the human body averages 55 square meters, or about 590 square feet.

Toxic or irritant dusts can be ingested in small amounts and may eventually cause damage in some form or other. Toxic materials can be absorbed in the digestive system and then picked up by the blood if they are not eliminated through the intestinal tract.

Another way in which toxic substances may enter the body is by absorption through the pores in the skin. Contact of toxic and irritant chemical agents with the skin may result in a case of skin irritation only; but many organic compounds such as TNT, cyanides, and most aromatic anamines and phenols can produce systemic poisoning by direct contact with the skin.

Potentially Hazardous Processes

Occupational hazards related to processes must be recognized before they can be controlled. Proper control can be ensured only by means of an industrial-hygiene survey of the environment or workplace, including the processes that go on in it. In many processes several hazards can exist together. Therefore, it is necessary to study an overall process carefully in order to identify all potentially hazardous conditions.

Certain types of processes should arouse suspicion of a health hazard unless it has been specifically established that they are properly controlled. Some of the most important processes of this nature are as follows:

Any process involving *combustion* should be inspected to determine what by-products of the combustion may be released to the environment and, possibly, how high the burner noise is.

Any process involving *high temperature,* with or without high combustion, should be examined to determine if workers are exposed to excessive heat and noise.

Any process involving *induction heating,* including microwave heating, should be examined with regard to effects of the heat on employees and also to the level of heat, if it is suspected ʹo be in a high range.

Any process involving the *melting of metal* should be studied to determine the toxicity of the metal fume and possibly of dust, if any is produced in the process.

Any process involving an *electrical discharge in the air* should be studied to determine whether ozones and oxides of nitrogen are produced.

Grinding, crushing, or combining materials involves the hazards of dust.

Conveying, sifting, sieving, screening, or bolting of any dry materials also present a dust hazard.

Mixing of dry materials presents a dust hazard.

Mixing of wet materials presents possible hazards from solvent vapors, mists, dust, and noise.

Dry grinding operations, including milling and sandblasting, should be examined for both dust and noise hazards.

Cold blending, forming, and cutting of metals or nonmetals should be examined for hazards arising from contact with a lubricant, inhalation of lubricant mist, and excessive noise.

Handling of small parts can present hazards from repeated motion and mechanical shock.

Coating operations, which generally use solvent degreasing, can be hazardous because the solvents can act as irritants if they are ingested or come into contact with the skin.

Electroplating should be examined to determine the toxicity of various metallic salts, acids, and alkalis and the hazards they present when skin contact or inhalation occurs.

Hot bending, forming, or cutting of metals or nonmetals may create hazards from lubricant mist, decomposition products of the lubricant, contact with the lubricant, heat, noise, and dust.

Painting operations should be examined for the possibility of hazards from inhalation of or contact with toxic and irritating solvents and inhalation of toxic pigments.

Ceramic coating presents the same hazards from toxic pigments as painting, plus hazards of heat from the furnaces and from the hot ware.

Mechanical coating with metals presents hazards from dusts and fumes of metals and fluxes in addition to heat and radiation.

Explosive processing produces gases (from the explosion itself; mainly carbon monoxide and oxides of nitrogen) and dust (from the materials being processed).

Warehousing should be checked for carbon monoxide and oxides of nitrogen.

EVALUATION

Obviously, each workplace is different, and conditions can change from day to day. But a general awareness of the fundamentals of evaluation can be an "early warning" system helping to identify possible health hazards. Safety engineers are sometimes asked to make decisions about the degree of health hazard arising out of an industrial system, operation, or process. The problem may be a matter of surveying general sanitation, or it may be highly complex in nature, requiring a complete study by a qualified industrial hygienist. To evaluate degree of exposure, the concentration of a contaminant is determined according to the terms, units, or percentages which appear in the standards on levels of exposure published by the American Conference of Governmental Industrial Hygienists (ACGIH).

Threshold Limit Values

Among the most widespread and serious of all hazardous conditions in the working environment are the airborne contaminants which exist as gases, dusts, fumes, mists, or vapors. In evaluating exposure, the concentration of an airborne contaminant is determined in the same terms or units as those in the published standards regarding levels of exposure, as has just been noted.

Definition   Annually, the American Conference of Governmental Industrial Hygienists adopts a list of "threshold limit values" (TLVs) for about 350 substances. The TLV is expressed in parts per million (ppm)— that is, parts of vapor or gas per million parts of air by volume—or as approximate milligrams (mg) of particulate per cubic meter of air (this is sometimes abbreviated mg/m^3). Mineral dusts are expressed as millions of particles per cubic foot (mppcf). For vapors or gases, another term used to express TLVs is percent of flammable vapors or gases in air, by volume. The lower flammable limit of carbon monoxide gas, for example, is 12.5 percent. This would correspond to 125,000 ppm, a fatal concentration if inhaled. Thus 1 part of carbon monoxide gas in 99 parts of air— a 1 percent mixture—would be in the safe range with regard to a fire hazard (10,000 ppm). The lesson here is that in most cases protecting against the health hazard eliminates the fire and explosion hazard.

Threshold limit values refer to time-weighted concentrations for a 7-hour or an 8-hour workday and a 40-hour week. They are based on the best available information from industrial experience or on experimental studies with human beings and animals, or, when possible, on a combination of both. The basis on which the values are established may differ from substance to substance.

Examples   Now that we understand the term "threshold limit value," let us apply it to a working condition. The TLV tells how many parts per million of vapor or gas can be present in the air of a room for a normal 8-hour workday without endangering workers' health. Carbon tetrachloride, a highly toxic solvent, for example, has a TLV of 10 ppm. Benzine has a TLV of 25 ppm. Less than 1 teaspoon of either of these substances in an unventilated room 10 by 10 by 10 feet would be a health hazard. On the other hand, acetone has a TLV of 1,000 ppm; in the same room it would take over 21 teaspoons of acetone to constitute a health hazard.

Uses   TLVs should be used as guides in the control of health hazards, not as fine lines between safe and dangerous concentrations.

Reasons for using TLVs vary. In some cases, protection against impairment of health may be the guiding factor; in others, it may be a reasonable degree of freedom from irritation, narcosis, nuisance, or other forms of stress. The American Conference of Governmental Industrial

Hygienists holds that limits based on physical irritation should be considered no less binding than those based on physical impairment. There is a growing body of evidence that physical irritation may promote and accelerate physical impairment.

These limits are intended for use in the field of industrial hygiene and should be interpreted and applied by persons trained in this field. They are not intended for use (or modification for use) (1) as ratios of two limits, (2) in the evaluation or control of community air-pollution nuisances, (3) in estimating the toxic potential of continuous uninterrupted exposures, (4) as proof or disproof of existing diseases or physical conditions, or (5) for adoption by countries whose working conditions differ from those in the United States and where substances and processes differ.

"Documentation of Threshold Limit Values" A separate companion piece to its list of TLVs is issued by ACGIH under this title. This publication gives the pertinent scientific information and data with reference to literature sources that were used to base each limit. Each documentation also contains a statement defining the type of response against which the limit is safeguarding the worker. For a better understanding of the TLVs, it is essential that the documentation be consulted when the TLVs are being used.

Ceiling versus Time-Weighted-Average Limits Although the time-weighted-average concentration provides the most satisfactory, practical way of monitoring airborne agents for compliance with the established limits, there are certain substances for which it is inappropriate. These include substances which are predominantly fast-acting. Threshold limits for such substances are appropriately based on the responses to them; and they are best controlled by a ceiling ("C") limit that should not be exceeded. Obviously, sampling to determine compliance with a "C" limit will differ from sampling to determine compliance with a time-weighted-average limit. A single brief sample is adequate for a "C" limit but not for a time-weighted limit, which requires a certain number of samples throughout a complete cycle of operations or throughout a work shift.

Whereas the ceiling limit establishes a definite level which concentrations should not be permitted to exceed, the time-weighted-average limit requires an explicit limit to excursions permissible above the listed values. The magnitudes of such excursions may be pegged to the magnitude of the threshold limit by an appropriate factor. It should be noted that the same factors are used by the ACGIH in deciding whether to include a substance in the "C" list.

"Skin" Some listed substances are followed by the designation "skin." This is a reference to a potential contribution to overall exposure by the cutaneous route, including mucous membranes and eyes, either by airborne substances or, more particularly, by substances with which workers come into direct contact. (It should be noted that vehicles can alter the degree of absorption by the skin.) This attention-calling designation is intended to suggest appropriate measures for the prevention of cutaneous absorption so that the threshold limit is not invalidated.

Review Threshold limit values are reviewed annually by the Committee on Threshold Limits of the ACGIH for revision or additions as further information becomes available.

Toxicity versus Hazard

The toxicity of a material is not synonymous with the health hazard it represents. Toxicity is the capacity of a material to produce injury or harm. Hazard is the possibility that a material will cause injury when a specific quantity of it is encountered under certain conditions.

The key elements to be considered when evaluating a health hazard are these: (1) how much of the material in contact with a body cell is required to produce injury; (2) the probability that the material will result in an injury if absorbed by the body; (3) the rate of generation of an airborne contaminant; (4) the total time of contact; (5) the control measures in use.

Toxicity is dependent on dose, rate, method and site of absorption, general state of health, individual differences, diet, and temperature.

Measurement and Monitoring

In order to evaluate the extent of exposure to various hazards, it is necessary to monitor the environment. All monitoring is done with two questions in mind:

1 What pollutants exist in the environment?
2 How much of each pollutant is there?

Since the primary emphasis of this chapter is airborne contaminants, let us examine measurement and monitoring of these.

General Requirements OSHA requirements in general have to do with exposure of workers, and therefore health monitoring is to be conducted at the breathing zone of the worker. This requirement differs from the requirements for fire protection or those of the Environmental Protection Agency, which are more concerned with areas. An exception is hazardous gases or vapors that may accumulate to cause an explosion and are covered by the OSH Act; here, OSHA calls for various area-monitoring devices.

Methods of Sampling A practical method of detecting airborne contaminants is to physically collect air samples from the breathing zone of a worker and then have them analyzed in a laboratory by various "wet chemistry" procedures or by advanced techniques, which use costly and complex instruments. But the preferred method is detection by a direct-reading instrument, which allows instantaneous or continuous monitoring of a contaminant in an employee's breathing zone.

Instruments Air-sampling instruments used in the evaluation of air-borne occupational health hazards may for the most part be classified as either direct-reading instruments, which provide an immediate indication of the level of a contaminant in a measured amount of air, or sample-collection devices, which collect a known amount of contaminant for subsequent laboratory analysis.

Unfortunately, there is not now, nor is there expected to be in the future, any one instrument that will indicate all known air contaminants and give a reading indicating what is present and in what concentration. However, there does seem to be a trend toward the development of specialized instruments for specific contaminants.

Three requisites in the selection of any instrument are:

1 Sensitivity to the particular contaminant
2 Rapid results
3 Accuracy

Sample-Collection Devices Often, a substance cannot be sampled by direct-reading instruments (because no appropriate instrument yet exists). For such a substance, samples must be collected and analyzed.

There are two basic methods for the collection of *gaseous samples.* The first involves the use of a gas-collecting device such as an evacuated flask or bottle to obtain a definite volume of air. This type is called simply a "grab" or "instantaneous" sample. In the second method, a known volume of air is passed through an absorbing medium to remove the desired contaminant from the sampled atmosphere. This technique provides a sample of the atmosphere over a recording time and is known as an "integrated" sample. Integrated sampling is required whenever compliance with a time-weighted average (TWA) must be established. In such sampling, the basic source of air is a battery-powered pump or blower used for prolonged periods. In personal air sampling, the complete collection device is worn by the worker near the breathing zone. A sampling system or sampling "train" may be used. This consists of a filter, an absorber (or adsorber), a flowmeter, a flow regulator, and an air mover.

The choice of grab or integrated sampling and the equipment to be used depends on several factors—including the type of contaminant to be sampled.

A sampling train for collecting *particulate samples* usually has the following critical elements: air-inlet orifices, particulate separator, airflow meter, flow-rate control valve, suction pump. The monitoring methods currently used for particulates include collecting particulates on a filter paper from air using an appropriate sampling device. The particulates collected are then analyzed further for size and, if necessary, chemical composition.

Although section 1910.93 of the Federal Register lists a number of hazardous particulates and permissible levels of exposure, only a few detailed methods are available for qualitative and quantitative analysis of samples collected. In theory, a standard may exist for dusts or fumes, but it is not an easy matter to collect a sample of the total particulate concentration in the air and then analyze it further to determine the specific concentrations.

What, then, is an employer required to do when faced with an exposure standard for particulates? The answer depends on the specific standard, but a few general steps may be considered:

1 Determine the total amount of airborne dusts by collecting samples.

2 Determine from the collected samples the proportion of respirable dust in the total airborne dusts.

3 Use "wet chemistry" or other techniques on the collected samples to determine the concentration of a specific particulate.

The employer may also use some type of a direct-reading instrument. Direct-reading instruments are discussed below.

Direct-Reading Instruments   Gases and vapors can be monitored by direct-reading instruments, but it is still difficult to design instruments to monitor insoluble materials. (A few exist, however; and many more will be designed and marketed as demand increases.)

The two most common types of read-out instruments are (1) *electrical devices,* which give a direct reading on a dial (examples are combustible-gas indicators, vapor testers, and explosimeters); and (2) *color-change and stain-length devices,* which produce a color change in a chemical through which air to be tested is drawn.

The ideal direct-reading instrument should be able to sample air in the worker's breathing zone and specify the concentration of the substance being measured, either as an instantaneous concentration or as a time-weighted average, as required. The reading could also be in terms of percentages of an appropriate standard.

It is important to recognize that no perfect instrument exists. But even a perfect instrument worn by a worker or placed in a location would not remain accurate without periodic observation (if for no other purpose than to note variation in operating conditions or work practices) and calibration. Both people and instruments malfunction, and environmental conditions can readily depart from normal. Malfunctions and environmental variations must be noted if sampling data are to be evaluated properly, and instruments must be calibrated accordingly. Calibration may be defined as determining the true values of the scale readings of a measuring instrument. Accurate calibration of a sampling device is essential to the correct interpretation of an instrument's indication.

Indirect-Reading Devices   Indirect-reading devices are used mostly to take samplings of gases and solvent vapors for which there may be no commercially available detector tubes. They are also used to sample for gas or vapor mixtures that cannot be separated by direct-reading instruments.

Personal Dosimeters   Personal dosimeters are specifically designed to monitor the air in the breathing zone of the wearer by sampling during a working cycle. These devices are usually self-contained and battery-operated. Personal air samples may be used to monitor particulates, gases, or vapors through the use of devices such as impingers, bubblers, filter paper, and adsorption columns.

Biological Monitoring   In addition to analysis of the quality of air in the workplace, biochemical tests on body tissues or excreta can be used to give an accurate measure of an employee's exposure to a hazard. Standard methods have been developed for analysis of expired air, blood, urine, fecal matter, and tissues such as hair and nails to detect absorption of a contaminant into the body. Two examples are testing for lead absorption and testing for benzine.

Biological (and other medical) monitoring provides information only after the fact of absorption of contaminant, of course, but it is necessary for adequate evaluation of an exposure.

CONTROL

General Methods and Principles

General methods of controlling harmful environmental stresses include the following:

Substitution of a less harmful material for a hazardous one
Change or alteration of a process to minimize a harmful condition or contact

Isolation or enclosure of a process or operation to reduce the number of persons exposed to a hazard

Wet methods to reduce generation of dust

Local exhaust at the point of generation and dispersion of contaminants (see Figure 8-1)

General or dilution ventilation with clean air

Personal protective equipment such as special clothing and eye and respiratory protection (see Chapter 15)

Good housekeeping, including cleanliness of the workplace; good waste disposal; adequate drinking, washing, and toilet water; clean eating facilities; and control of insects and rodents

Special control methods for specific hazards such as reduction of exposure time, film badges and similar monitoring devices, continuous sampling with preset alarms, and medical programs to detect intake of toxic materials

Training and education to supplement engineering controls

FIGURE 8-1
A good local exhaust system at the point where contaminants are generated and dispersed should be flexible and easy to move about.

The general principles of control involve, first, the identification of industrial exposures and, second, the evaluation of such exposures by proper scientific methods to determine whether or not they represent actual health hazards. These two steps have just been discussed. For example, to deal with airborne hazards the air is sampled, the amount and quality of contaminants is determined for each type of occupation or process involved, and the results are compared with the accepted standards of maximum allowable concentration that have been adopted by OSHA or NIOSH. Third, after it has been determined that an actual health hazard which cannot be entirely eliminated is represented by an occupation or process, control measures—both engineering and medical—are instituted.

For example, to protect the health of employees who work where air contaminants created by a manufacturing process are released into the atmosphere, three steps would have to be taken:

The properties of the specific chemical agent and its possible physiological effects on employees would have to be ascertained.

The particular exposure might then be evaluated by chemical analysis or air samples; and a step-by-step analysis of the operations would be made to find the areas where employees were exposed to hazardous amounts of the material. The operational analysis would also determine how the material became airborne and was dispersed.

The type and extent of controls would depend upon the toxic properties of the air contaminant and on the evaluation made of the exposure and the operation dispersing the contaminant. (The extensive controls needed for lead oxide dust, for example, would not be needed for limestone dust, since much greater quantities of limestone dust can be tolerated.)

Red 0-4
Lft
Tox

Engineering Control

Basic Principles   In discussing the methods of engineering control, we are concerned with three basic principles:

1 Investigation of occupational conditions
2 Measurement of occupational exposures and rating of hazards
3 Institution of specific measures of protection

Approaches   Among the methods of approach for engineering control measures are the following:

Making a physical and hygienic *survey* of plant conditions
Drawing up an occupational *history and analysis*
Collecting adequate *information* concerning materials and processes used

Taking atmospheric and gross *samples* of materials and making *quantitative determinations* (physical and chemical)

Arriving at a *rating* of the hazard by comparing the findings in the given instance with accepted standards

Making an *appraisal of protective apparatus* and devices to indicate their efficiency

Providing an adequate and proper *interpretation* of the information so that it can be applied practically to the problem involved

Methods   When the foregoing approaches to engineering control have been carried out and information has been secured from them, the information can indicate the necessity of using specific methods of protection. The following have been found useful in different combinations:

Mechanical devices such as exhaust ventilation, personnel respirators, and filter-type breathing masks

Alterations of processes such as segregation, enclosure, and wetting

Substitution of nontoxic materials for those of proved toxicity

Educational programs for management, employees, engineers, and plant physicians

Continuous *monitoring and maintenance of protective equipment* to ensure efficiency

Medical Control

It is, of course, obvious that after a disease has been contracted, medical control measures are useful only to prevent additional cases. However, before diseases are contracted, there are a number of preventive controls that can be instituted by plant physicians:

Intelligent use of *preemployment physical examinations,* including chest x-rays and appropriate clinical laboratory tests

Use of *periodic physical examinations* after employment, including chest x-rays and clinical laboratory tests, where typical hazards have been proved to exist by industrial-hygiene investigations

Taking responsibility for seeing that efficient *protections* are used for hazardous exposures, including supplying personal respiratory protection where industrial conditions make it impossible to control a hazard by other means

Prompt *transference to nonhazardous occupations* of employees found to have symptoms of an occupational disease associated with exposure to a hazard

Prompt *removal from work* of all persons suffering from active occupational diseases when such persons could become a menace to themselves, their fellow workers, or the employer

Adequate provision for *treatment, rehabilitation,* and *prompt return to work* of employees who have disabilities resulting from occupational diseases.

EXAMPLE 1: DUSTS

Airborne particles are usually classified as dusts, fumes, or mists. As an example of the principles of industrial hygiene, this section will examine dusts.

Recognition

Dusts are mechanically generated particles resulting from such operations as drilling, blasting, crushing, sandblasting, stone and granite cutting, grinding, and polishing. Most particles of airborne dusts are below 1 micrometer in diameter: the average diameter is 0.5 to 0.7 micrometer. (One micrometer = one-millionth of a meter, or about 0.000039 inches; this unit was formerly called a "micron.") Particles less than 5 micrometers in diameter are of most importance to this discussion, as larger particles settle out of the air very quickly and are not likely to be breathed in.

Types of Dust Hazards Dusts usually fall into three groups when considered as hazards: (1) toxic, (2) nuisance, and (3) fibrosis-producing, including asbestos, a carcinogenic (cancer-producing) dust.

Toxic Dusts Toxic dusts are usually considered to be the dusts of poisonous materials such as lead and arsenic or their compounds that produce injury to organs and systems such as the circulatory system, heart, liver, kidneys, brain, and nervous system. A toxic dust may also be deposited in the mouth and throat by inhalation or contamination by dirty hands and later swallowed, entering the body through the digestive tract.

Nuisance Dusts A nuisance dust is one that is often associated with an increase in respiratory disease such as head colds, bronchitis, pneumonia, and tuberculosis. Many such dusts are uncomfortable and irritating to the worker and may be harmful; but they do not produce toxic effects or form fibrous tissue in the lungs.

Fibrosis-Producing Dusts A fibrosis-producing dust is one that causes or contributes to abnormal increases in fibrous tissue in an organ or region of the body, or to a degeneration of fibrous tissue. Asbestos is the most important of the fibrosis-producing dusts.

A carcinogenic dust: asbestos. Because of its strong fibers, which are (unlike cotton) almost indestructible, heatproof, fireproof, and resistant to most chemicals, asbestos has found its way into more than 3,000 products ranging from potholders and children's toys to welding rods and insulation materials. No factory, store, or home is without asbestos and very few groups of workers are not (potentially, at least) exposed to it. Extensive studies have revealed that inhaled asbestos causes body cells to turn cancerous. Mesothelioma, a cancer of the membranous lining of the chest or abdomen, can be caused by asbestos, which can also cause cancer of the tubes in the lungs, the intestinal tract, the stomach, the large intestine (colon), and the rectum.

Asbestos fibers are extremely fine; fibers that can be seen with the naked eye actually consist of thousands of smaller fibers, called "fibrils," so small that there are 1 million of them to 1 inch. (For comparison, there are just over 600 human hairs side by side to 1 inch.) When such small fibers get into the air, they float and never settle. Occupational groups that may have generally high exposures to asbestos dust include shipyard workers and (obviously) asbestos-manufacturing workers and asbestos-textile workers.

OSHA has proposed for asbestos an 8-hour time-weighted limit of 0.5 fibers per cubic centimeter of air. The British Threshold Standard for asbestos is two fibers per centimeter.

Incidence and Etiology   Dusts are the most common type of air contaminants encountered in industrial processes and can occur whenever various materials are subjected to processes which break them up into fine particles or when powered or pulverized materials are handled or transported. Inorganic dusts are most frequent, but organic dusts are also found in different industrial operations.

Metals can produce both dusts and fumes; this depends on the nature of the operation and the state of the materials when processed. For example, among the heavy metals (such as lead), dust may have as much etiologic importance as fumes, or more. (Fumes are usually found whenever molten metals are being processed or used in industrial operations. Welding, an operation whose use has greatly increased, can produce fumes of the metals which are being welded or fumes and gases from the variegated coatings of welding rods.)

Numerous dusts and metallic dusts are considered of etiologic importance, as the following outline shows.

1 Organic dusts
   a Nonliving organic dusts
      (1) Toxic and irritant dusts (various organic compounds such as tetryl, mercury fulminate, and trinitrotoluene)
      (2) Allergy-producing dusts (woods, pollen, etc.)
   b Living organic dusts

    (1) Bacteria
    (2) Fungi
2 Inorganic dusts (mineral dusts)
  a Toxic and irritant dusts (mainly the dusts from heavy metals and their salts, such as arsenic, lead, and manganese)
  b Fibrosis-producing dusts (examples are free silica and asbestos)
  c Non-fibrosis-producing dusts (the so-called "inert" dusts, such as alundum, corundum, emery, limestone, magnesite, marble, plaster of paris, and polisher's rouge)

An important matter related to understanding the etiology of diseases caused by dusts (and fumes) is the use of the terms "exposure" and "hazard." They are not synonymous. "Exposure" merely means contact with some kind of substance which is a *potential* health hazard. Whether the substance is an *actual* health hazard can be determined only by an industrial-hygiene survey in which the amount and nature of the exposure are rated according to accepted standards of concentration, particle size, and quality. It is often erroneously stated, for example, that there is a "lead hazard," when it is known only that *contact* is being made with lead and there are no data to support the contention that there is an actual hazard present.

From the etiologic point of view, particularly with reference to pathological effects, the following list shows how different types of substances (among the dusts and fumes) act on various systems in the body to produce pathological effects:

General systemic poisoning (Examples: arsenic, lead, manganese, mercury)

Substances acting on the *circulatory system* (Examples: arsenic, lead, mercury, vanadium)

Substances acting on the *respiratory system* (Examples: asbestos, free silica)

Substances acting on the *gastrointestinal system* (Examples: antimony, cadmium, lead, mercury, zinc)

Substances acting on the *skeleton and joints* (Examples: lead, mercury, mesothorium, phosphorus, radium)

Substances acting on *organs of special sense* (Examples: arsenic, lead, manganese, mercury)

Substances acting as *fatal poisons* (Examples: lead, mercury)

Substances acting on the *genitourinary system* (Examples: arsenic, lead, mercury, vanadium)

Substances acting on the *brain and nervous system* (Examples: arsenic, lead, manganese)

**Evaluation: Principal Factors Determining the Health
Hazard from Airborne Dust**

The following factors are the most important:

1 Chemical composition of the dust
2 Amount of dust in the air
3 Size of the dust particles
4 Length of time of exposure

Materials such as rocks and minerals are not what might be termed
"pure" chemical compounds but rather are mixtures of several chemical
compounds. "Chemical composition" refers to the kinds and amounts of
these various components. (In the case of silica-bearing substances, for
example, the amount of silica present may vary from almost nothing to
100 percent. The amount of free silica present is the important factor in
the silicosis-producing properties of a dust.)

Regardless of the amount or type of dust, the more dust is retained,
the more severe the resulting illness will be. Dust particles may be so
small that they are visible only by microscopic analysis; particles of this
size behave essentially like air, passing practically unimpeded through
the protective barriers of hair and mucus in the upper airways. The air
sacs and small air tubes, therefore, receive a large dose, and the intensity
of lung reaction is greater than if the same quantity of larger particles had
been inhaled.

Control: Ventilation

Noxious dusts (and also noxious mists, vapors, and fumes) can be effectively controlled or eliminated through the use of systematic ventilation.
Exhaust systems play a very important part in eliminating contaminated
air and dusts. If air sampling shows contaminants in the atmosphere, or if
city and state codes or OSHA standards require it, then exhaust systems
must be installed. Air-cleaning devices may be required to prevent an exhaust system from throwing out contaminants that could represent a
community health hazard, pollute the outside atmosphere, or create a
general nuisance. Increased awareness of the environment makes such
community regulations increasingly likely.

A good local exhaust system consist of four basic parts:

1 *Hood*—into which the airborne contaminant is drawn
2 *Ducts*—for carrying the contaminated air to some central point in
the system
3 *Air-cleaning device*—such as an arrestor for purifying the air before
it is discharged

4 *Fan and motor*—to provide the suction for keeping the air moving through the system

Some typical processes requiring ventilation are:

Spray booths
Welding operations
Dust control in foundries
Grinding operations
Woodworking machines
Fog or steam removal at dip tanks
Cast-iron machining
Melting furnaces
Oil mists
Materials conveying

EXAMPLE 2: PREVENTION OF DERMATITIS FROM EPOXY RESINS

Epoxy resins were introduced to American industry and American workers relatively recently. Epoxy resins have many uses because they are strong, resistant, and lightweight and join firmly to other materials. In manufacturing, such as the aircraft, automobile, and electrical industries, as well as in painting and many other operations, the epoxy resins are being used very widely.

But working with epoxy resins can spell trouble for the employee who does not know what the hazards are and how to guard against them. Employees working with the resins may develop a disabling skin condition, or dermatitis, within a few weeks.

Recognition and Evaluation

Epoxy resins are harmful:

- When improperly handled
- When their vapors are inhaled

When completely cured, epoxy resins are inactive and have little potential danger for workers. But wet or uncured resins and chemicals used to harden the resinous mixture, thin it, strengthen it, or make it flexible should be regarded and handled as hazardous materials.

Potentially Hazardous Operations   Operations in which hazards are found include the following:

Mixing: Employees who mix resins and hardeners without safeguards will have direct contact with the basic components or breathe the vapors from the mixture.

Molding and casting: Workers often mold or brush the mixture onto a form, cast it in a mold by hand, or laminate it with fiberglass, using no protection from direct contact. Sometimes workers wear adequate protection on the hands and forearms but leave the neck and face unprotected.

Tooling: Sanding, grinding, or drilling the cast of hardened epoxy resins or laminates (made by alternating layers of epoxy resins with layers of other materials) may cause tiny particles of materials to become embedded in the skin. Often the embedded material will lead to mechanical irritation of the skin, with possible infection from scratching the irritated areas.

FIGURE 8-2 and FIGURE 8-3
Two different types of designs for exhaust systems.

What the Hazard Can Produce   Dermatitis, an inflammation of the skin, is the disease that most often attacks workers handling epoxy resins and the chemicals used to manufacture them. Redness, itching, swelling, blisters, oozing, crusting, and scaling can result from primary irritation or allergic sensitization dermatitis:

Primary irritation: A direct reaction experienced by most people from skin contact with strong chemicals such as acids, alkalis, solvents, and other materials. The face, neck, forearms, and hands are most commonly affected.

Allergic sensitization: A severe type of dermatitis often occurring over widespread areas of the body about a week or more after the first contact. Swelling, redness, and oozing are present, and the employee is usually not able to work.

Mechanical dermatitis: A reaction of the skin to an irritating foreign body. Tiny bits of fiberglass or other materials become embedded in the skin and lead to irritation, itching, and sometimes secondary infection.

Other conditions: Respiratory, nose, and throat irritation; headache; nausea; intestinal upsets; and other conditions may result from breathing the vapors or other materials used in the manufacturing processes. The eyes are also affected by the fumes or by direct contact.

Control

Meticulous workroom housekeeping and constant awareness on the part of employees will guard against the two chief dangers in epoxy plant operations: (1) skin contact with the resin or other materials, and (2) exposure to their vapors.

General Basic Rules Outbreaks of dermatitis and other diseases can be avoided by following these basic rules:

Inform employees of possible hazards.

Provide ventilation to control vapors produced while mixing the resins and hardener as well as to control the glass and epoxy particulates during tooling.

Maintain industrial and personal hygiene by means of good plant housekeeping procedures, appropriate industrial cleansers, protective clothing, and (where needed) protective creams.

Specific Rules The following specific rules can protect workers from the effects of epoxy resins:

Learn what the hazards of epoxy resin operations are and how to avoid direct contact. Handling the materials or breathing their fumes should be avoided.

Avoid unnecessary contamination in the plant to cut down on exposure. Use special isolated areas of the plant for mixing, molding, curing, casting, and tooling of the resins.

A well-ventilated hood should be used when mixing batches of resins and hardener to prevent the escape of any possible hazardous vapors. Limit the mixing to only a few employees.

Use ventilated hoods for grinding, sawing, drilling, or polishing the molded laminate. The ventilation will remove dusts from the air and lessen the chances of skin contact.

Protective sleeves and cotton liners under rubber gloves should be worn during the molding operations to protect the skin.

Water-soluble skin-protective gels should be provided for the protection of skin against the action of the solvents.

When washing the hands, neutral or acid soaps should be used instead of alkaline, powdered, or abrasive soaps.

Avoid using acetone or similar solvents to cleanse the skin. These solvents may cause dermatitis or other skin irritations.

Use disposable paper towels instead of clean-up rags to avoid contact with materials that would ordinarily remain on the rags.

Avoid spilling resins and hardening agents or letting them drip. Immediately wash up spills or drips that do occur, using warm water with a mild soap.

Keep tables, machinery, tools, floors, walls, and windows free of fiberglass particles and dusts. Use heavy disposable paper on the tables and change it as often as possible.

To the Supervisor This has been a capsule review of the hazards involved in working with epoxy resins and ways of preventing possible illness.

Safe handling can be achieved only with the cooperation of every individual involved. Observance of proper methods of handling, housekeeping, and personal hygiene will prevent dermatitis and other health problems. Failure to follow precautions can cause troublesome dermatitis and other complaints.

IN CONCLUSION: PURCHASING CHEMICALS

In view of the increased emphasis on toxicity, it is strongly recommended that when chemicals are being purchased for plant use, processes, and manufacturing, the manufacturer should supply a ''Material Safety Data Sheet'' (OSHA Form 20).

The primary information shown in the data sheet for any chemical will include trade and chemical names and synonyms; chemical family, and possibly the formula; a list of hazardous ingredients; physical data; data on fire and explosion hazards; data on reactivity; proper procedures for cleaning up spills or leaks; special protection needed; special precautions that should be followed when using it; and first-aid procedures in the event of an accident.

Under the Toxic Substances Control Act, purchasers should get this information from the supplier upon request. It is in the best interest of employees to ask the supplier for the Material Safety Data Sheet *before* the materials are actually delivered into your plant.

The company or plant itself must have personnel (or at least one person) who can understand and interpret the data and be able to recognize any gaps where additional information or technical expertise is required.

OSHA HAZARD COMMUNICATION STANDARD

The Hazard Communication Standard is a ''performance standard,'' which means that it describes objectives that must be met, but without specifying the method for accomplishing those objectives. The method is up to the individual organization to choose.

The standard requires chemical manufacturers and importers to assess the chemicals which they produce or import, and all employers to use hazard communication programs to provide information—to their employees—concerning hazardous chemicals. These programs must include the labeling of chemicals, materials safety data sheets (MSDS), training, and access to written records. In addition, distributors of hazardous chemicals are required to ensure that containers they distribute are properly labeled, and that a material safety data sheet is provided to their customers in one segment of industry, the manufacturing division. Eventually, the federal standard may be expanded to include other employees, either by rule-making or by court order.

Chemical manufacturers and importers are required to evaluate the hazards of the chemicals they produce or import and to transmit this information to downstream employers, who in turn must pass on the information to their employees through a comprehensive hazard communication program designed by the employer. This shall be done by means of labels on containers, material safety data sheets, and employee training. The intent of this standard is to ensure that the employees will receive as much information as needed concerning the hazards in their workplaces, and that this information be presented to them in a usable, readily accessible form.

WORD POWER

etiology   The study of the causes of diseases.

fumes   Solid particles created by the condensation of a substance from a liquid to a vapor or a gas.

gases   Formless fluids that will expand to occupy whatever space is available to them.

pel   Permissible Exposure Level. Its intent is the same as TLV, but it is a legal limit, which—when exceeded—is punishable by a penalty as set by OSHA regulations.

ppm.   The atmospheric concentration of gases and vapors, often expressed as parts of contamination per million parts of air by volume.

toxicity   The ability of a substance to produce an unwanted effect when the chemical involved has reached a sufficient concentration within the body.

toxicology   The study of the nature and actions of poisons.

QUESTIONS AND EXERCISES

1 What are the key elements that should be considered when evaluating a health hazard?

2 How can the use of the preemployment physical examination be used to reduce the chances for contracting occupational diseases in the workplace?

3 What is the purpose of the Material Safety Data Sheet?

4 How is dermatitis caused? How can it be eliminated?

5 What is the difference between a TLV and a PEL?

6 How can atmospheric contaminants be detected and measured?

7 What are the requisites for the selection of air-sampling instruments?

8 What are the three key concepts which must be applied for an effective industrial hygiene program?

9 What is the difference between ceiling limit and the time-weighted-average limit?

10 What are the two basic methods for collecting "gaseous" samples?

11 Hazardous dusts fall into three major groups. Name them.

12 What are the four major components of a good local exhaust system?

13 Name five processes that require ventilation.

14 What are the two factors that generally make epoxy resins harmful to workers?

15 Name some good personal hygiene factors that would prevent or control dermatitis.

BIBLIOGRAPHY

Accident Prevention Manual for Industrial Operations, 5th ed., National Safety Council, Chicago, Ill., 1964, p. 28–4.

Compliance Operations Manual, U.S. Department of Labor, Occupational Safety and Health Administration, January 1972, pp. xiii–6.

Hazard Communication Standard: Final Rule, Federal Register, Part iv, Occupational Safety and Health Administration, November 26, 1983, p. 53280.

The Health Hazard Evaluation Program of the National Institute for Occupational Safety and Health (NIOSH), Cincinnati, Oh.

James, Dan: "OSHA's Communication Standard," *Professional Safety,* May 1986, p. 31.

National Safety News, vol. 110, no. 4, October 1974, p. 95; vol. 111, no. 5, March 1975, p. 190; vol. 112, no. 4, October 1975, pp. 50, 51; vol. 115, no. 4, April 1977, p. 58; vol. 116, no. 2, August 1975, TOSCA.

Preventing Dermatitis If You Work with Epoxy Resins, U.S. Department of Health, Education, and Welfare Public Health Service Publication no. 1040, May 1963.

CHAPTER 9

ELEMENTS OF WORKERS' COMPENSATION LAWS

HISTORY

Prior to the enactment of Workers' Compensation laws, and to a considerable extent after these laws were passed, it was extremely difficult for employees to recover damages for injuries received from accidents occurring during the course of employment. It was necessary for the employee to prove that negligence on the part of his or her employer resulted in an accident causing bodily injury. Naturally, there was great reluctance to bring suit or claim against an employer for fear of job loss and a possible long period of unemployment. It was a sort of a game in which the employer held the high cards. About this time, many countries in Europe and, of course, the United States were directing their efforts toward industrialization. It was the age of the "Industrial Revolution" and workers on the farm, in local trades, and men on the high seas turned their attention to the financial benefits to be gained from working in manufacturing plants and related industries. The relationship between employee (servant) and employer (master) became more acute as a rapid increase in accidents occurred due to very poor working conditions and mechanical equipment. The accidents often resulted in serious bodily injury or death.

The needs and demands of the people involved, both servant and master, brought about the adoption of state workers' compensation laws in the United States between the years 1910 and 1920. These laws placed upon the employer the responsibility for accidents, and proof of negligence was no longer necessary by the employee as long as the accident occurred in the course of his or her employment. Most European countries, especially England and Germany, were going through the same industrial revolution and had adopted laws similar to workers' compensation before such action was taken in America. The first modern workers' compensation law was enacted in Germany about 1884 and was followed by the English plan about 1887.

The German plan was a comprehensive one which covered sickness, accidents, disability, and old age. Negligence by the employer was not a factor in the German plan. Actually, there were three funds in this plan. The sickness fund was made up of two-thirds worker's contributions and one-third employer's contributions. The accident fund was supported entirely by the employers, who also managed the fund. The disability fund was supported equally by both the worker and the employer. The German plan was compulsory for both employers and employees.

The English plan was enacted about 1897 and came closer to modern-day workers' compensation legislation. It served as a pattern for the American acts (laws) which came later. In the English plan, the employer carried the entire burden of the cost—either by paying benefits out of profit and loss, or by covering the risk with an insurance company. The administration of workers' compensation insurance by industry gave a tremendous impetus to their overall business and operations. The English plan provided no form of compensation for sickness. Just as in the German plan, the English plan did not include legislation covering negligence by the employer.

The first workers' compensation law in the United States was enacted by Congress in 1908 and covered only government employees. It was called the Federal Employers' Liability Act of 1906, but was held unconstitutional soon after it was enacted into law. However, the second Federal Employers' Liability Act of 1908 was sustained, thereby giving compensation benefits to employees of the United States government.

In 1910, New York passed the first Workers' Compensation Act and was soon followed by ten other states in 1911. Three more states followed the same route in 1912, and, in the following year, eight more states joined in the procession. The majority of all of the acts were passed between the years 1911 and 1920. Today, each of the 50 states and American Samoa, Guam, Puerto Rico, and the United States Virgin Islands have a workers' compensation law.

Employers' Obligations Prior to Workers' Compensation Laws

Prior to the advent of workers' compensation laws, employers had but three general obligations under both common and statute law. They were as follows:

- To safeguard the general public
- To safeguard their customers
- To safeguard their employees

The employer had no further obligations relating to personal injuries in the workplace under the then existing laws. The employer's obligation to employees under common law was very limited. They were as follows:

1 Provide a safe place for employees to work in.
2 Provide safe tools and equipment for them to work with.
3 Provide employees with a knowledge of the hazards of the workplace.
4 Provide competent fellow employees and supervisors.
5 Provide safety rules for employees and insist on their enforcement.

All of these were fine rules, except that at that time there was no enforcement agency to ensure that employers were complying with their responsibilities.

Employees' Rights Prior to Workers' Compensation Laws

Before workers' compensation laws were enacted, an employee who suffered loss of income while recuperating from an industrial injury received very little or nothing from the employer in recompense for his or her injuries. To receive damages, the only "right" that the employee had was to file suit and prove in a court of law that the incurred injury was the result of the employer's negligence.

The employee faced great difficulty in obtaining satisfaction under this system. First, there were long delays in securing court action; second, there was uncertainty as to the results of the trial; third, there was a high cost involved in negligence suits; and, fourth, very often the injured worker was still denied any compensation for his incurred injury.

Employers' Rights Prior to Workers' Compensation Laws

Employers, even though negligent, could defend themselves successfully by the use of three common law defenses. These defenses were:

Contributory negligence
Negligent fellow worker
Assumption of risk

Contributory Negligence The employer was not held liable if the employee by his own negligence contributed to the accident which subsequently caused his or her injury.

Negligent Fellow Worker The employer was not held liable for a work-related injury caused by the negligence of a fellow worker.

Assumption of Risk If the employee was injured as a result of certain risks and hazards associated with the employment, it was presumed that the employee assumed those risks when he or she accepted the employment.

In rare circumstances today, where the employer is subject to the act and the employee rejects it, the employer can use the three standard common law defenses. On the other hand, if the employee accepts the act, and the employer rejects it, he is deprived of the three common defenses; and the same would be true if both the employer and the employee rejected the act.

MAJOR STIPULATIONS IN THE NEW WORKERS' COMPENSATION LAWS

At the close of the 19th century, it became apparent that the accepted common law defenses used by the employer operated too harshly on claims submitted by disabled workers. The situation led to demands for reform and changes in the system. As a result, the so-called "employers liability laws" were adopted by many states. Eventually, the employers liability laws became the workers' compensation laws as we know them today. The major changes in the laws were:

The employer must pay benefits to injured workers without their having to go to court to prove negligence.

Workers gave up the right to sue the employer in a court of law.

The employer was required to pay medical costs for employees injured in work-related accidents.

The employer had to give up the use of the three common law defenses.

In short, workers' compensation laws hold that employers should assume the cost of occupational injuries without regard to fault. The law also serves to relieve employers of liability from common-law suits involving negligence.

The Basic Objectives Underlying Workers' Compensation Laws

The six basic objectives that underlie workers' compensation laws are:

1 To provide sure, prompt, and reasonable income and medical benefits to work-related accident victims, or income benefits to their dependents, regardless of fault.

2 To provide a single remedy, and reduce court delays, costs, and workloads arising out of personal injury litigation.

3 To relieve public and private charities of financial drain incident to uncompensated industrial accidents.

4 To eliminate payment of fees to lawyers and witnesses, as well as time-consuming trials and appeals.

5 To encourage maximum employer interest in safety and rehabilitation through an appropriate experience-rating mechanism.

6 To promote frank study of causes of accidents (rather than concealment of fault), thus reducing preventable accidents and human suffering.

REQUIREMENTS FOR BENEFITS UNDER WORKERS' COMPENSATION

Compensation will be paid "for personal injury or death by accidents arising out of and in the course of employment." To be within the course of employment, a worker need not confine himself to the core matter of his or her work. Workers may be—during their employment—performing any act incidental to the work, without departing from the scope of their employment. The phrase, "for personal injury or death by accident arising out of and in the course of employment," has been the source of more than half of the personal injury claims under the Workers' Compensation Act.

Conditions Required for Establishing a Claim

In order for the injured worker or surviving dependents to establish a claim for benefits there are four conditions or criteria that must be established under the law. These conditions are as follows:

- There has to be an injury.
- The injury must have resulted from an accident.
- The injury must have arisen out of the worker's employment.
- The injury must have occurred during the course of employment.

At times, there seems to be some confusion as to the meaning of the words "accident" and "injury." Let's separate the two words and give them some meaning.

Accident An "accident" may be defined as a sudden and unexpected event that is unplanned, uncontrollable, undesirable, and causes near-injury or injury at a definite time and place.

Injury An "injury" denotes the effects or the results of an accident.

The injury must have resulted from an accident This condition implies that there has had to have been an accident and there must have been an injury arising out of the accident.

The injury must have arisen out of the worker's employment The worker must be doing work assigned to him or her by a supervisor, or, the worker must be carrying out a duty or a work activity that is normally expected of an employee by the employer.

The injury must have occurred during the course of employment There has to be an agreement as to when the course of employment is, and the conditions as to what is considered employment.

Conditions Differing from the Above

There are many types of employment that are distinct from the day-to-day type of employment. There are those employees whose job may take them from place to place during the course of their workday. There are also jobs that have no defined set of working hours. There are other employment situations whose conditions do not fall within the normal scope of workplace employment, but yet fall within the scope of reasonable conditions required for establishing a claim for a workplace injury. Listed here are but a few examples of the conditions that create this particular category of employment.

1 Traveling employees (employees whose work takes them from one particular work location to another).

2 Traveling salespersons.

3 Stand-by employees subject to call at any hour by the employer.

4 Employees living on the employer's premises.

The word employment includes any act which may reasonably be said to be incidental to the work. Thus, it will include any act which the employment by its very nature seems to call for. However, since the nature and purpose of various employments are so different, it can be difficult to try to define exactly what may or may not be incidental to a particular employment.

COVERAGE

Approximately 80 percent of all workers in the United States are covered by workers' compensation laws. The following list gives the categories of persons and employments that are partially covered or are not covered in many states. Since changes in workers' compensation laws occur almost daily, the laws covering the workers and conditions involved should be consulted.

1 *Agricultural workers:* At one time, these workers were not covered; now 37 states and Puerto Rico have laws that provide at least some coverage in varying degrees.

2 *Domestic workers:* Twenty-four states permit employers to provide voluntary coverage and the rest of the states have some form of compulsory coverage. Louisiana and Wyoming exclude domestic workers from any form of coverage.

3 *Casual employment:* Employment which is occasional, incidental, and not occurring at regular intervals.

4 *Hazardous employment:* Eleven states indicate that the workers' compensation laws cover only employments listed as "hazardous" or "extra-hazardous." However, the lists have broadened in many of these 11 states, so that many occupations not ordinarily considered especially hazardous are included.

5 *Employees of charitable or religious organizations:* Persons whose work is irregular, for short-term periods, or temporarily outside the scope of the regular activities of the employer are covered in about a dozen states.

6 *Small organizations:* In many states workers' compensation laws do not apply where there are fewer than a stipulated number of employees.

7 *Railroad and maritime workers:* These workers are not covered by workers' compensation laws because railroad employees who are engaged in interstate commerce and maritime workers are generally covered by the Federal Employers' Liability Act.

8 *Contractors and subcontractors:* Independent contractors are not covered by the insurance of one for whom they are performing services. They must provide their own coverage for their own employees.

9 *Minors:* Minors are entitled to the same workers' compensation and coverage other legally employed persons would receive under similar circumstances. In a few states, compensation to minors who are illegally employed is increased if they are injured. In some states, the amount for a permanent injury is based on considerations of future earning capacity.

10 *Extraterritoriality:* A company may hire a worker in one state but give him or her work in another. Generally, if a contract of hire is made within one state, and if the employee is a resident of or the employer's place of business is within that state, its workers' compensation laws will govern even if an accident occurs outside the state.

In addition to the above employment relationships, there can be any number of other employments that can or cannot be covered under workers' compensation in some states. These can also include persons in the armed forces reserves and national guardsmen, prisoners of federal, state, and local institutions, members of the employer's family, and discharged employees.

Actions or Conditions Disallowing Workers' Compensation Benefits

There are certain actions or inactions that could result in the loss of benefits to injured employees. They may differ in language from state to state, but generally all of the acts cover these actions. They are:

1 Intoxication or the use of a controlled substance
2 An intentional self-inflicted injury
3 The commission of a felony or a misdemeanor
4 The willful failure, or refusal, to obey a reasonably written or printed safety rule of the employer
5 Horseplay or altercation with a fellow employee
6 The willful failure, or refusal, to perform a duty as required by the acts
7 Falsification of application for employment

Benefits Provided under Workers' Compensation

Because workers' compensation imposes an absolute, but limited, liability upon the employer for employee disabilities caused by the employment, benefits payable to the injured employee attempt to cover most of the worker's economic loss. This includes both loss of earnings and extra expenses associated with the injury. Specifically, the benefits are:

Cash benefits including both impairment benefits and disability benefits. The former are paid for certain specific physical impairments, while the latter are available whenever there is an impairment and loss of wages.

Medical benefits amounting to almost 30 percent of all workers' compensation benefits paid. In almost all of the acts, unlimited medical benefits are provided either specifically by statute or by administrative discretion.

Choice of physician Practices vary from state to state with respect to the choice of an attending physician. States are divided nearly evenly between those that give this decision to the employer and those that give the decision to the employee. In some states, selection must be made from an approved list. The employer normally has the right to have his own physician conduct examinations.

Rehabilitation benefits including medical rehabilitation and vocational benefits for those cases involving severe disabilities. Most states have time and monetary limitations for rehabilitation.

CASH BENEFITS

In considering workers' compensation income or cash benefits—which replace employee loss of income or earning capacity due to occupational injury or disease—four classifications of disability are used:

- Temporary total disability
- Permanent total disability
- Temporary partial disability
- Permanent partial disability ("permanent partial" is divided into "nonscheduled" and "scheduled" disabilities)

Temporary total disability implies that the disabled employee is incapable of doing any physical work for a limited time; however, full recovery and return to employment is expected. (The majority of workers' compensation cases fall into this category.)

Permanent total disability generally indicates that the disabled worker is regarded as totally and permanently unable to perform gainful employment.

Temporary partial disability implies that the disabled worker can perform some work but not full or regular duties, and that full recovery is anticipated.

Permanent partial disability implies that the disabled employee suffers from an injury from which he or she may not recover, but is able to perform some other type of work.

Income Benefits for Permanent and Total Disability

Income or cash benefits are payable under either temporary total or permanent total disability. For computing weekly benefit payments, a formula expressed as a percentage of weekly wages is used. In most states, limitations are placed on maximum and minimum benefits payable weekly; some states also limit the total number of weeks and total dollar amount of benefit eligibility. Where there is permanent total disability, most states provide payments extending through the employee's lifetime. Almost all of the states use 66⅔ percent of an employee's average weekly wage to compute weekly benefit payments.

For temporary total or permanent total disability, the wage replacement percentage in each jurisdiction is the same. However, in permanent total disability cases, the time limits tend to be longer and the total dollar amounts higher than in cases of temporary total disability. Some states provide additional amounts for dependents, as well as other benefits.

Income Benefits for Partial Disability

Most awards and dollars paid out as income benefits are either for temporary total or permanent partial disability. As partial disabilities involve current earnings or wage-earning ability, in many states weekly benefit payments for temporary or permanent partial disabilities of the "nonscheduled" type are based on a wage-loss replacement percentage. The percentage applies to the difference between wages earned before and after injury. In some states, "nonscheduled" permanent partial disabilities are compensated as a percentage of the total disability cases.

FIGURE 9-1
Estimated premium costs of Workers' Compensation to employers in the United States, selected years, 1940-82. Source: Social Security Bulletin, December 1984/Vol. 47, No. 12.

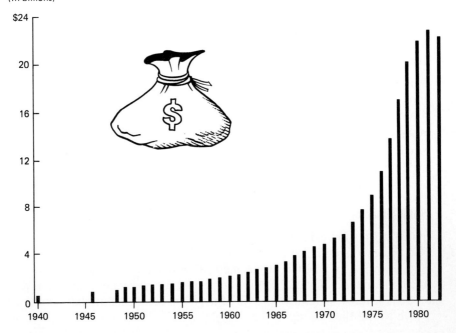

"Scheduled" Income Benefits

"Scheduled" benefits cover injuries involving loss—or loss of use of—specific body members, where wage loss based on the nature of the impairment is assumed. In most jurisdictions, the actual amount payable is a specific number of weeks of benefits (based on the member involved) multiplied by the weekly benefit amount (based on earnings at time of injury). In some states, the "scheduled" award is in addition to any payment otherwise payable to the employee while he or she may be temporarily totally disabled (the healing period). Furthermore, some states limit the amount payable for such periods of temporary total disability.

Survivors' Benefits for Fatal Injuries

Benefits payable in the event of fatal injuries comprise more than 14 percent of all total income benefits. The benefits provided include a burial allowance, as well as proportion of the worker's former weekly wages.

Although death is the ultimate work-related tragedy, the economic loss associated with death cases is often less than that of a permanent total disability. Because of these considerations, death benefits are generally paid to the spouse until remarriage and to the children until a specific age. In addition, some of the acts provide a maximum benefit total, expressed as a maximum period for the payment of benefits.

Limitations on Filing Claims

The right to compensation under the act is barred unless a claim for compensation is filed within the stipulated time requirement after the occurrence of an accident or occupational disease. Each state has its own statutory provision regarding the limitations on filing a claim. The statutes also provide a time period for filing a claim if death results from an accidental injury, occupational disease, or radiation exposure.

Waiting Period

State statutes provide that a waiting period must elapse, during which period income benefits are not payable. This waiting period affects only compensation payments, while medical and hospital care are provided for immediately. If the disability continues for a certain number of days or weeks, most state laws provide for payment of income benefits retroactive to the date of injury.

Rehabilitation Benefits

The mutual interest of disabled employees and employers generally favor starting rehabilitation as soon as possible. Although rehabilitation is considered an integral part of complete medical treatment, its use may extend beyond this consideration (for example, where it includes vocational rehabilitation and retraining). However, rehabilitation is provided in all states even if unspecified in the law. Maintenance allowances and special fund sources to finance rehabilitation are also provided for in many of the state statutes.

Insurance carriers and many employers with medical departments are leaders in carrying on rehabilitation for the industrially injured. Likewise, many major industries have comprehensive programs for employment of the physically handicapped. Smaller industries maintain modified programs for the placement of disabled workers in congenial tasks.

The Federal Vocational Rehabilitation Act is now effective in all states, and includes federal funds to aid states in vocational rehabilitation of the industrially disabled.

Second Injury Funds

Second injury funds were developed within the framework of the workers' compensation system to compensate an employee with a preexisting injury who incurs a second injury that produces a disability greater than that of the first injury. Under such a system, the employer pays only the benefits that are due for the second injury. The employee is fully protected because the fund pays the difference between what he or she would have received for the resulting condition if there had been no prior disability.

Second injury funds encourage the hiring of the physically handicapped. By removing an employer's fear of increased workers' compensation costs, second injury funds enhance the employment opportunities of disabled workers.

Where there is no second injury fund provided by law, an employer whose employee sustains a second injury usually is liable for compensation due for the total resulting injury. Second Injury Fund legislation is not intended to relieve an employer of all liability when a handicapped worker is further injured on the job. Its purpose, rather, is to divide or apportion, liability between the employer and the fund—the employer paying only for any injury that occurred while the worker was in his or her employment, the fund paying the balance of benefits payable for the combined effect of the first disability and second injury.

GENERAL PROVISIONS AFFECTING COMPENSATION

As originally envisioned, the workers' compensation system was to be self-administered by the individual states. Generally, the states have moved to administer their workers' compensation laws through their court system, a special commission or board, or a combination of both. State legislatures began to enact workers' compensation laws that created a new legal relationship between the employer and the law. Today, every state has a Workers' Compensation Act, and insurance premiums total over several billion dollars annually.

INSURING THE RISK

Since workers' compensation has become compulsory under the law, most employers have turned to insurance companies for assistance. This line of insurance has made a very substantial contribution to the tremendous growth of casualty insurance. Employers today can insure their risks by several different methods. They are as follows:

Self-insurance

Private insurance carrier

Employers' mutual and reciprocal associations (companies joined together to insure their risks jointly)

Independent system (employers administer their own systems; benefits must be equal to, or better than, those provided by state laws)

METHODS OF PREMIUM CHARGES

"Premium charges" refers to money paid out by a policyholder to an insurance company for protection, benefits, or services to be rendered, based upon the insuring agreements and provisions defined in the policy contract. Obviously, premium income is the lifeblood of the insurance industry.

The cost of workers' compensation insurance is paid for by the employer. The cost of any type of insurance is dependent on the risk the insurer feels that the policy entails—a risk being "an exposure to a hazard." This is also true in workers' compensation insurance. However, since there are differences in exposure and types of benefits from state to state, the cost of workers' compensation insurance will vary from state to state. The four most common methods of determining insurance premiums are:

- Premium discounting
- Experience rating
- Retrospective rating
- Manual rating

Premium discounting This plan is mandatory in most states. It was devised to serve large companies that pay high premiums for the workers' compensation insurance. The discounts are graded in accordance with the size of the policyholder's premium.

Experience rating The premiums are based on estimates of expected losses and are adjusted from the loss experience over the previous three years. A good safety performance record by a company would therefore reduce the premium, while a poor safety performance record would increase the premium.

Retrospective rating Since experience rating relates premiums to past experience, retrospective rating relates premiums to the current policy period. The insured will pay the expected premium at the beginning of the period for which he or she is insured, and then an adjustment is made at the end of the period if there is a reduction in the injury-loss experience.

Manual rates These rates are based on information reported to the rate-making organizations in the insurance industry. Business operations, and certain categories of workers, are classified according to a manual which cites rates for the different classifications of jobs.

The National Council on Compensation Insurance is the largest rate-making organization for workers' compensation coverage. A number of states have set up their own bureaus for rate-making purposes for risks operating in their particular state.

ACCIDENT REPORTING REQUIREMENTS

Almost all of the states require that employers report work-related injuries to their Workers' Compensation Boards or Commissions. In some states, the employer's insurance carrier is permitted by law to relieve the employer of this burden. The time element for reporting such injuries ranges from 48 hours in some states to as much as 28 days in others. In almost all of the states, acts provide penalties for failure to report. The reporting requirements also specify the types and the nature of illnesses and injuries to be reported.

HEARINGS

Where there has been an inability for which the employer and the employee must agree on compensation, either party may then request a hearing on the dispute. The hearing on a disputed claim may be held before the full industrial board or commission or by a single member of the board. The board will hear the parties at issue, including their representatives and witnesses. After hearing the evidence of both parties, the hearing member makes a determination of the dispute and decides the award.

The claimant in a compensation hearing must establish every fact necessary to support an award. The claimant must show the time of the accident, his proof of employment, that the harm for which he or she seeks compensation was accidental, that it arose out of the employment, and that the harm occurred during the course of the employment. Since the industrial board or commission is an administrative body rather than a court, should a decision of this body be in dispute with either party, that party may appeal to the appellate court of his or her state.

THIRD PARTY LIABILITY

Where the worker's injury or death arises during the course of his or her employment, but is caused by the fault of some third party outside the employment circle, two remedial avenues exist. The injured employee may obtain compensation from the employer, or the worker has the right to bring suit for damages. If a monetary recovery is made by way of a court action or settlement, the employer is entitled to a reimbursement equal to the benefits already paid.

WORKERS' OCCUPATIONAL DISEASE ACTS

The term "occupational disease" is defined as a disease arising out of, and in the course of, employment. Contrary to the Compensation Acts, the Disease Acts require that there be a special hazard in the employment. Ordinary ailments found in the general public which bear no direct connection to the employment, cannot be considered as occupational diseases. They must be "peculiar or characteristic of" the employee's occupation.

Therefore, it is obvious that the worker who breathes or coughs his or her health away in a dusty environment has contributed as much to the manufacture of a product as has the worker who has lost a hand or any eye making the same product.

Although workers' compensation laws initially had no specific provisions for occupational disease, all states now recognize responsibility for them. Coverage extends to all diseases arising out of and in the course of employment, with unlimited medical care provided. In most states, compensation benefits are the same as for accidental injuries. The time limit for filing a claim averages from 1 to 3 years after the last exposure or first manifestation of the disease or death.

The burden of proving all elements of recovery of compensation is on the worker, and the worker's discomfort or irritation is not enough; there must be a loss of wage-earning ability.

SELF-INSURANCE

A self-insurer is a company which has been granted the privilege of assuming the payment of workers' compensation benefits from general company operating funds by the state.

There are two major advantages of self-insurance: direct control over losses and cash flow. The incidence of accidents may be much lower for the company in particular than the industry in general. This could make the actual cost for coverage less than the premium the company would have to pay through private insurance. In addition, because losses are paid as they occur rather than through one insurance premium payment, the company has the advantage of a better cash flow.

EMPLOYER'S ACCIDENT REPORTING REQUIREMENTS

Every employer must keep a record of all injuries, fatal or otherwise, alleged by an employee to have been sustained during his or her course of employment. The acts have a time limit requirement which obligates employers or their insurance carrier to submit a "first report" of an accident to the State's Workers' Compensation Board or Commission. The acts also provide penalty provisions against employers who fail to comply.

QUESTIONS AND ANSWERS RELATED TO WORKERS' RIGHTS UNDER WORKERS' COMPENSATION LAWS

1 Q. What must I do if I become injured in an accident on my job?

A. Report the injury immediately to your supervisor or to the medical department. In either case, be sure to follow company policy in respect to this reporting procedure.

2 Q. Must I make a written statement or a report to my employer after the occurrence of an accident?

A. No. This is generally a management function and should be written by your supervisor.

3 Q. If any supervisor asks me to sign the accident report, must I do so?

A. The law does not require that you sign the accident report unless there is a specific requirement to do so under the act in your state.

4 Q. If the injury that I have incurred in an industrial accident is the result of my own negligence, am I entitled to workers' compensation benefits?

A. Yes. Negligence or carelessness itself will not prevent you from receiving benefits under workers' compensation because it is constructed as no-fault insurance.

5 Q. For what injuries am I entitled to receive workers' compensation benefits?

A. You are entitled to receive benefits for any accidental injury or occupational disease arising out of, and in the course of, employment at the workplace.

6 Q. What constitutes an accidental injury?

A. It is an injury incurred by an employee performing a task for an employer at a designated place of employment. The injury may be a laceration, a slip, or a fall, or could have been incurred by being struck by a piece of equipment or a moving vehicle. It must be a disabling injury.

7 Q. What medical benefits am I entitled to after the occurrence of a disabling injury?

A. The act requires that the employer furnish injured employees medical, surgical, nursing, and hospital services as needed for recuperation.

8 Q. Who pays for the medical and hospital expenses?

A. The employer (if self-insured) or the employer's insurance carrier.

9 Q. Do I have the right to choose a physician if I am disabled on the job?

A. As of 1984, the employer could choose the physician in 13 states, the employee could do so in 21 states, and the balance of the states differ in the manner of "choice of physician."

10 Q. If my employer asks me to report to the company doctor for an examination or treatment, must I do so?

A. Yes, providing that the employer's request as to where and when to report is reasonable and will not jeopardize your life.

11 Q. Can I refuse to report to my employer's doctor for such an examination or treatment?

A. No, because your compensation may be suspended or entirely lost if you refuse.

12 Q. what is the statute of limitation for filing a workers' compensation claim?

A. In some states the statute of limitation is 1 year, in others it is 2 years, while in others it can be 3 years.

13 Q. Assuming that I have a disabling injury, when would my weekly compensation benefits start?

A. It depends on acts of the state in which you live. In many states, if your disability period is longer than 7 days but less than 21 days, your weekly benefits would begin from the eighth day. If your disability is longer than 21 days, then compensation benefits are retroactive from the date of injury.

14 Q. What monetary compensation benefits will I receive if I lose a specific body part such as a finger, a hand, or an eye?

A. It varies from state to state and can be found under "Income Benefits For Scheduled Injuries" in the workers' compensation laws of your state.

15 Q. If I lose a body part such as the outer phalange of a finger or thumb and can continue to work after medical treatment, will I be compensated for that loss?

A. Yes. You do not have to incur a loss of wages in order to be compensated for a specific loss.

16 Q. If my employer refuses my claim for workers' compensation benefits, do I have any other recourse?

A. Yes, you may then proceed to file a claim with your state's Workers' Compensation Board or Commission.

17 Q. What happens next?

A. Your case will most probably be assigned for a hearing before a member of the board.

18 Q. If I win my case, can my employer appeal?

A. Yes. He has the same right of appeal as you do.

19 Q. Is there a time limit for filing a claim for an occupational disease?

A. Yes. The time limit for filing a claim for an occupational disease varies from state to state.

20 Q. What should I do if I have been fired by my employer for filing a claim for workers' compensation?

A. The Supreme Court has ruled that you can file suit against your employer for damages.

21 Q. Do workers' compensation laws cover occupational disease?

A. All states recognize the employer's responsibility for occupational disease. In some states, there is coverage under the existing workers' compensation laws, while in others there is some form of an Occupational Disease Act in its statutes.

22 Q. Are civilian workers of the federal government covered under workers' compensation?

A. No. They are covered under the Federal Employees' Compensation Act.

QUESTIONS AND EXERCISES

1 What is workers' compensation?
2 Determine if workers under workers' compensation in your state have the "employee's choice of physician" clause in the act.
3 For what injuries may an employee *not* recover workers' compensation benefits?
4 Define the following degrees of disability: temporary total disability, temporary partial disability, permanent partial disability, and permanent total disability.
5 What are the criteria for recovering benefits under workers' compensation?
6 If an employer is subject to the workers' compensation laws of his or her

state, may an injured worker sue the employer for negligence in a court of law?

7 Before workers' compensation, what were employers' three common defenses that they used to protect their interests against injured employees?

8 May a worker sue an outside contractor or the manufacturer of a machine or tool that caused a personal injury to her or to him while performing work for the employer? Explain.

9 Is work-related stress covered under workers' compensation in your state?

10 Does your state have a Workers' Compensation Disease Act? If it does, what occupational diseases are covered by it?

11 What are the three methods by which employers may insure their coverage under workers' compensation?

12 An employee asked for, and received, permission from his supervisor to use the company's radial saw to do some personal work for his home. He was granted permission, and in doing so, lost two fingers while operating the saw. In your opinion, is this employee entitled to workers' compensation benefits? Explain the reasoning for your answer.

13 Have workers' compensation costs and other costs of industrial accidents imposed a burden on industry, workers, and society in general? Why?

14 Self-insurance is becoming popular among those who qualify. Can you explain why?

15 What is a statutory law in workers' compensation? How does it affect an employee?

16 Are you in favor of a federal workers' compensation law in the United States, which would cover all workers under one law? Explain the reasons for your answer.

17 Would business and industry favor a federal workers' compensation law to cover all workers under one law? Why?

18 Are workers' compensation laws a detriment for handicapped workers trying to enter into occupations in industry? Explain.

19 Before the enactment of workers' compensation laws, was the employer held responsible under contributory negligence?

20 Have workers' compensation laws become a factor in reducing workplace injuries?

BIBLIOGRAPHY

"Analysis of Workers' Compensation Laws," U. S. Chamber of Commerce, Washington, D.C., 1988, pp. vii, 16.

Hammer, Willie: *Occupational Safety Management and Engineering,* 3d ed., Prentice-Hall, Englewood Cliffs, N.J., 1985, p. 34.

Small, Ben F.: *Workmen's Compensation Law of Indiana,* The Michie Law Publishing Company, Charlottesville, Va., 1950, pp. 4, 5.

CHAPTER 10

OCCUPATIONAL SAFETY AND HEALTH ACT

CONGRESSIONAL ACT

The Occupational Safety and Health Act (OSHA) of 1970 is a landmark piece of legislation because it made safety and health on the job a matter of law for 4 million American businesses and their 57 million employees. During the year prior to the passage of this law, there were approximately 14,000 work-related fatalities and nearly 2.5 million disabling injuries. In light of these statistics, Congress recognized that action was necessary.

Purpose of OSHA

The purpose of OSHA is "to assure so far as possible every working man and woman in the nation safe and healthful working conditions and to preserve our human resources." This means that employers must provide their employees a place of employment free from recognized hazards that might cause serious injury or death. The act also requires that all employees must comply with the safety and health rules, standards, regulations, and orders issued under the occupational safety and health act.

Coverage of the Act

The authority of OSHA is extensive in its scope. In general, coverage of the act applies to all employers engaged in any kind of business—industrial, institutional, or residential—and their employees in all 50 states, the

District of Columbia, Puerto Rico, and all other territories and posses-
sions of the United States. Coverage of the act is provided either directly
by federal OSHA or through an OSHA-approved state program.

As defined by the act, an employer is any "person engaged in a business affecting commerce who has employees, but does not include the
United States or any State or political division of a State." Therefore, the
act applies to employers and employees in such varied fields as manufacturing, construction, longshoring, agriculture, law and medicine, charity
and disaster relief, organized labor, and private education. Such coverage includes religious groups to the extent that they employ workers for
secular purposes.

The following are not covered under the act:

Self-employed persons

Farms at which only immediate members of the farm employer's family are employed

Workplaces already protected by other federal agencies under other
federal statutes, such as the Atomic Energy Act and the Coal Mine
Health and Safety Act

ADMINISTRATION OF THE LAW

When the Occupational Safety and Health Act became law, there were
three federal agencies created in its administration. They were the Occupational Safety and Health Administration (OSHA), the Occupational
Safety and Health Review Commission (OSHRC), and the National Institute for Occupational Safety and Health (NIOSH).

Occupational Safety and Health Administration (OSHA)

This agency, under the Department of Labor, is responsible for:

1 Establishing mandatory safety and health standards.
2 Inspecting workplaces for compliance with the standards.
3 Proposing penalties for noncompliance with the standards.
4 Investigating fatalities and catastrophes resulting in the hospitalization of five or more employees. (Such situations must be reported by the
employer to OSHA within 48 hours.)
5 Providing free safety training for employers and employees.
6 Providing free consultation to employers to enable them to comply
with the standards.

To assist in carrying out its responsibilities, OSHA has established ten regional offices throughout the United States (including Alaska and Hawaii), as shown in Table 10-1.

Occupational Safety and Health Review Commission (OSHRC)

This commission, which is an independent agency of the government, was established to conduct hearings when citations of noncompliance with the OSHA standards and proposed monetary standards are contested by the affected employers or by their employees. The commission's actions are limited to contested cases. It can also access penalties and, when necessary, may conduct its own investigation and may affirm, modify, or vacate the findings of OSHA.

TABLE 10-1
UNITED STATES DEPARTMENT OF LABOR REGIONAL OFFICES

Region 1—Boston	Region 2—New York	Region 3—Philadelphia
Connecticut	New Jersey	Delaware
Maine	New York	District of Columbia
Massachusetts	Puerto Rico	Maryland
New Hampshire	Virgin Islands	Pennsylvania
Rhode Island		Virginia
Vermont		West Virginia

Region 4—Atlanta	Region 5—Chicago	Region 6—Dallas
Alabama	Illinois	Arkansas
Florida	Indiana	Louisiana
Georgia	Michigan	Oklahoma
Kentucky	Minnesota	Texas
Mississippi	Ohio	New Mexico
North Carolina	Wisconsin	
South Carolina		
Tennessee		

Region 7—Kansas City	Region 8—Denver	Region 9—San Francisco	Region 10—Seattle
Iowa	Colorado	Arizona	Alaska
Kansas	Montana	California	Idaho
Missouri	North Dakota	Hawaii	Oregon
Nebraska	South Dakota	Nevada	Washington
	Utah		
	Wyoming		

National Institute for Occupational Safety and Health (NIOSH)

This agency is the principal federal agency engaged in safety and health research. Its primary duties include:

1 Recommending new safety and health standards to OSHA for adoption.

2 Conducting research on various safety and health problems.

3 Investigating toxic substances and developing criteria for the use of such substances in the workplace.

4 Making workplace investigations for toxic substances when called upon to do so.

5 Conducting educational programs to provide an adequate supply of qualified personnel to carry out the purposes of the OSH Act.

6 Publishing an annual listing of all known toxic substances and the concentrations at which toxicity is known to occur.

Bureau of Labor Statistics

Also within the Department of Labor, the Bureau of Labor Statistics establishes the occupational injury and illness record-keeping requirements for employers having 11 or more employees. It also conducts statistical surveys and establishes methods for acquiring data on injuries and illnesses. Questions about record-keeping requirements and related reporting procedures can be directed to any of the OSHA regional or area director's offices or to the regional office of the Bureau of Labor Statistics.

OSHA'S OBJECTIVE

Under the OSH Act, OSHA was created to:

1 Encourage employers and employees to reduce workplace hazards and to implement new or improved existing safety and health programs.

2 Provide for research in occupational safety and health and develop innovative ways of dealing with occupational safety and health problems.

3 Establish "separate but dependent responsibilities and rights" for employers and employees for the achievement of better safety and health conditions.

4 Maintain a reporting and record-keeping system to monitor job-related injuries and illness.

5 Establish training programs to increase the number of competent occupational safety and health personnel.

6 Develop mandatory job safety and health standards, and enforce them effectively.

7 Provide for the development, analysis, evaluation, and approval of state occupational safety and health programs.

While OSHA continually reviews and redefines specific standards and practices, its basic purpose remains constant—to implement its Congressional mandate fully and fairly to all concerned.

EMPLOYER RESPONSIBILITIES UNDER OSHA

Employers have certain responsibilities under OSHA in addition to those cited in Chapter 2. An employer must "furnish to each of his employees employment and a place of employment free from recognized hazards that are causing or likely to cause death or serious physical harm to his employees." The employer must:

1 Ensure compliance with OSHA standards.

2 Keep records of work-related injuries and illnesses.

3 Maintain records of employee exposure to toxic materials and harmful physical agents.

4 Notify employees of the provisions of, their rights under, and obligations to, OSHA.

5 Minimize or reduce accidents.

6 Provide medical examinations when required by OSHA standards.

7 Report to the nearest OSHA office the prescribed fatality and accident reporting requirements within 48 hours.

8 Abate cited violations of the OSHA standards within the prescribed period.

EMPLOYEE RESPONSIBILITIES UNDER OSHA

Although OSHA does not cite employees for violations of their safety responsibilities, each employee "shall comply with all occupational safety and health standards, rules, and regulations, and orders issued under the act," that are applicable. These responsibilities include the following:

1 Complying with all applicable OSHA standards

2 Complying with employers' safety and health rules and regulations

3 Reporting hazardous conditions to the supervisor

4 Reporting job-related injuries or illnesses to the supervisor, and seeking medical treatment promptly

5 Cooperating with OSHA compliance officers during inspections by them, when asked to do so

ENFORCEMENT PROCEDURES UNDER THE OCCUPATIONAL SAFETY AND HEALTH ACT

Inspection Priorities

OSHA has established the following priorities for workplace inspections:

1 Top priority is given to imminent danger situations. An imminent danger is any condition where there is reasonable certainty that a danger exists that can be expected to cause death or serious physical harm immediately, or before the danger can be eliminated through normal enforcement procedures.

2 Second priority is given to the investigation of fatalities or catastrophic events requiring the hospitalization of five or more employees.

3 Third priority is given to employee complaints. (The complaint must describe in sufficient detail any hazard—real or alleged—that exists in the plant or on the jobsite.)

4 Next in priority are programs aimed at inspecting specified high-hazard industries, occupations, or health substances. Industries are selected for inspection in this category on the basis of statistical information published by the Bureau of Labor Statistics.

5 The last priority is the follow-up inspection designed to determine if the previously cited violations have been corrected within the specified required period. Generally, unabated violations found during the follow-up inspection incur additional daily penalties while they continue to exist.

Inspection Procedure

Entry into the Establishment   Workplace inspections are to be made during the regular business hours of the establishment, except when special circumstances may require otherwise. In preparing for the inspection, the compliance officer becomes familiar with as many relevant facts as possible about the workplace designated for inspection. This preparation may take into account such things as the history of the establishment, the nature of the business, a history of previous citations and related correspondence, and the particular standards that are likely to apply. Other preparations would include a camera and instrumentation to detect and measure noise levels and toxic substances, fumes, and gases.

In 1978, the rules governing OSHA entry into a business establishment were modified by the United States Supreme Court. In a court ruling (*Marshall v. Barlow's, Inc.*), the Supreme Court ruled that OSHA may not conduct warrantless inspections without an employer's consent. It may, however, inspect after acquiring a judicially authorized search warrant.

Opening Conference   After entering the plant, the compliance officer will meet with the employer, or a representative, to explain why the investigation is being conducted, and to have the employer designate the management representative for the walk-around inspection of the plant or jobsite. A union representative (or an employee, if there is no union) can also accompany the compliance officer during the inspection tour. Compliance regulations require that a representative of the employer and a representative of the employees shall be given the opportunity to accompany the compliance officer during the inspection. Section 8(e) of the act provides that, if there is no authorized employee representative, the compliance officer must consult with a reasonable number of employees concerning matters of safety and health in the plant or on the jobsite.

Inspection Tour   Before starting on the inspection tour, the compliance officer will inspect the plant's injury and illness records, which should be kept up-to-date by the employer.

The compliance officer must take the time necessary to inspect all aspects of the operations at the establishment being inspected. He or she may start in any area of the plant, and may take hours, days, or longer to complete the inspection, depending on the size of the plant.

The purpose of the OSHA inspection is to inspect the operation for compliance with the OSHA standards. The compliance officer will inspect machines, electrical systems, mechanical equipment, housekeeping conditions, and so on, for possible violations of the standards. Violations fall into one of six categories. These categories and the resultant penalties will be discussed later in this chapter.

Numerous apparent violations—such as blocked aisles, unsafe floor surfaces, hazardous projections, unclean toilets, and deficiencies of similar nature—can generally be corrected immediately. If corrections under these conditions can be made immediately by the management, it would display to the compliance officer an act of good faith and a desire to comply with the OSHA standards. Although corrected, an apparent violation could be the basis for a citation or proposed penalty by the compliance officer.

Closing Conference   The compliance officer notes any violations of the OSHA standards found during the inspection of the plant. A closing conference is held by the compliance officer with management officials and a separate conference is arranged with employee representatives when requested. Generally, there will be no closing conference with the employees or their representatives, since they will have communicated and participated with the compliance officer during the inspection.

Closing Conference with the Employer The compliance officer will confer with the employer or representative and will note all conditions and practices disclosed by the inspection which may constitute safety or health violations. The officer will also indicate, where possible, the applicable section or sections of the standards which have been violated. He or she will also request from the employer a time period or a promised date (abatement period) for the correction of the alleged violations found during the walk-around inspection tour.

Citations If the Area Director, on the basis of the compliance officer's report, determines that alleged violations have existed during the plant inspections, he will issue a written citation with the proposed penalties for each violation with their abatement dates, and send it to the employer by registered mail. This document is known as the *Notification of Proposed Penalty*. The employer then has 15 working days from the date of receipt within which to notify the area director's office that the citation and/or the proposed penalty will be contested. If the employer does not notify the area director that the citation and/or proposed penalty will be contested, the assessment of the penalty is final.

Strikes or Labor Disputes No inspection activities are to be initiated by a compliance officer, either at a plant or elsewhere, while that place of business, plant, or work site is involved in a strike or a labor dispute involving a work stoppage or picketing, without the approval of the area director.

Advance Notice of Inspection

Advance notices of inspection are never given by OSHA, except in special situations and under certain conditions. The exceptions for an advance notice of inspection are as follows:

1 In cases of imminent danger.

2 When the inspection can be conducted most effectively after regular business hours.

3 When it is necessary to assure the presence of employer and employee representatives.

4 When the Area Director determines that an advance notice would enhance the probability of a more thorough and effective inspection.

Confidentiality of Trade Secrets

All information reported to or otherwise obtained by the Secretary of Labor or his representative (compliance officer) with any inspection under the OSH Act, shall be considered as confidential and shall be protected by the secretary, the Occupational Safety and Health Review Commission, and the court.

Types of Violations

Imminent Danger A condition or practice that causes a danger which could be expected to cause death or serious physical harm immediately.

Serious Violation A condition or practice which has the probability of causing death or serious physical harm.

Nonserious Violation A violation of the standards which would not cause death or serious physical harm. Usually, in a nonserious violation, there is no permanent injury caused. Tripping on a level surface would be an example of such a violation. Poor housekeeping might also be an example of a nonserious violation.

De Minimis Violation A de minimis violation occurs when an employer complies with the clear intent of a standard, but deviates "in a manner which has no direct bearing on safety and health." Such a violation exists when: (1) the employer complies with a proposed amendment to a standard that provides protection equal to or greater than the original regulation, or (2) the workplace is technically advanced beyond the requirements of a particular standard and provides equal or better protection. No penalty is assessed for a de minimis violation, and OSHA may not issue a citation for failure to correct the alleged citation.

Willful Violation A violation of the standards where the employer committed an intentional and known violation of the standards and made no reasonable effort to eliminate it.

Repeated Violation A violation of the standard, where upon reinspection, a similar violation of the standards is found.

See Table 10-2 for a schedule of OSHA citations and proposed penalties assessed for the various violations. Citation and penalty procedures may differ somewhat in states with their own safety and health programs.

Additional Violations and Penalties

Additional violations for which citations and proposed penalties may be issued are as follows:

Violations of posting requirements can bring a civil penalty of up to $1,000.

TABLE 10-2
OSHA CITATIONS AND PENALTIES

Violation	Maximum penalty under law
De minimis	No penalty
nonserious	$1,000 for each violation
Serious	$1,000 for each violation
Willful, no death	$10,000 fine
Willful repeated violations	$10,000 fine and/or six months jail term
Willful, death results	$20,000 fine
Willful, death results, second violation	$20,000 fine and/or one year jail term
Faillure to correct cited violation	$1,000 per day fine
Failure to post official documents	$1,000 fine
Falsification of documents	$10,000 fine and/or six-month jail term

Assaulting a compliance officer, resisting, opposing, intimidating or interfering with compliance officers in the performance of their duties is a criminal offense, subject to a fine of not more than $5,000 and a jail term of not more than three years.

Failure to maintain the "Log and Summary of Occupational Injuries and Illnesses" (OSHA Form 200) or the Supplemental Record (OSHA Form 101 or equivalent): $100 for each form not maintained.

Failure to report to OSHA within 48 hours an on-the-job fatality or a catastrophic accident involving the hospitalization of five or more employees: $400.

Failure to notify the authorized employee representative about an advance notice of inspection: $200.

Penalty Reductions

Reductions in proposed penalties may be given for the following conditions:

- Evidence of good faith
- Size of business
- History of previous violations

What Constitutes Evidence of Good Faith?

1 Evidence of an ongoing safety program by the employer
2 The employer's desire to comply with the OSHA standards
3 The employer's record and display of safety accomplishments

Penalty Reductions for Evidence of Good Faith

1 There is a 20 percent penalty reduction if the employer has an effective ongoing safety program.

2 There is a 10 percent penalty reduction if the employer has an average safety program and corrects unsafe conditions—but with reservations.

3 There will be no reduction in penalties if the employer has no safety program.

Penalty Reductions for the Size of the Business

1 There is a 10 percent penalty reduction if the organization employs fewer than 20 employees.

2 There is a 5 percent penalty reduction for businesses employing 20 to 100 employees.

3 There are no penalty reductions for businesses employing over 100 employees.

Penalty Reductions for a History of Previous Violations

1 There is a 20 percent penalty reduction on an initial OSHA inspection having original violations.

2 There is a 10 percent penalty reduction if there are no repeat violations found on any subsequent inspections.

PROCEDURE FOR APPEALS BY EMPLOYERS

When an employer receives an OSHA citation and wishes to contest it or the proposed penalties, the employer may request an informal meeting with the OSHA area director to discuss the case. The area director has the authority to enter into settlement agreements for the purpose of revising citations and penalties in order to avoid prolonged legal disputes.

Notice of Contest

An employer who decides to contest the proposed penalty or the abatement date has 15 working days from the date of the receipt of the citation in which to do so. The notification must be made in writing to the Area Director and is called the Notice of Contest. A copy of the Notice of Contest must also be given to the employees' authorized representative. And, if the employees affected are not represented by a union, a copy of the notice must be posted in a prominent location in the workplace.

Review Process

Within 7 days of the receipt of the employer's Notice of Contest, the area director forwards the case to the Occupational Safety and Health Review Commission, which will assign the case to an administrative law judge. The commission notifies the employer and the Secretary of Labor of the commission's receipt of the notice of contest. The Secretary of Labor then becomes the *complainant* and the employer is the *respondent*. In

FIGURE 10-1
Contest (appeal) and review procedure.

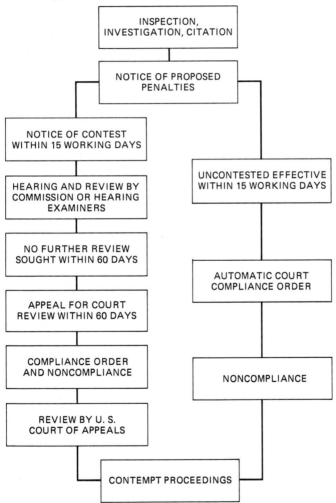

- INSPECTION, INVESTIGATION, CITATION
- NOTICE OF PROPOSED PENALTIES
- NOTICE OF CONTEST WITHIN 15 WORKING DAYS
- HEARING AND REVIEW BY COMMISSION OR HEARING EXAMINERS
- NO FURTHER REVIEW SOUGHT WITHIN 60 DAYS
- APPEAL FOR COURT REVIEW WITHIN 60 DAYS
- COMPLIANCE ORDER AND NONCOMPLIANCE
- REVIEW BY U. S. COURT OF APPEALS
- UNCONTESTED EFFECTIVE WITHIN 15 WORKING DAYS
- AUTOMATIC COURT COMPLIANCE ORDER
- NONCOMPLIANCE
- CONTEMPT PROCEEDINGS

order that the affected employees be informed of the hearing date, a copy of the commission's notice must be given to the employees' authorized representative; and if there is no union representative, a copy of the notice must be posted in a prominent location in the workplace.

After hearing the evidence presented, and after considering any written brief submitted by either party, the judge will prepare a written decision and file it with OSHRC. Each party will receive a copy of the decision, which becomes final 30 days after it is received by OSHRC.

APPEALS BY EMPLOYEES

Employees may not contest:

1 The employer's citations
2 Employer's amendments to citations
3 Penalties in the employer's citations
4 The lack of penalties

Employees may contest:

1 The time element in the citation for the abatement of a hazardous condition by the employer.
2 The employer's Petition for Modification of Abatement which requests an extension of time for correcting a hazardous condition. Employees must contest the PMA within 10 days of its posting or within 10 working days after the receipt of a copy of it by the employees' representative.

Employees have the right to request an informal conference with the area director regarding an inspection, citation, notice of proposed penalty, or the employer's Notice of Contest.

PETITION FOR MODIFICATION OF ABATEMENT

If an employer has made a sincere effort to comply with the promised abatement date, and cannot meet this date because of factors beyond his or her control, the employer can file a "Petition for Modification of Abatement," or PMA. The petition should show a request, and the reasons, for the additional time needed to complete the abatement of the hazard. Affected employees and their union representatives, if any, must also be notified of the petition. The PMA will trigger a hearing, similar to one for a Notice of Contest, before OSHRC.

WITHDRAWAL OF CONTEST

An employer who has filed a notice of contest, and at some later time decides to withdraw the case before the hearing date, may do so by:

1 Showing that the alleged violation has been abated or will be abated
2 Paying the fine for the proposed penalty
3 Informing the affected employees and their authorized representatives of the withdrawal of contest

VARIANCES FROM THE STANDARDS

Employers may request a variance from a standard or regulation from OSHA if they cannot fully comply with a standard by the effective date, due to shortages of materials, equipment, or professional or technical personnel, or can prove their facilities or methods of operation provide employee protection "at least as effective as" that required by the standard. This is a *temporary* variance. A temporary variance may be granted for the period needed to achieve compliance or for 1 year, whichever is shorter. It is renewable twice, each time for 6 months.

An employer may also ask for a *permanent* variance from a standard if she or he can prove by a preponderance of evidence that the conditions, practices, means, methods, operations, or processes proposed will provide employment and a place of employment to the employees as safe as those provided in the standards.

RECORD-KEEPING AND REPORTING REQUIREMENTS

Employers with 11 or more employees must maintain records of occupational injuries and illnesses as they occur. Employers with 10 or fewer employees are exempt from keeping such records, unless they are selected by the Bureau of Labor Statistics (BLS) to participate in the Annual Survey of Occupational Injuries and Illnesses.

The purpose of keeping occupational injury and illness records is to permit BLS survey material to be compiled, to help define high-hazard industries, and to inform employees of the status of their employer's safety record. Employers in state-plan states are required to keep the same records as employers in other states.

An *occupational injury* is an injury such as a cut, fracture, sprain, or amputation, which results from a work-related accident or from exposure involving a single incident in the work environment.

An *occupational illness* is any abnormal condition or disorder, other than one resulting from an occupational injury, caused by exposure to environmental factors associated with employment. Included are acute and chronic illnesses or diseases that may be caused by inhalation, absorption, ingestion, or direct contact with toxic substances or harmful agents.

All occupational illnesses must be recorded, regardless of severity. All occupational injuries must be recorded, if they result in:

1 Death (must be recorded, regardless of the length of time between the injury and death)
2 One or more lost workdays
3 Restriction of work or motion
4 Loss of consciousness
5 Transfer to another job
6 Medical treatment beyond first aid

RECORD-KEEPING FORMS

Record-keeping forms are maintained on a calendar year basis. They must be maintained for 5 years at the establishment and must be made available to representatives of the Occupational Safety and Health Administration, Health and Human Services, and the Bureau of Labor Statistics. Only two forms are needed for record-keeping as explained below.

Forms

OSHA Form No. 200, Log and Summary of Occupational Injuries and Illnesses Each recordable occupational injury and illness must be logged on this form within 6 working days from the time the employer learns of it. If the log is prepared at a central location by automatic data processing equipment, a copy current to within 45 calendar days must be present at all times in the establishment.

OSHA Form No. 101, Supplemental Record of Occupational Injuries and Illnesses The form OSHA No. 101 contains much more detail about each injury or illness. It must be completed within 6 working days from the time the employer learns of the work-related injury or illness. A substitute for the form OSHA No. 101 (such as insurance or workers' compensation) may be used if it contains all of the required information.

Annual Survey

Employers selected to participate in the annual statistical survey receive in the mail, soon after the close of the year, OSHA Form No. 2005 for this purpose. Each employer selected must complete this report, using form No. 200 as the source of information, and return it to the Bureau of Labor Statistics. Small business employers (employers with 10 or fewer

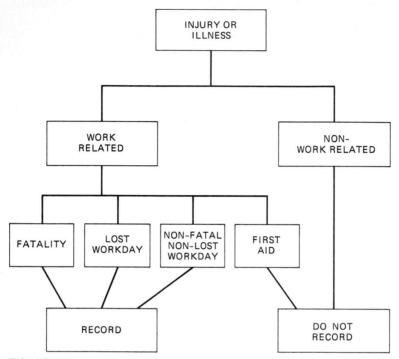

FIGURE 10-2
Conditions requiring the recording of occupational injuries and illnesses on OSHA form no. 200 by employers of 11 or more employees.

employees) selected for the survey are notified at the beginning of the year and are supplied with an OSHA Form No. 200.

A copy of the totals and the information following the fold line of the last page of OSHA Form No. 200 for the year must be posted for employees to read no later than February 1, and must be kept in place until March 1.

Many specific OSHA standards have additional record-keeping and reporting requirements.

ADDITIONAL RECORD-KEEPING REQUIREMENTS

OSHA record-keeping requirements also stipulate that certain specific standards show written records, while others may imply that a record of a specific inspection or training activity be noted. The standards themselves must be studied for more information. Table 10-3 shows OSHA standards that *require* record keeping, recording dates, and the name or

TABLE 10-3
INDEX OF STANDARDS REQUIRING RECORD KEEPING

Subject	Standards source	Activity
Injuries and illnesses	1904	Recording occupational injuries and illnesses
Exposure and medical records	1910.20	Access to employee exposure and medical records
Outrigger scaffolds	1910.28(e)(3)	Designed by a licensed PE
Powered platforms	1910.66(e)(3)	Maintenance, inspection, testing
Manlifts	1910.68(e)(3)	Inspections
Emergencies	1910.38(a)(1)	Emergency action plan and fire-prevention plan
Noise	1910.95(c)	Written hearing conservation program
Noise	1910.95(d)	Employee noise exposures

initials of the person responsible for the recording. Table 10-4 shows OSHA standards that *imply* record keeping, recording dates, and the person responsible for the recording.

There are approximately 74 general industry and construction standards that require records, and at least 140 standards that imply records should be kept. Employers should develop their own checklists by re-

TABLE 10-4
INDEX OF STANDARDS THAT IMPLY THAT RECORDS BE KEPT

Subject	Standards source	Activity
Floors	1910.22(d)	Floor loading capacities
Ladders	1910.15(d)(1)(x)	Inspection of wooden ladders
Ladders	1910.26(c)(2)(iv)	Inspection of metal ladders
Scaffolds	1910.28(d)(14)	Inspection of welded frames on scaffolds
Scaffolds	1910.28(f)(2)	Certification of hoisting machines on mason's adjustable multiple-point suspension scaffold
Scaffolds	1910.28(f)(11)	Inspection of mason's adjustable multiple-point suspension scaffold
Scaffolds	1910.28(g)(3)	Two-point suspension scaffold hoisting machines certification
Scaffolds	1910.28(g)(8)	Inspection of two-point suspension scaffold

searching each general industry, construction, agriculture, or maritime OSHA standard to determine if a written record is required.

POSTING REQUIREMENTS FOR EMPLOYERS

Employers are responsible for keeping employees informed about OSHA and about the various safety and health matters with which they and the employer are involved. Federal OSHA regulations and those states with their own occupational safety and health programs, require employers to post certain materials at a prominent location or locations in the workplace. These materials include:

1 "Job Safety and Health Protection" workplace poster (OSHA Form No. 2203), or the state's equivalent poster.

2 Copies of OSHA citations for violations of the OSHA standards. They must be posted at or near the location of the alleged violations for three days, or until the violations are abated, whichever is longer.

3 Copies of summaries of petitions for variances from any of the standards, including record-keeping procedures.

4 The "Log and Summary of Occupational Injuries and Illnesses" (OSHA Form No. 200). The summary portion of the log must be posted annually no later than February 1 of the succeeding calendar year and remain posted until March 1.

TABLE 10-5
OSHA ENFORCEMENT ACTIVITY

Inspections*	1980	1987	Percent change
Total	63,363	59,071	↓ 7%
Health	11,871	9,722	↓ 18
Accident	2,286	1,458	↓ 36
Manufacturing	27,224	17,384	↓ 36
After a complaint	16,093	9,743	↓ 39
Violations*			
Willful	1,238	529	↓ 57%
Serious	44,645	36,982	↓ 13
Workers covered by inspections†			
Workers	3.67	2.46	↓ 33%

*In number for fiscal years ending September 30.
†In millions for fiscal years ending September 30.
Source: U.S. Senate Labor and Human Resources Committee, April, 1988.

STATE PLANS

One reason that the Occupational Safety and Health Act was passed was the inadequacy of existing state plans. In many instances, state safety codes and regulations were lacking in their provisions and enforcement because of scanty funding and the lack of personnel empowered to enforce the laws. Although the new law is a federal law, it permits the states to gain sole authority to administer safety, if the specific conditions are met.

Any state can assume responsibility for its own occupational safety and health program by submitting its plan to the Secretary of Labor. Such plans are approved by the Secretary of Labor under the following conditions:

1 The state must create an agency to carry out the plan.

2 The state's plan must include safety standards and regulations. Their enforcement must be at least as effective as the federal plan.

3 The state's plan must include provisions for rights of entry and inspection of the workplace, including a prohibition on advance notice of inspections.

4 The state's plan must also cover state and local government employees.

The U. S. Department of Labor–OSHA will pay for half the cost of the approved state program. It will also monitor state programs for 3 years prior to certification of the program. States with their own occupational safety and health programs also have a state system for review and appeal of citations, penalties, and abatement periods. The procedures are generally similar to the federal OSHA system, but cases are heard by a state review board or its equivalent authority.

Monitoring State Plans

Immediately after approving a state plan, OSHA implements a monitoring system to determine (1) whether the state is meeting the commitments contained in this plan and (2) whether its program will be "as effective as" the federal OSHA program. When OSHA monitors a state's plan, it uses the following methods:

Quarterly Reports On a quarterly basis, the state submits a summary of its enforcement and standards activities. OSHA analyzes the state's progress towards standards and enforcement goals.

Semi-Annual Evaluations OSHA conducts investigations of the state's performance and summarizes its findings in a comprehensive report submitted to the state each 6 months. The state has the opportunity to respond to this report and to the program recommendations contained in it. The investigation covers:

State inspections
State Compliance Officer's qualifications
Case file reviews of inspections
Proper equipment
Development procedures
Variances
State OSHRC procedures

The regional OSHA office will forward any recommendations regarding a state plan to the Assistant Secretary of Labor.

CONSULTATIVE ASSISTANCE SERVICES

OSHA makes free consultative services available to employers requesting advice and help for the purpose of establishing and maintaining a safe and healthful workplace.

The process begins with the employer's request for consultation and a commitment to correct any serious job safety and health hazards identified by the OSHA consultant, who is usually employed by OSHA on a contract basis for this purpose. No penalties are proposed or citations issued for hazards identified by the consultant.

Besides helping employers to identify and correct specific hazards, consultation can also include assistance in developing and implementing safety and health programs. Any conditions of "imminent danger" identified by the consultant must be resolved immediately. If immediate action is not taken, OSHA requires that it be immediately notified by the consultant, so that appropriate enforcement action can be taken as prescribed by regulations.

WORD POWER

act or OSH act   The Williams-Steiger Occupational Safety and Health Act of 1970.
Assistant Secretary of Labor   The Assistant Secretary of Labor for Occupational Safety and Health.
commerce   Any trade, traffic, commerce, transportation, or communication among several states or between a state and any place outside or thereof.
establishment   A single physical location where business is conducted or where services or industrial operations are performed. Examples are a factory, hotel, farm, movie theater, warehouse, or a gasoline service station.
hazardous substance   A substance or material which can be explosive, flammable, poisonous, corrosive, or otherwise harmful to the body.

shall Indicates provisions in the standards which are mandatory.

standard A pattern of guidance by which the quality, the construction of, or the safety of a product or place of employment may be adopted as uniform.

QUESTIONS AND EXERCISES

1 Name the 3 federal agencies that were created under the OSH Act.

2 How many regional offices does OSHA have?

3 Name the required equipment that a compliance officer may need in the course of an official plant inspection.

4 An employer has _____ days in which to contest a citation, its penalties, or the abatement date.

5 Within the review process, the Secretary of Labor is called the _____ and the employer is called the _____ .

6 Employers with _____ or more employees are required to maintain records of occupational injuries and illnesses.

7 OSHA must forward a Notice of Contest to OSHRC within how many days after its receipt from the employer?

8 Who has the authority to approve state plans for OSHA certification?

9 In your opinion, what would be considered as a demonstration of good faith by an employer? What are some of the factors that should be taken into consideration when making such a determination?

10 Explain the functions of OSHA, OSHRC, and NIOSH.

11 In the order of their importance, what are OSHA's inspection priorities?

12 What was the Supreme Court's decision in *Marshall v. Barlow?*

13 Is the employer required to give a copy of a Notice of Contest to the employee's authorized representative?

14 Is an area director authorized to enter into a settlement agreement with employers when they are issued a citation showing proposed penalties and abatement dates?

15 What is a Petition for Modification of Abatement?

16 If employees are not allowed to contest the employer's citation, amendments to citations, penalties, or lack of penalties, what can they contest?

17 Once a state plan has been approved by OSHA, it will fund up to _____ of the program's operating costs.

18 The workplace poster informing employees of their rights and responsibilities under OSHA is called the _____ .

19 What is considered to be an occupational injury or illness?

20 The Annual Survey of Occupational Injuries and Illnesses is conducted by the _____ .

21 An employer may ask OSHA for a _____ from a standard if he or she can prove that his or her method of operation provides employee protection "as good as" that provided by OSHA.

22 Employers may ask OSHA for a _____ from a standard if they cannot abate a hazardous condition by the effective date.

23 Under what conditions may OSHA give an employer an Advance Notice of Inspection?

24 Are workplaces protected by other federal agencies under other federal statutes covered under OSHA?

25 Name at least three examples of notices or records that are required to be posted in a prominent place by OSHA.

26 How are employee representatives to accompany the compliance officer during an inspection chosen?

27 What should the compliance officer discuss with the employer during the closing conference?

28 What differences should be taken into account when determining the difference between an imminent danger violation and a serious violation?

29 Name at least five employer responsibilities under OSHA.

30 Are compliance officers required to make inspections while the establishment is involved in a labor problem or a work stoppage?

BIBLIOGRAPHY

All About OSHA, rev. eds., U. S. Department of Labor, Occupational Health and Safety Administration, OSHA 2056, 1982, p. 2; 1985, p. 3.

Coble, David F.: "OSHA Recordkeeping—Are You Ready?," *Professional Safety,* December 1987, p. 23.

Hammer, Willie: *Occupational Safety Management and Engineering,* 3d ed., Prentice-Hall, Englewood Cliffs, N.J., p. 53.

How OSHA Monitors State Plans, Programs and Policy Series, U. S. Department of Labor, Occupational Safety and Health Administration, OSHA 2221, January 1980, pp. 2–3.

The Occupational Safety and Health Act, Public Law 91–596, 1970, sec. 6 (A), p. 5; sec. 8(D), p. 7; sec. 15, p. 17.

Recordkeeping Guidelines for Occupational Injuries and Illnesses, U. S. Department of Labor, Bureau of Labor Statistics, OMB 1220–0029.

Vincoli, Jeffrey W.: "OSHA Investigation Refresher," *Professional Safety,* March 1988, p. 26.

CHAPTER 11

OCCUPATIONAL SAFETY AND HEALTH

GENERAL INDUSTRY STANDARDS

A standard is a pattern of guidance by which the quality, the construction, or the safety of a product or place of employment may be adopted as uniform for the manufacturer, the consumer, and the public in general.

In carrying out its duties, OSHA is responsible for promulgating legally enforceable standards. OSHA standards are safety standards which may require conditions, or the adoption or use of one or more practices, means, methods, operations, or processes necessary to protect workers on the job.

OSHA standards fall into four major categories: General Industry, Maritime, Construction, and Agriculture. In this chapter, we will be working with the General Industry Standards. One of the best sources of information on the standards is the *Federal Register,* where all OSHA standards are published when adopted, as are all amendments, corrections, insertions, or deletions. It is published daily, Monday through Friday, and provides a system to make publicly available the regulations and legal notices issued by all federal agencies.

The Code of Federal Regulations (CFR) contains the OSHA general and permanent rules and regulations previously released in the *Federal Register.* The code is published annually in paperback volumes, available from the Superintendent of Documents, U. S. Government Printing Office, Washington, D.C. 20402.

A major factor that affected the preparation of the first OSHA standards was the lack of personnel knowledgeable in safety and health standards. Therefore, OSHA gathered the consensus standards issued by such organizations as the American National Standards Institute, National Fire Protection Association, American Society of Mechanical Engineers, National Electrical Code, National Safety Council, and many others. These standards were originally known as consensus standards, or voluntary standards. Then, when the OSH Act became the law, these standards became regulatory or mandatory standards. The words "shall" and "required" in the standards indicate provisions which are mandatory. Other safety standards previously issued under other federal safety laws had already been mandatory.

The greatest complaints about OSHA concerned many of the so-called "nitpicking" rules and regulations considered by both industry and employees as more a nuisance than a help in protecting employees on the job. Over the span of the past 10 years, OSHA has eliminated many of those rules, thereby concentrating on reducing or eliminating the more serious safety and health hazards.

Although this chapter does not cover all of the OSHA General Industry safety and health standards, the selection of those presented is based on those standards generally applicable to almost all places of employment. They are laid out in alphabetical order with the appropriate standards number placed after each title.

To find information about a hazard in the *Code of Federal Regulations,* the user should first read the introductory statement to each category. Then, the user should analyze the hazard and classify it under the following categories: the workplace, machines and equipment, materials, the employee, special process, or power source. By looking up the applicable category, the location of the appropriate standard in 29 CFR (Code of Federal Regulation) can be found. Since hazards may be logically classified in more than one category, look under each category which might be appropriate to a given situation.

It should be recognized that the information contained in these standards is only a digest and should in no way be considered as a complete substitute for any provision of the OSH Act or for any standards promulgated under the act.

General Industry Safety and Health Regulations, Part 1910

1 *Abrasive blasting* (Standards Source 1910.344)
   a Blast-cleaning nozzles shall be equipped with an operating valve which must be held open manually (deadman control). A support shall be provided on which the nozzle may be mounted when not in use.

   b The concentration of respirable dust or fumes in the breathing zone of the abrasive-blasting operator or any other worker shall be below the levels specified in 1910.93.

   c Blast-cleaning enclosures shall be exhaust-ventilated in such a way that a continuous inward flow of air will be maintained at all openings in the enclosure during the blasting operation.

   d The air for abrasive-blasting respirators shall be free of harmful quantities of contaminants.

 2 *Abrasive grinding* (Standards Source 1910.215)

   a Abrasive wheels shall be used only on machines provided with safety guards, with the following exceptions:

    (1) Wheels used for internal work while within the work being grounded.

FIGURE 11-1
Abrasive grinders, like the one shown here, must conform to Standards Source 1910.215 of the OSHA General Industry Standards.

(2) Mounted wheels used in portable operations, 2 inches and smaller in diameter.

(3) Type 16, 17, 18, 18R, and 19 cones, plugs, and threaded-hole pot balls where the work offers protection.

b Abrasive wheel safety guards shall cover the spindle ends and nut and flange projections, except the following:

(1) Safety guards on all operations where the work provides a suitable measure of protection to the operator may be so constructed that the spindle end, nut, and outer flange are exposed.

(2) Where the nature of the work is such as to entirely cover the side of the wheel, the side covers of the guard may be omitted.

(3) The spindle end, nut, and outer flanges may be exposed on machines designed as portable saws.

c An adjustable work rest of rigid construction shall be used to support the work on offhand grinding machines. Work rests shall be kept adjusted closely to the wheel, with a maximum clearance of one-eighth inch.

d Dry-grinding machines shall be provided with a suitable hood or closure connected to an exhaust system if airborne contaminants exceed allowable levels.

e Machines shall be securely anchored to prevent movement or designed so that in normal operation they will not move.

3 *Air receivers, compressed* (Standards Source 1910.169)

a Air receivers should be supported with sufficient clearance to permit a complete external inspection and to avoid corrosion of external surfaces.

b Air receivers shall be installed so that drains, handholes, and manholes are easily accessible.

c Every air receiver shall be equipped with an indicating pressure gage located so as to be readily visible and with one or more spring-loaded safety valves.

4 *Air tools* (Standards Source 1910.243)

a For portable tools, a tool retainer shall be installed on each piece of utilization equipment which without such a retainer may eject the tool.

b Hose and hose connections used for conducting compressed air to utilization equipment shall be designed for the pressure and service to which they are subjected.

5 *Aisles and passageways* (Standards Source 1910.22)

a Where mechanical handling equipment is used, sufficient safe clearance shall be allowed for aisles at loading docks, through doorways, and whenever turns or passage must be made.

FIGURE 11-2
A permanent aisle which is well marked, is kept clear, and has no obstructions in or across it that could create a hazard. This is a typical example of the requirements in Standards Source 1910.22.

   b Aisles and passageways shall be kept clear and in good repair, with no obstructions across or in aisles that could create hazards.

   c Permanent aisles and passageways shall be appropriately marked.

6 *Belt sanding machines* (Standards Source 1910.213)

   a Belt sanding machines shall be provided with guards at each nip point where the sanding belt runs onto the pulley.

   b The unused run of the sanding belt shall be guarded against accidental contact.

7 *Calenders, mills, and rolls* (Standards Source 1910.216)

   a A safety trip-type bar, rod, cable to activate an emergency stop switch shall be installed on calenders, rolls, or mills to prevent persons or parts of the body from being caught between the rolls.

   b A fixed guard across the front and one across the back of the mill, approximately 40 inches vertically above the working level and 20 inches horizontally from the crown face of the roll, should be used where applicable.

8 *Chains, Cables, Ropes, Hooks* (Standards Source 1910.179)

   a Chains, cables, ropes, slings, etc., shall be inspected daily, and defective gear shall be removed and repaired or replaced.

   b Hoist chains and hoist ropes shall be free from kinks or twists and shall not be wrapped around the load.

c All U-bolt wire clips or hoist ropes shall be installed so that the U-bolt is in contact with the dead end (the short or non-load-carrying end) of the rope. Clips shall be installed in accordance with the clip manufacturer's recommendation. All nuts or newly installed clips shall be retightened after one hour of use.

9 *Chipguards* (Standards Source 1910.242)

In operations involving cleaning with compressed air, protective shields and barriers shall be provided to protect personnel against flying chips or other such hazards.

10 *Chlorinated hydrocarbons* (Standards Source 1910.93)

a Carbon tetrachloride or other chlorinated (halogenated) hydrocarbons shall not be used where the airborne concentration exceeds the threshold limit value (TLV) listed.

b Degreasing and other cleaning operations involving chlorinated hydrocarbons shall be located so that vapors from these operations will not reach or be drawn into the atmosphere surrounding any welding operations.

11 *Compressed air, use of* (Standards Source 1910.242)

Compressed air used for cleaning purposes shall not exceed 30 psi when the nozzle end is obstructed or dead-ended, and must be used with effective chip guarding and personal protective equipment.

12 *Conveyors* (Standards Source 1910.265)

a Conveyors installed within 7 feet of a floor or walkway shall be provided with crossovers at aisleways or other passageways (sawmills only).

b Where conveyors 7 feet or more above the floor pass over working areas, aisles, or thoroughfares, suitable guards shall be provided to protect personnel from the hazard of falling materials (sawmills only).

c Open hoppers and chutes shall be guarded by standard railings and toeboards or by some other comparable safety device.

13 *Cranes and hoists (overhead and gantry)* (Standards Source 1910.179)

a All functional operating mechanisms, air and hydraulic systems, chains, rope slings, hooks, and other lifting equipment shall be visually inspected daily.

b Complete inspection of the crane shall be performed frequently (daily to monthly) and periodically (at 1- to 12-month intervals).

c Overhead cranes shall have stops at the limit of travel of the trolley. Bridge and trolley bumpers or equivalent automatic devices shall be provided. Bridge trucks shall have rail sweeps.

d The rated load of the crane shall be plainly marked on each side of the crane. If the crane has more than one hoisting unit, each hoist shall have its rated load marked on it or its load block, and this marking shall be clearly legible from the ground or floor.

14 *Cylinders, compressed gas, used in welding* (Standards Source 1910.252)

    a Compressed gas cylinders shall be kept away from excessive heat, shall not be stored where they might be damaged or knocked over by passing or falling objects, and shall be stored at least 20 feet away from highly combustible materials.

    b Where a cylinder is designed to accept a valve protection cap, caps shall be in place except when the cylinder is in use or is connected for use.

    c Acetylene cylinders shall be stored and used in a vertical valve-end-up position only.

    d Oxygen cylinders in storage shall be separated from fuel-gas cylinders or combustible materials (especially oil or grease) by a minimum distance of 20 feet or by a noncombustible barrier at least 5 feet high having a fire-resistance rating of at least one half-hour.

15 *Dip tanks containing flammable or combustible liquid* (Standards Source 1910.108)

    a Dip tanks of over 150 gallons capacity or 10 square feet in liquid surface area shall be equipped with a properly trapped overflow pipe leading to a safe location outside the buildings.

    b There shall be no open flames, spark-producing devices, or heated surfaces having a temperature sufficient to ignite vapors in or within 20 feet of any vapor area.

    c All dip tanks, except hardening and tempering tanks exceeding 150 gallons liquid capacity or having a liquid surface area exceeding 4 square feet, shall be protected with at least one of the following automatic extinguishing facilities:

        (1) Water spray system

        (2) Foam system

        (3) Carbon dioxide system

        (4) Dry chemical system

        (5) Automatic dip-tank cover

16 *Dockboards* (Standards Source 1910.30)

    a Dockboards shall be strong enough to carry the load imposed on them.

    b Portable dockboards shall be anchored or equipped with devices which will prevent them from slipping. They shall have handholds or other effective devices to allow safe handling.

    c Positive means shall be provided to prevent railroad cars from being moved while dockboards are in position.

17 *Drains for flammable and combustible liquids* (Standards Source 1910.106)

a Emergency drainage systems shall be provided to direct any leakage of flammable liquid or fire-protection water to a safe location.

b Emergency drainage systems for flammable liquids, if connected to public sewers or discharged into public waterways, shall be equipped with traps or separators.

18 *Drill-presses* (Standards Source 1910.219)

The V-belt drive of all drill presses, including the usual front and rear pulleys, shall be guarded to protect the operator from contact or breakage.

19 *Drinking water* (Standards Source 1910.141)

a Portable water shall be provided in all places of employment.

b The nozzle of a drinking fountain shall be set at such an angle that the jet of water will not splash back down on the nozzle; and the end of the nozzle shall be protected by a guard to prevent a person's mouth or nose from coming into contact with the nozzle.

c Portable drinking-water dispensers shall be designed and serviced to ensure sanitary conditions, shall be capable of being closed, and shall have a tap. Unused disposable cups shall be kept in a sanitary container, and a receptacle shall be provided for used cups. The common drinking cup is prohibited.

20 *Electrical installations* (Standards Source 1910.309)

a Every new electrical installation and all new equipment installed, replaced, modified, repaired, or rehabilitated after March 15, 1972, shall comply with the provisions of the 1971 National Electrical Code, NFPA 70-1971; ANSI C1-1971 (Revision of 1968).

b Electrical installations not covered by the preceding paragraph shall comply with the articles and sections of the 1971 National Electrical Code that are set forth in standards source 1910.309(a).

21 *Emergency flushing, eyes and body* (Standards Source 1910.151)

Where the eyes or body of any person may be exposed to injurious corrosive materials, suitable facilities for quick drenching or flushing of the eyes and body shall be provided within the work area for immediate emergency use.

22 *Exits* (Standards Source 1910.36)

a Every building designed for human occupancy shall be provided with exits sufficient to permit the prompt escape of occupants in case of emergency.

b In hazardous areas or where employees may be endangered by the blocking of any single means of egress by fire or smoke, there shall be at least two means of egress remote from each other.

c Exits and the way of approach to and travel from exits shall be maintained so that they are unobstructed and are accessible at all times.

FIGURE 11-3
Covers of electrical boxes should never be left open. Leaving them open creates the
danger of electrical shock and violates OSHA regulations.

d All exits shall discharge directly onto the street or some other open
space that gives safe access to a public way.
e Exit doors serving more than 50 people or at high-hazard areas
shall swing in the direction of travel.
f Exits shall be marked by readily visible, suitably illuminated signs.
Exit signs shall be distinctive in color and provide a contrast with

FIGURE 11-4
Electrical installations should not be obstructed in any way.

their surroundings. The word "EXIT" shall consist of plainly legible letters, not less than 6 inches high.

23 *Explosives and blasting agents* (Standards Source 1910.109)

   a All explosives shall be kept in approved magazines.

   b Stored packages of explosives shall be laid flat with top side up. Black powder, when stored in magazines with other explosives, shall be stored separately.

   c Smoking, matches, open flames, spark-producing devices, and fire-arms (except firearms carried by guards) shall not be permitted inside or within 50 feet of magazines. The land surrounding a magazine shall be kept clear of all combustible materials for a distance of at least 25 feet. Combustible materials shall not be stored within 50 feet of magazines.

24 *Eye and face protection* (Standards Source 1910.133)

   a Protective eye and face equipment shall be required where there is reasonable probability of injury that can be prevented by such equipment.

b Eye- and face-protection equipment shall be in compliance with ANSI 287.1—1968, "Practice for Occupational and Educational Eye and Face Protection."

25 *Fan blades* (Standards Source 1910.212)

When the periphery of the blades of a fan is less than 7 feet above the floor on working level, the blades shall be guarded. The guard shall have the openings no larger than ½ inch. The use of fabric nets with ½-inch maximum openings to modify existing substandard guards is acceptable.

26 *Fire protection* (Standards Source 1910.157)

a Portable fire extinguishers suitable to the conditions and hazards involved shall be provided and maintained in an effective operating condition.

b Portable fire extinguishers shall be conspicuously located and mounted where they will be readily accessible. Extinguishers shall not be obstructed or obscured from view.

c Portable fire extinguishers shall be given maintenance service at least once a year. A durable tag must be securely attached to show the maintenance or recharge date.

d In storage areas, clearance between sprinkler systems deflectors and the top of storage varies with the type of storage. For combustible material stored over 15 feet but not more than 21 feet high in solid piles, or over 12 feet but not more than 21 feet high in piles that contain horizontal channels, the minimum clearance shall be 36 inches. The minimum clearance for smaller piles or for noncombustible materials shall be 18 inches.

27 *Flammable liquids incidental to principal business* (Standards Source 1910.106)

a Flammable liquids shall be kept in covered containers when not actually in use.

b Flammable and combustible liquids shall be drawn from or transferred into containers within a building only through a closed piping system, from safety cans, by means of a device drawing through the top, or by gravity through an approved self-closing valve. Transferring by means of pressure shall be prohibited.

c Inside storage rooms for flammable and combustible liquids shall be of fire-resistant construction and shall have self-closing fire doors at all openings, 4-inch sills or depressed floors, a ventilation system that provides at least six air changes within the room per hour, and—in areas used for storage of class I liquids—electrical wiring approved for use in hazardous locations.

FIGURE 11-5
Portable fire extinguishers should be conspicuously mounted where they will be readily accessible, not obstructed (as shown here), or obscured from view.

    d Outside storage areas shall be graded so as to divert spills away from buildings or other exposures, or be surrounded with curbs or dikes at least 6 inches high with appropriate drainage to a safe location for accumulated liquids. The area shall be protected against tampering or trespassing, where necessary, and shall be kept free of weeds, debris, and other combustible material not necessary to the storage.

    e Areas where flammable liquids with flashpoints below 100° Fahrenheit are used shall be ventilated at a rate of not less than 1 cubic foot per minute per square foot of solid area.

28 *Floors, general conditions* (Standards Source 1910.22)

    a All floor surfaces shall be kept clean, dry, and free from protruding nails, splinters, loose boards, holes, and projections.

    b Where wet processes are used, drainage shall be maintained and base floors, platforms, mats, or other dry standing places should be provided where practicable.

29 *Floor loading limit* (Standards Source 1910.22)

    a Floors in buildings used for mercantile, business, industrial, or storage purposes, other than those resting directly on solid ground, shall be posted to show maximum safe floor loads.

30 *Floor openings, hatchways, open sides, etc.* (Standards Source 1910.23)

    a Floor openings shall be guarded by a standard railing on all exposed sides or be protected by a suitable cover.

    b Open-sided floors, platforms, etc., 4 feet or more above the adjacent floor or ground level shall be guarded by a standard railing on all open sides, except where there is an entrance to a ramp, stairway, or fixed ladder.

31 *Forklift trucks* (Standards Source 1910.178)

    a All new forklift trucks acquired and used after February 15, 1972, shall comply with ANSI B56.1—1969, "Power Industrial Trucks, Part II." Approved trucks shall bear a label indicating approval.

    b High-lift trucks shall be equipped with a substantial overhead guard unless operating conditions do not permit.

    c Fork trucks shall be equipped with a vertical-load back-rest extension when the type of load presents a hazard to the operator.

    d The brakes of highway trucks shall be set, and wheel shocks placed under the rear wheels, to prevent the truck from rolling while being boarded with forklift trucks.

    e Wheel stops or other recognized protection shall be provided to prevent railroad cars from moving while they are boarded with forklift trucks.

32 *General-duty clause* [PL91-596, Section 5(a)(1)]

Hazardous conditions or practices not covered in an OSHA standard may be covered under Section 5(a)(1) of the act, which states: "Each employer shall furnish to each of his employees employment and a place of employment which are free from recognized hazards that are causing or are likely to cause death or serious physical harm to his employees."

33 *Guards, construction of* (Standards Source 1910.219)

Guards for mechanical power-transmission equipment shall be made of metal or other suitable material. Wood guards may be used in the woodworking and chemical industries, in industries where atmospheric conditions would rapidly deteriorate metal guards, or where temperature extremes make metal guards undesirable.

34 *Head protection* (Standards Source 1910.132)

Head-protection equipment (helmets) shall be worn when there is a

FIGURE 11-6
A mechanical power press with all guards in place and constructed in accordance with Standards Source 1910.219.

possible danger of head injuries from impact, flying or falling objects, or electrical shock and burns.

35 *Hand tools* (Standards Source 1910.242)

a Each employer shall be responsible for the safe condition of tools and equipment used by employees, including tools and equipment which may be furnished by employees.

b All hand tools shall be kept in safe condition. Handles of tools shall be kept tight in the tool, and wooden handles shall be free of splinters or cracks. Wedges, chisels, etc., shall not have mushroomed heads. Wrenches shall not be used when sprung to the point that slippage occurs.

36 *Housekeeping* (Standards Source 1910.22)

All places of employment, passageways, storerooms, and service rooms shall be kept clean and orderly and in a sanitary condition.

37 *Jointers* (Standards Source 1910.213)

a Each hand-fed planer and jointer with a horizontal head shall be equipped with a cylindrical cutting head. The opening in the table shall be kept as small as possible.

b Each hand-fed jointer with a horizontal cutting head shall have an automatic guard which will cover the section of the head on the working side of the fence or gage.

c A jointer guard shall automatically adjust itself to cover the unused portion of the head and shall remain in contact with the material at all times.

d Each hand-fed jointer with a horizontal cutting head shall have a guard which will cover the section of the head back of the gage or fence.

38 *Ladders, fixed* (Standards Source 1910.27)

a All fixed ladders shall be designed for a minimum concentrated live load of 200 pounds.

b All rungs shall have a minimum diameter of ¾ inch, if metal, or 1⅛ inches, if wood. They shall be a minimum of 16 inches in clear length and be spaced uniformly no more than 12 inches apart.

c Metal ladders shall be painted or treated to resist corrosion or rusting when the location demands.

d Cages, wells, or safety devices for ladders affixed to towers, watertanks or chimneys shall be provided on all ladders more than 20 feet long. Landing platforms shall be provided each 30 feet of length; but where no cage is provided, landing platforms shall be provided for every 20 feet of length.

e Tops of cages on fixed ladders shall extend 42 inches above top of landing, unless other acceptable protection is provided, and the bottom of the cage shall be not less than 7 feet or more than 8 feet above the base of the ladder.

f Side rails shall extend 3½ feet above the landing.

39 *Ladders, portable* (Standards Source 1910.25)

   a The maximum length for portable wood ladders shall be: step, 20 feet; single straight ladders, 30 feet; two-section extension ladders, 60 feet; sectional ladders, 60 feet; trestle ladders, 20 feet; platform stepladders, 20 feet; painters' stepladders, 12 feet; masons' ladders, 40 feet.

   b Stepladders shall be equipped with a metal spreader or locking device of sufficient size and strength to securely hold the front and back sections in open position.

   c Non-self-supporting ladders shall be erected on a sound base at a 4-1 pitch and placed to prevent slipping.

   d The top of the ladder used to gain access to a roof should extend at least 3 feet above the point of contact.

   e Wooden ladders should be kept coated with a suitable protective material.

40 *Lunchrooms* (Standards Source 1910.141)

   a Employees shall not consume food or beverages in toilet rooms or any area exposed to a toxic material.

   b Covered receptacles of corrosion-resistant or disposable material shall be provided in lunch areas for disposal of waste food. The cover may be omitted where sanitary conditions can be maintained without the use of a cover.

41 *Machine guarding* (Standards Source 1910.212)

One or more methods of machine guarding shall be provided to protect the operator and other employees in the machine area from hazards such as those created by point of operation, ingoing nip points, rotating parts, and flying chips or sparks.

42 *Machinery, fixed* (Standards Source 1910.212)

Machines designed for a fixed location shall be securely anchored to prevent ''walking'' or moving or designed in such a manner that they will not move in normal operation.

43 *Mats, insulating* (Standards Source 1910.309)

Where motors or controllers operating at more than 150 volts to ground are guarded against accidental contact only by location, and where adjustment or other attendance may be necessary during operations, suitable insulating mats or platforms shall be provided.

44 *Medical services and first aid* (Standards Source 1910.151)

   a The employer shall ensure the ready availability of medical personnel for advice and consultation on matters of plant health.

   b When a medical facility for treatment of injured employees is not available near the workplace, a person or persons shall be trained to render first aid.

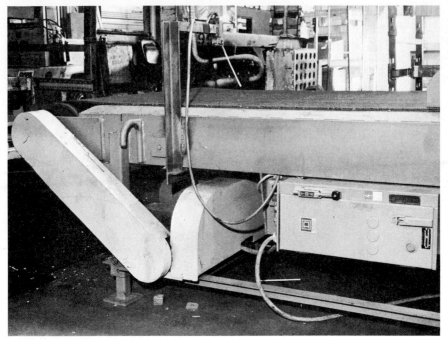

FIGURE 11-7
This conveyor is fully guarded against all possible ingoing nip points and rotating parts.

   c First-aid supplies approved by the consulting physician shall be readily available.

45 *Noise exposure* (Standards Source 1910.95)

   a Protection against the effects of occupational noise exposure shall be provided when the sound levels exceed those shown in Table G-16 of the Safety and Health Standards. Feasible engineering or administrative controls (or both) shall be utilized to keep exposure below the allowable limit.

   b When engineering or administrative controls fail to reduce the noise level to within the levels of Table G-16 of the Safety and Health Standards, personal protective equipment shall be provided and used to reduce the noise to an acceptable level.

   c Exposure to impulsive or impact noise should not exceed 140 decibels (dB) peak sound-pressure level.

   d In all cases where the sound levels exceed the values shown in Table G-16 of the Safety and Health Standards, a continuing, effective hearing-conservation program shall be administered.

46 *Personal protective equipment* (Standards Source 1910.132)

   a Proper personal protective equipment, including shields and bar-riers, shall be provided, used, and maintained in a sanitary and reliable condition where there is a hazard from processes or the environment that may cause injury or illness to the employee.

   b Where employees furnish their own personal protective equipment, the employer shall be responsible for its adequacy and its proper maintenance in a sanitary condition.

47 *Power transmission, mechanical* (Standards Source 1910.219)

   a All belts, pulleys, chains, flywheels, shafting, and shaft projections or other rotating or reciprocating parts within 7 feet of the floor or working platform shall be effectively guarded.

   b Belts, pulleys, and shafting located in rooms used exclusively for power-transmission apparatus need not be guarded when the following requirements are met:

     (1) The basement, tower, or room occupied by transmission equipment is locked against unauthorized entrance.

     (2) The vertical clearance in passageways between the floor and power-transmission beams, ceiling, or any other objects is not less than 5 feet 6 inches.

     (3) The intensity of illumination conforms to the requirements of ANSI A11.1—1965 (R1970).

     (4) The footing is dry, firm, and level.

     (5) The route followed by the oiler is protected in a manner that will prevent accidents.

48 *Punch presses* (Standards Source 1910.217)

   a It shall be the responsibility of the employer to provide and ensure the usage of "point-of-operation" guards or properly applied and adjusted point-of-operation devices on every operation performed on a mechanical power press. This requirement shall not apply when the point-of-operation opening is ¼ inch or less.

   b A substantial guard shall be placed over the treadle of foot-operated presses.

   c Pedal counterweights, if provided on foot-operated presses, shall have the path of travel of the weight enclosed.

49 *Radiation* (Standards Source 1910.96, 1910.97)

   a Employers shall be responsible for proper controls to prevent any employee from being exposed to radiation, either ionizing or electromagnetic, in excess of acceptable limits.

   b Each radiation area shall be conspicuously posted with appropriate signs, barriers, or both. Employers shall maintain records of the ra-

FIGURE 11-8
Guarded foot treadle.

    diation exposure of all employees for whom personnel monitoring is required.

    c Employers shall maintain records of the radiation exposure of all employees for whom personnel monitoring is requested.

50 *Railings* (Standards Source 1910.23)

    a A standard railing shall consist of a top rail, an intermediate rail, and posts; it shall have a vertical height of 42 inches from upper surfaces of top rail to floor, platform, etc.

    b A railing for open-sided floors, platforms, and runways shall have a toeboard whenever, beneath the open side, persons can pass, there is moving machinery, or there is equipment with which falling materials could cause a hazard.

    c Railings shall be of such construction that the complete structure shall be capable of withstanding a load of at least 200 pounds in any direction on any point on the top rail.

    d A stair railing shall be of construction similar to a standard railing,

FIGURE 11-9
Unguarded foot treadle.

but the vertical height shall be not more than 34 inches nor less than 30 inches from upper surface of top rail to the surface of the tread in line with the face of the riser at the forward edge of the tread.

e See Toeboards, No. 63.

51 *Revolving drums* (Standards Source 1910.212)

Revolving drums, barrels, or containers shall be guarded by an interlocked enclosure that will prevent the drum, etc., from revolving unless the guard enclosure is in place.

52 *Saws, band* (Standards Source 1910.213)

a All portions of bandsaw blades shall be enclosed or guarded except for the working portion of the blade between the bottom of the guide rolls and the table.

b Bandsaw wheels shall be fully encased. The outside periphery of the enclosure shall be solid. The front and back shall be either solid or wire mesh or perforated metal.

53 *Saws, portable circular* (Standards Source 1910.243)

All portable power-driven circular saws having a blade diameter greater than 2 inches shall be equipped with guards above and below the base plate or shoe. The lower guards shall cover the saw to the depth of the teeth, except for the minimum arc required to permit the base plate to be tilted for bevel cuts, and shall automatically return to the covering position when the blade is withdrawn from the work. (This provision does not apply to circular saws used in the meat industry for cutting meat.)

54 *Saws, radial* (Standards Source 1910.213)

   a Radial saws shall have an upper guard which completely encloses the upper half of the saw blade. The sides of the lower exposed portion of the blade shall be guarded by a device that will automatically adjust to the thickness of, and remain in contact with, the material being cut.

   b Radial saws used for ripping shall have non-kickback fingers or dogs.

   c Radial saws shall be installed so that the cutting head will return to the starting position when released by the operator.

55 *Saws, swing or sliding cutoff* (Standards Source 1910.213)

   a All swing saws or sliding cutoff saws shall be provided with a hood that will completely enclose the upper half of the saw.

   b Limit stops shall be provided to prevent swing- or sliding-type cutoff saws from extending beyond the front or back edges of the table.

   c Each swing or sliding cutoff saw shall be provided with an effective device to return the saw automatically to the back of the table when released at any point of its travel.

   d Inverted sawing or sliding cutoff saws shall be provided with a hood that will cover the part of the saw that protrudes above the top of the table or material being cut.

56 *Saws, table* (Standards Source 1910.213)

   a Circular table saws shall have a hood over the portion of the saw above the table, mounted so that the hood will automatically adjust itself to the thickness of, and remain in contact with, the material being cut.

   b Circular table saws shall have a spreader aligned with the blade, spaced no more than ½ inch behind the largest blade mounted in the saw. (The provision of a spreader in connection with grooving, dadoing, or rabbiting is not required.)

   c Circular table saws used for ripping shall have non-kickback fingers or dogs.

   d Feed rolls and blades of self-feed circular saws shall be protected by a hood or guard to prevent the hands of the operator from coming into contact with the in-running rolls at any point.

57 *Scaffolds* (Standards Source 1910.28)

    a All scaffolds and their supports shall be capable of supporting the load they are designed to carry with a safety factor of at least 4.

    b All planking shall be Scaffold Grade, as recognized by grading rules for the species of wood used. The maximum permissible spans for 2 × 9 inch or wider planks are shown in Table 11-1 below. The maximum permissible span for 1¼ × 9 inch or wider planks of full thickness is 4 feet, with medium loading of 50 pounds per square foot.

TABLE 11-1

	Material				
	Full-thickness undressed lumber			Nominal-thickness lumber	
Working load, pounds per square foot	25	50	75	25	50
Permissible span, feet	10	8	6	8	6

    c Scaffold planking shall be overlapped a minimum of 12 inches or secured from movement.

    d Scaffold planks shall extend over their end supports not less than 6 inches or more than 18 inches. Railings and toeboards shall be installed on all open sides and ends of platforms more than 10 feet above the floor. There shall be a screen with ½-inch maximum openings between the toeboard and the midrail where persons are required to work or pass under the scaffold.

58 *Spray-finishing operations* (Standards Source 1910.94)

    a All spray-finishing shall be conducted in spray booths or in spray rooms.

    b There shall be no open-flame or spark-producing equipment in any spraying areas or within 20 feet of spraying areas, unless separated by a partition.

    c Spray booths and spray rooms shall be constructed of metal or other substantial noncombustible material.

    d Electrical wiring and equipment located in any spraying area shall be of explosion-proof type, approved for class I, group D locations.

    e All spraying areas shall be provided with mechanical ventilation adequate to remove flammable vapors, mists, or powders to a safe location and to confine and control combustible residues so that life is not endangered.

f Electric motors driving exhaust fans shall not be placed inside flammable-materials spray booths or ducts. Belts or pulleys within the booth or duct shall be thoroughly enclosed.

g The quantity of flammable or combustible liquid kept in the vicinity of spraying operations shall be the minimum required for operations and should ordinarily not exceed a supply for one day or one shift.

h Conspicuous "NO SMOKING" signs shall be posted at all flammable-materials spraying areas and storage rooms.

59 *Stairs, fixed industrial* (Standards Source 1910.23, 1910.24)

a Every flight of stairs having four or more risers shall be provided with a standard railing on all open sides. Handrails shall be provided on at least one side of closed stairways, preferably on the right side descending.

b Stairs shall be constructed so that rise height and tread width are uniform throughout.

c Fixed stairways shall have a minimum width of 22 inches.

d See Railings, No. 50(d), for stair-railing design requirements.

60 *Stationary electrical devices* (Standards Source 1910.309)

All stationary electrically powered equipment, tools, and devices located within reach of a person who can make contact with any grounded surface or object shall be grounded.

61 *Storage* (Standards Source 1910.176)

a All storage shall be stacked, blocked, interlocked, and limited in height so that it is secure against sliding or collapse.

b Storage areas shall be kept free from accumulation of materials that constitute hazards or harbor pests. Vegetation control will be exercised when necessary.

c Where mechanical handling equipment is used, sufficient safe clearance shall be allowed for aisles, at loading docks, through doorways, etc.

62 *Tanks, open-surface* (Standards Source 1910.94)

Where ventilation is used to control potential exposures to employees, it shall be adequate to reduce the concentration of the air contaminated to the degree that a hazard to employees does not exist.

63 *Toeboards* (Standards Source 1910.23)

a Railings protecting floor openings, platforms, scaffolds, etc., shall be equipped with toeboards whenever, beneath the open side, a person can pass, there is moving machinery, or there is equipment with which falling material could cause a hazard.

b A standard toeboard shall be at least 4 inches in height and may be of any substantial material, either solid or open, with openings not to exceed 1 inch in greatest dimension.

c See Railings, No. 5.

64 *Toilets* (Standards Source 1910.141)

    a Every place of employment shall be provided with adequate toilet facilities which are separate for each sex. Water closets shall be provided according to the following: 1 to 15 persons, one facility; 16 to 35 persons, two facilities; 36 to 55 persons, three facilities; 56 to 80 persons, four facilities; 81 to 110 persons, five facilities; 111 to 150 persons, six facilities; over 150 persons, one more facility for each additional 40 persons.

    b Separate toilet rooms for each sex need not be provided where toilet rooms will be occupied by no more than one person at a time, can be locked from the inside, and contain at least one water closet.

    c Each water closet shall occupy a separate compartment with a door and walls or partitions between fixtures sufficiently high to ensure privacy.

    d Requirements (a) and (b) above do not apply to mobile crews or normally unattended locations, as long as employees have transportation immediately available to nearby toilet facilities.

    e Adequate washing facilities shall be provided in every toilet room or be adjacent thereto.

    f Covered receptacles shall be kept in all toilet rooms used by women.

65 *Toxic vapors, gases, mists, and dusts* (Standards Source 1910.93)

    a Exposure to toxic vapors, gases, mists, or dusts at a concentration above the threshold limit values contained or referred to in Safety and Health Standards, shall be avoided.

    b To achieve compliance with paragraph (a) above, administrative or engineering controls must first be determined and implemented whenever feasible. When such controls are not feasible to achieve full compliance, protective equipment or any other protective measures shall be used to keep the exposure of employees to air contaminants within the limits prescribed. Any equipment or technical measures used for this purpose must be approved for each particular use by a competent industrial hygienist or other technically qualified person.

66 *Trash* (Standards Source 1910.141)

Trash and rubbish shall be collected and removed so as to avoid creating a menace to health and as often as necessary to maintain good sanitary conditions.

67 *Washing facilities* (Standards Source 1910.141)

    a Adequate washing facilities shall be provided in every place of employment and maintained in a sanitary condition. For industrial establishments, at least one lavatory with adequate hot and cold water shall be provided for every 10 employees up to 100 persons, and one lavatory for each 15 persons over 100 persons.

b A suitable cleansing agent, individual hand towels, or some other approved apparatus for drying the hands, and receptacles for disposing of hand towels, shall be provided at washing facilities.

68 *Welding* (Standards Source 1910.252; see also Cylinders, Compressed Gas, No. 14)

a Welding equipment shall be chosen for safe application to the work and shall be installed properly. Employees designated to operate welding equipment shall be properly instructed and qualified to operate it.

b Mechanical ventilation shall be provided when welding or cutting:

(1) Beryllium, cadmium, lead, zinc, or mercury.

(2) Fluxes, metal coatings, or other material containing fluorine compounds.

(3) Where there is less than 10,000 cubic feet of space per welder.

(4) Where the overhead height is less than 16 feet.

(5) In confined spaces.

c Proper shielding and eye protection to prevent exposure of personnel from welding hazards shall be provided.

d Proper precautions (isolating welding and cutting, removing fire hazards from the vicinity, providing a fire watch, etc.) for fire prevention shall be taken in areas where welding or other "hot work" is done.

e Work and electrode lead cables shall be frequently inspected. Cables with damaged insulation or exposed bare conductors shall be replaced.

69 *Woodworking machinery* (Standards Source 1910.213)

a All woodworking machinery such as table saws, swing saws, radial saws, bandsaws, jointers, tenoning machines, boring and mortising machines, shapers, planers, lathes, sanders, veneer cutters, and other miscellaneous woodworking machinery shall be effectively guarded to protect the operator and other employees from hazards inherent to their operation.

b A power-control device shall be provided on each machine to make it possible for the operator to shut off the power to the machine without leaving his or her position at the point of operation.

c Power controls and operating controls should be located within easy reach of an operator while at the regular work location, making it unnecessary for the operator to reach over the cutter to make adjustments. (This does not apply to constant-pressure controls used only for setup purposes.)

d Each operating treadle shall be protected against unexpected or accidental tripping.

QUESTIONS AND EXERCISES

The purpose of these questions and exercises, some of which are hypothetical situations, is to teach the student how to research specific items in the Standards and relate them to facilities, processes, or equipment in the workplace. A copy of the Code of Federal Regulations, Part 29, General Industry Standards would be most helpful for these exercises.

1 *Situation:* Assume that your plant has no medical or first aid services available to its employees. (a) Do the OSHA standards cover medical services? (b) What are the requirements for this situation under the regulations?
2 Do the standards have any requirements for exit signs? If so, what is the standards number and what are its requirements?
3 In 1910.176 (a), what are the requirements for aisles and passageways?
4 *Situation:* Your plant has 150 employees. What are management's responsibilities and requirements regarding Employee Emergency Plans and Fire Prevention Plans?
5 Which standards source requires medical examinations for employees using respirators? How often must respirators be inspected?
6 What are the requirements for using compressed air for cleaning purposes?
7 *Situation:* Management has installed a machine it feels is impossible to install a guard on, to prevent injury to the operator and to other employees. Under the standards, can management legally operate the machine?
8 Can an employer comply with 1910.95, Occupational Noise Exposure, by requiring employees to wear hearing protection, or must other control methods be utilized first?
9 Under OSHA requirements, the maximum sound level to which an employee can be exposed for a 6-hour day is: (a) 90dB(A), (b) 92dB(A), (c) 95dB(A), or (d) 97dB(A)?
10 Where did OSHA get its standard for Occupational Head Protection? Name the source, source number, and title.
11 *Situation:* A plant manager has been notified that a compliance officer has arrived at the plant for the purpose of making an inspection. The plant manager feels that the plant is not ready for an inspection. May she or he refuse entry to the compliance officer, and if so, what are the legal rights for doing so?
12 Where did the standards requirements for rubber protective equipment for electrical workers originate? Name the source.
13 Which standards source contains the requirement for mechanical ventilation for welders, when welding in confined spaces?
14 Which standards source requires that all woodworking machinery be guarded to protect operators and other employees from the hazards inherent to their operation?
15 Do the standards require that employers furnish a lunchroom for their employees?
16 Is the employer required to furnish drinking water at the workplace or work site?
17 What is the General Duty Clause? When can it be used by the compliance officer?

18 In 1910.212(b), why must a machine designed for a fixed location be securely anchored?

19 What is a toeboard, and what are the physical requirements for its usage?

20 May spray-painting operations be conducted in an open area of the plant in order to obtain better ventilation for the operator?

21 When are handrails required on stairs?

22 What is the minimum permissible width for stairways?

23 What are the housekeeping requirements for a factory or plant?

24 Is the employer responsible for the safe condition of personal tools owned and used by employees in the workplace?

25 Under what conditions are machine guards constructed of wood more desirable than those fabricated of metal?

26 When must eye protection be utilized?

27 What is the requirement for exit doors serving more than 50 people?

28 When is it required for a workplace to have a person or persons trained in first aid?

29 Do the standards allow for wooden ladders to be painted?

30 *Situation:* If a Japanese manufacturer moved his plant to the United States, and brought Japanese workers from Japan to work in it, must the company abide by the OSH Act?

BIBLIOGRAPHY

Code of Federal Regulations, Part 29: *General Industry Standards,* U. S. Department of Labor, Occupational Safety and Health Administration, July 1, 1985.

General Industry, OSHA Safety and Health Standards Digest, rev., U. S. Department of Safety and Health Administration, OSHA 2201, September, 1983.

12

CONSTRUCTION INDUSTRY SAFETY AND HEALTH STANDARDS

According to the Bureau of Labor Statistics, the construction industry—which employs over a million workers—was considered the most dangerous industry in the United States, with 15.2 occupational injuries and illnesses per every 100 full-time employees in 1985.

The construction industry abounds with many skills and trades—each with its own inherent dangers. They include ironworkers, riggers, carpenters, welders, plumbers, concrete workers, and many others. The three major parts of the body affected in injuries in the construction industry are the trunk, the lower extremities, and hands and fingers. Fractures, strains and sprains, and cuts and abrasions are the leading kinds of injuries incurred.

Excavation and trenching cave-ins result in an estimated 100 fatalities annually in the United States. With little or no warning, an unsupported, improperly shored or sloped trench or excavation wall can collapse, trapping workers below in seconds. For each fatality, there are an estimated 50 related, serious injuries annually. In addition to human losses due to excavating and trenching accidents, the financial costs can be staggering—property damage, work stoppage, and workers' compensation, among others.

PLANNING FOR SAFETY

Most on-the-job problems and accidents are a direct result of inadequate planning when preparing the bid. After work has begun, correcting mistakes in construction slows down the operation, adds to the cost, and increases construction business failure. The contractor must build safety into the prebid planning in the same way that all of the other prebid factors are considered.

Although this chapter does not contain all of the Construction Industry Safety and Health Standards, those selected cover approximately 90 percent of the basic applicable standards. They are laid out in alphabetical order, with the standards reference number placed after each title.

CONSTRUCTION INDUSTRY SAFETY AND HEALTH STANDARDS, PART 1926

1 *Abrasive Grinding* (Standards Source 1926.303)
   a All abrasive wheel bench and stand grinders shall be provided with safety guards which cover the spindle ends, nut and flange projections, and are strong enough to withstand the effects of a bursting wheel.
   b An adjustable work rest of rig construction shall be used on floor and bench-mounted grinders, with the work rest kept adjusted to a clearance not to exceed one-eighth inch between the work rest and the surface of the wheel.
   c All abrasive wheels shall be closely inspected and ring-tested before mounting to ensure that they are free from defect.
2 *Air tools* (Standards Source 1926.302)
   a Pneumatic power tools shall be secured to the hose or whip so that they cannot be accidentally disconnected.
   b Safety clips or retainers shall be securely installed and maintained on pneumatic impact tools to prevent attachments from being accidentally expelled.
   c The manufacturer's safe operating pressure for all fittings shall not be exceeded.
   d All hoses exceeding ½-inch inside diameter shall have a safety device at the source of supply or branch line to reduce pressure in case of hose failure.
3 *Belt sanding machines* (Standards Source 1926.304)
   a Belt sanding machines shall be provided with guards at each nip point where the sanding belt runs onto a pulley.
   b The unused run of the sanding belt shall be guarded against accidental contact.
4 *Compressed air, use of* (Standards Source 1926.302)
   a Compressed air used for cleaning purposes shall not exceed 30 psi, and then only with effective chip guarding and personal protective equipment.

b This requirement does not apply to concrete form, mill scale, and similar operations.

5 *Compressed gas cylinders* (Standards Source 1926.350)

  a Valve protection caps shall be in place when compressed gas cylinders are transported, moved, or stored.

  b Cylinder valves shall be closed when work is finished and when cylinders are empty, or are moved.

  c Compressed gas cylinders shall be secured in an upright position at all times, except if necessary for short periods of time when cylinders are actually being hoisted or carried.

  d Cylinders shall be kept at a safe distance or shielded from welding or cutting operations. Cylinders shall be placed where they cannot become part of an electrical circuit.

6 *Concrete, concrete forms, and shoring* (Standards Source 1926.700)

  a All equipment and material used shall comply with American National Standards Institute A10.9-1970, ''Safety Requirements for Concrete Construction and Masonry Work.''

  b Employees shall not be permitted to work above vertically protruding reinforcing steel, unless it has been protected to eliminate the hazard of impalement.

  c Powered and rotating-type concrete troweling machines that are manually guided shall be equipped with a deadman-type operating control.

  d Formwork and shoring shall safely support all loads imposed during concrete placement. Drawings or plans of jack layout, formwork, shoring, working docks, and scaffolding systems shall be available at the jobsite.

7 *Cranes and derricks* (Standards Source 1926.550)

  a The employer shall comply with the manufacturer's specifications and limitations where available.

  b Rated load capacities, recommended operating speeds, and special hazard warnings or instructions shall be conspicuously posted on all equipment. Instructions or warnings shall be visible from the operator's station.

  c Equipment shall be inspected by a competent person before each use and during use, and all deficiencies corrected before further use.

  d Accessible areas within the swing radius of the rear of the rotating superstructure shall be properly barricaded to prevent employees from being struck or crushed by the crane.

  e Except where electrical distribution and transmission lines have been de-energized and visibly grounded at the point of work, or where insulating barriers not a part of, or an attachment to, the equipment or machinery have been erected to prevent physical contact with the lines, no part of a crane or its load shall be operated within 10 feet of a line rated 50 kv or below; 10 feet + 0.4 inches for each 1 kv over 50 kv for lines rated over 50 kv, or twice the length of the line insulation, but never less than 10 feet.

f An annual inspection of the hoisting machinery shall be made by a competent person or by a government or private agency recognized by the U. S. Department of Labor. Records shall be kept of the dates and results of each inspection.

g All crawlers, truck, or locomotive cranes in use shall meet the requirements as prescribed in the American National Standards Institute B30.5-1968, "Safety Code for Crawler, Locomotive, and Truck Cranes."

8 *Disposable chutes* (Standards Source 1926.252)

a Whenever materials are dropped more than 20 feet to any exterior point of a building, an enclosed chute shall be used.

b When debris is dropped through holes in the floor without the use of chutes, the area where the material is dropped shall be enclosed with barricades not less than 42 inches high and not less than 6 feet back from the projected edges of the opening above. Warning signs of the hazard of falling material shall be posted at each level.

9 *Drinking water* (Standards Source 1926.51)

a An adequate supply of potable water shall be provided in all places of employment.

b Portable drinking water containers shall be capable of being tightly closed and be equipped with a tap.

c The common drinking cup is prohibited.

d Unused disposable cups shall be kept in sanitary containers, and a receptacle shall be provided for used cups.

10 *Electrical—general* (Standards Source 1926.400)

a All electrical work shall be in compliance with the 1971 National Electrical Code, unless provided by OSHA regulations.

b No employer shall permit an employee to work in such proximity to any part of an electric power circuit that the worker may contact the same in the course of his or her work unless the employee is protected against electric shock by de-energizing the circuit and grounding it or by guarding it by effective insulation or other means. In work areas where the exact location of underground electric power lines is unknown, workers using jack-hammers, bars, or other hand tools which may contact an energized line shall be provided with insulated protective gloves.

11 *Electrical grounding* (Standards Source 1926.400)

For 15-ampere and 20-ampere receptacle outlets on single-phase, 120-

volt circuits for construction sites which are not a part of the permanent wiring of the building or structure, the employer shall use either ground-fault circuit interrupters or an assured equipment grounding conductor program for employee protection.

12 *Excavating and trenching* (Standards Source 1926.651)

   a Before opening any excavation, efforts shall be made including utility company contact to determine if there are underground utilities in the area, and they shall be located and supported during the excavation process.

   b The walls and faces of trenches 5 feet or more deep and all excavations in which employees are exposed to danger from moving ground or cave-in shall be guarded by a shoring system, sloping of the ground, or some other equivalent means.

   c In excavations which employees may be required to enter, excavated or other material shall be effectively stored and retained at least 2 feet or more from the edge of the excavation.

13 *Excavating the trenching inspections* (Standards Source 1926.650)

Daily inspections of excavations shall be made by a competent person. If evidence of possible cave-ins or slides is apparent, all work in the excavation shall cease until the necessary precautions have been taken to safeguard the employees.

14 *Explosives and blasting* (Standards Source 1926.904)

Explosive material shall be stored in approved facilities as required by provisions of the Internal Revenue Service regulations published in 27 Code of Federal Regulations 181, "Commerce in Explosives."

15 *Eye and face protection* (Standards Source 1926.102)

   a Eye and face protection shall be provided when machines or operations present potential eye or face injury.

   b Eye and face protective equipment shall meet the requirements of ANSI 287.1-1986, "Practice for Occupational and Educational Eye and Face Protection."

   c Employees involved in welding operations shall be furnished with filter lenses or plates of at least the proper shade number.

16 *Fire protection* (Standards Source 1926.150)

   a A fire fighting program is to be followed throughout all phases of the construction and demolition work involved. It shall provide for effective fire fighting equipment to be available without delay, and designed to effectively meet all fire hazards as they occur.

   b Fire fighting equipment shall be conspicuously located and readily accessible at all times, shall be periodically inspected, and be maintained in operating condition.

   c Carbon tetrachloride and other toxic vaporizing liquid fire extinguishers are prohibited.

   d If the building under construction includes the installation of automatic sprinkler protection, the installation shall closely follow the construction and be placed in service, as soon as applicable laws permit, following completion of each story.

e A fire extinguisher, rated not less than 2A, shall be provided for each 3,000 square feet of the protected building area, or major fraction thereof. Travel distance from any point of the protected area to the nearest fire extinguisher shall not exceed 100 feet.

f One or more fire extinguishers, rated not less than 2A, shall be provided on each floor. In multistory buildings, at least one fire extinguisher shall be located adjacent to stairways.

g The employer shall establish an alarm system at the work site so that employees and the local fire department can be alerted for an emergency.

17 *Flagmen* (Standards Source 1926.201)

a When signs, signals, and barricades do not provide necessary protection on or adjacent to a highway or street, flagmen or other appropriate traffic controls shall be provided.

b Flagmen shall be provided with and shall wear a red or orange warning garment while flagging. Warning garments worn at night shall be of reflectorized material.

18 *Flammable and combustible liquids* (Standards Source 1926.152)

a Only approved containers and portable tanks shall be used for storage and handling of flammable and combustible liquids.

b No more than 25 gallons of flammable or combustible liquids shall be stored in a room outside of an approved storage cabinet. No more than 60 gallons of flammable or 120 gallons of combustible liquids shall be stored in any one storage cabinet. No more than three storage cabinets may be located in a single storage area.

c Inside storage rooms for flammable and combustible liquids shall be of fire-resistive construction, have self-closing fire doors at all openings, 4-inch sills or depressed floors, a ventilation system that provides at least six air changes within the room per hour, and have electrical wiring at equipment approved for Class I, Division I locations.

d Storage in containers outside buildings shall not exceed 1,100 gallons in any one pile or area. The storage area shall be graded to divert possible spills away from building or other exposures, or shall be surrounded by a curb or dike. Storage areas shall be located at least 20 feet from any building and shall be free from weeds, debris, and other combustible materials not necessary to the storage.

e Flammable liquids shall be kept in closed containers when not actually in use.

f Conspicuous and legible signs prohibiting smoking shall be posted in service and refueling areas.

19 *Floor openings, open sides, hatchways, etc.* (Standards Source 1926.500)

a Floor openings shall be guarded by a standard railing and toeboards or cover. In general, the railing shall be provided on all exposed sides, except at entrance to stairways.

b Every open-sided floor or platform, 6 feet or more above adjacent floor or ground level, shall be guarded by a standard railing, or the equivalent, on all open sides except where there is entrance to a ramp, stairway, or fixed ladder.

c Runways 4 feet or more high shall have standard railings on all sides, except runways 18 inches or more wide used exclusively for special purposes may have the railing on one side omitted where operating conditions necessitate.

d Ladderway floor openings or platforms shall be guarded by standard railings with standard toeboards on all exposed sides, except at entrance to opening, with the passage through the railing either provided with a swinging gate or so offset that a person cannot walk directly into the opening.

e Temporary floor openings shall have standard railings.

f Floor holes into which persons can accidentally walk shall be guarded by either a standard railing with standard toeboard on all exposed sides, or a standard floor hole cover. When the cover is not in place, the floor hole shall be protected by a standard railing.

20 *Gases, vapors, fumes, dusts, and mists* (Standards Source 1926.500)

a Exposure to toxic gases, vapors, fumes, dusts, and mists at concentrations above those specified in the "Threshold Limit Values of Airborne Contaminants for 1970" of the American Conference of Governmental Industrial Hygienists.

b Administrative or engineering controls must be implemented whenever feasible to comply with Threshold Limit Values.

c When engineering and administrative controls are not feasible to achieve full compliance, protective equipment or other protective measures shall be used to keep the exposure of employees to air contaminants within the limits prescribed. Any equipment and technical measures used for this purpose must first be approved for each particular use by a competent industrial hygienist or other technically qualified person.

21 *General duty clause* (Public Law 91-596)

Hazardous conditions or practices not covered in an OSHA standard may be covered under section 5(a)(1) of the Occupational Safety and Health Act of 1970 which states: "Each employer shall furnish to each of his employees employment and a place of employment which are free from recognized hazards that are causing or are likely to cause death or serious physical harm to his employees.''

22 *General requirements* (Standards Source 1926.20)

a The employer shall initiate and maintain such programs as may be necessary to provide for frequent and regular inspections of the job site, materials, and equipment.

b The employer shall instruct each employee in the recognition and avoidance of unsafe conditions and in the regulations applicable to his or her work environment to control or eliminate any hazards or other exposure to illness or injury.

c The use of machinery, tools, material, or equipment which is not in compliance with any applicable requirements of Part 1926 is prohibited.

23 *Hand tools* (Standards Source 1926.301)

a Employers shall not issue or permit the use of unsafe hand tools.

b Wrenches shall not be used when jaws are sprung to the point that slippage occurs. Impact tools shall be kept free of mushroomed heads. The wooden handles of tools shall be kept free of splinters or cracks and shall be kept tight in the tool.

24 *Head protection* (Standards Source 1926.100)

a Head protective equipment (helmets) shall be worn in areas where there is a possible danger of head injuries from impact, flying or falling objects, or electrical shock and burns.

b Helmets for protection against impact and penetration of falling and flying objects shall meet the requirements of ANSI 289.1-1969.

c Helmets for protection against electrical shock and burns shall meet the requirements of ANSI 289.2-1971.

25 *Hearing protection* (Standards Source 1926.52)

a Feasible engineering or administrative controls shall be utilized to protect employees against sound levels in excess of those shown in Table D-1. (For OSH Act Table D-2, see Table 12-1.)

b When engineering or administrative controls fail to reduce sound levels within the limits of Table D-2, ear protective devices shall be provided and used.

c Exposure to impulse or impact noise should not exceed 140 decibels peak sound pressure level.

d In all cases where the sound levels exceed the values shown in Table D-2 of the Safety and Health Standards, a continuing, effective hearing conservation program shall be administered.

26 *Heating devices, temporary* (Standards Source 1926.154)

Solid fuel salamanders used for heating purposes are prohibited in buildings and on scaffolds.

TABLE 12-1
PERMISSIBLE NOISE EXPOSURES

(OSH Act Table D-2)	
Duration/day, hours	Sound level, dB(A)
8	90
6	92
4	95
3	97
2	100
1½	102
1	105
½	110
¼ or less	115

27 *Hoists, materials and personnel* (Standards Source 1926.552)
   a The employer shall comply with the manufacturer's specifications and limitations.
   b Rated load capacities, recommended operating speeds, and special hazard warnings or instructions shall be posted on hoist cars and platforms.
   c Hoistway entrances of material hoists shall be protected by substantial full-width gates or bars.
   d Hoistway doors or gates of personnel hoists shall be not less than 6 feet 6 inches tall, and be protected with mechanical locks which cannot be operated from the landing side and are accessible only to persons on the car.
   e Overhead protective coverings shall be provided on the top of the hoists's cage or platform.
   f All material hoists shall conform to the requirements of ANSI A10.5-1969, "Safety Requirements for Material Hoists."
28 *Housekeeping* (Standards Source 1926.25)
   a Forms and scrap lumber with protruding nails and all other debris, shall be kept clear from all work areas.
   b Combustible scrap and debris shall be removed at regular intervals.
   c Containers shall be provided for collection and separation of all refuse. Covers shall be provided on containers used for flammable or harmful substances.
   d Wastes shall be disposed of at frequent intervals.
29 *Illumination* (Standards Source 1926.56)
   Construction areas, ramps, runways, corridors, offices, shops, and storage areas shall be lighted not less than the minimum illumination

intensities listed in Table D-3 while any work is in progress. (For OSH Act Table D-3, see Table 12-2 below.)

30 *Jointers* (Standards Source 1926.304)

    a Each hand-fed planer and jointer with a horizontal head shall be equipped with a cylindrical cutting head. The opening in the table shall be kept as small as possible.

    b Each hand-fed jointer with a horizontal cutting head shall have an automatic guard which will cover the section of the head on the working side of the fence or cage.

    c A jointer guard shall automatically adjust itself to cover the unused portion of the head, and shall remain in contact with the material at all times.

    d Each hand-fed jointer with a horizontal cutting head shall have a guard which will cover the section of the head back of the cage or fence.

31 *Ladders* (Standards Source 1926.450)

    a The use of ladders with broken or missing rungs or steps, broken or split side rails, or with other faulty or defective construction is prohibited. When ladders with such defects are discovered, they shall immediately be withdrawn from service.

    b Portable ladders shall be placed on a substantial base at a 4–1 pitch, have clear access at top and bottom, extend a minimum of 36 inches above the landing, or where not practical, be provided with grab rails and be secured against movement while in use.

    c Portable metal ladders shall not be used for electrical work or where they may contact electrical conductors.

TABLE 12-2
MINIMUM ILLUMINATION INTENSITIES IN FOOT-CANDLES

Foot-candles	(OSH Act Table D-3) Area or operation
5	General construction area lighting
3	General construction areas, concrete placement, active storage areas, loading platforms, refueling and field maintenance areas
5	Indoor: warehouses, corridors, hallways, and exitways
5	Tunnels, shafts, and general underground work areas
10	General construction plant and shops (e.g., batch plants, screening plants, mechanical and electrical equipment rooms, carpenter shops, rigging lofts and active storerooms, mess halls, indoor toilets, and workrooms)
30	First aid stations, infirmaries, and offices

d Job-made ladders shall be constructed for their intended use. Cleats shall be inset into side rails ½ inch, or filler blocks used. Cleats shall be uniformly spaced, 12 inches, top-to-top.

e Except where either permanent or temporary stairways or suitable ramps or runways are provided, ladders shall be used to give safe access to all elevations.

32 *Liquefied petroleum gas* (Standards Source 1926.153)

a Each system shall have containers, valves, connectors, manifold valve assemblies, and regulators of an approved type.

b All cylinders shall meet Department of Transportation specifications.

c Every container and vaporizer shall be provided with one or more approved safety relief valves or devices.

d Containers shall be placed firmly upright on firm foundations or otherwise firmly secured.

e Portable heaters shall be equipped with an approved automatic device to shut off the flow of gas in the event of flame failure.

f Storage of LPG within buildings is prohibited.

g Storage locations shall have at least one approved portable fire-extinguisher, rated not less than 20–B:C.

33 *Medical services and first aid* (Standards Source 1926.50)

a The employer shall ensure the availability of medical personnel for advice and consultation on matters of occupational health.

b When a medical facility is not reasonably accessible for the treatment of injured employees, a person trained to render first aid shall be available at the work site.

c First aid supplies approved by the consulting physician shall be readily available.

d The telephone numbers of the physicians, hospitals, or ambulances shall be conspicuously posted.

34 *Motor vehicles and mechanized equipment* (Standards Source 1926.601)

a All vehicles in use shall be checked at the beginning of each shift to assure that all parts, equipment, and accessories that affect safe operation are in proper operating condition and free from defects. All defects shall be corrected before the vehicle is placed in service.

b No employer shall use any motor vehicle, earthmoving, or compacting equipment having an obstructed view to the rear unless: The vehicle has a reverse signal alarm distinguishable from the surrounding noise level, or, the vehicle is backed up only when an observer signals that it is safe to do so.

35 *Personal protective equipment* (Standards Sources 1926.28, 1926.104, 1926.106)

a The employer is responsible for requiring the wearing of appropriate personal protective equipment in all operations where there is an exposure to hazardous conditions or where the need is indicated for using such equipment to reduce the hazard to employees.

b Lifelines, safety belts, and lanyards shall be used only for employee safeguarding.

c Employees working over or near water, where the danger of drowning exists, shall be provided with U. S. Coast Guard-approved life jackets or buoyant work vests.

36 *Powder-actuated tools* (Standards Source 1926.302)

a Only trained employees shall be allowed to operate powder-actuated tools.

b All powder-actuated tools shall be tested daily before use and all defects discovered before or during use shall be corrected.

c Tools shall not be loaded until immediately before use. Loaded tools shall not be left unattended.

37 *Power transmission and distribution* (Standards Source 1926.950)

a Existing conditions of the equipment shall be determined before starting work, by an inspection or by a test.

b Electric equipment and lines shall be considered energized until determined otherwise by testing or until grounding.

c Operating voltage of equipment and lines shall be determined before working on or near energized parts.

d Rubber protective equipment shall comply with the provisions of the ANSI J6 series, and shall be visually inspected before use.

38 *Power transmission, mechanical* (Standards Source 1926.300)

a Belts, gears, shafts, pulleys, sprockets, spindles, drums, flywheels, chains, or other reciprocating, rotating, or moving parts of equipment shall be guarded if such parts are exposed to contact by employees or otherwise constitute a hazard.

b Guarding shall meet the requirements of ANSI B15.1–1953 (Revised 1985), ''Safety Code for Mechanical Power Transmission Apparatus.''

39 *Railings* (Standards Source 1926.500)

a A standard railing shall consist of top rail, intermediate rail, toeboard, and posts, and have a vertical height of approximately 42 inches from the upper surface of the top rail to the floor or platform.

b The top rail of a railing shall be smooth-surfaced, with a strength to withstand at least 200 pounds. The intermediate rail shall be approximately halfway between the top rail and the floor.

c A stair railing shall be of construction similar to a standard railing, but the vertical height shall be not more than 34 inches nor less than

30 inches from the upper surface of the top rail to the surface of the tread in line with the face of the riser at the forward edge of the tread.

40 *Respiratory protection* (Standards Source 1926.103)

    a In emergencies, or when feasible engineering or administrative controls are not effective in controlling toxic substances, appropriate respiratory protective equipment shall be provided by the employer and shall be used.

    b Respiratory and protective devices shall be approved by the Mine Safety and Health Administration, the National Institute for Occupational Safety and Health, or be acceptable to the U. S. Department of Labor for the specific contaminant to which the employee is exposed.

    c Respiratory protective devices shall be appropriate for the hazardous material involved and the extent and nature of the work requirements and conditions.

    d Employees required to use respiratory protective devices shall be thoroughly trained in their use.

    e Respiratory protective equipment shall be inspected regularly and maintained in good condition.

41 *Rollover protective structures (ROPS)* (Standards Source 1926.1000)
Rollover protective structures (ROPS) applies to the following types of materials handling equipment: To all rubber-tired, self-propelled scrapers, rubber-tired front-end loaders, rubber-tired dozers, wheel-type agricultural and industrial tractors, crawler tractors, crawler-type loaders, and motor graders, with or without attachments, that are used in construction work. This requirement does not apply to sideboom pipelaying tractors.

42 *Safety nets* (Standards Source 1926.105)

    a Safety nets shall be provided when workplaces are more than 25 feet above the surface where the use of ladders, scaffolds, catch platforms, temporary floors, safety lines, or safety belts is impractical.

    b Where nets are required, operations shall not be undertaken until the net is in place and has been tested.

43 *Saws, band* (Standards Source 1926.304)

    a All portions of bandsaw blades shall be enclosed or guarded, except for the working portion of the blade between the bottom of the guide rolls and the table.

    b Bandsaw wheels shall be fully enclosed.

44 *Saws, portable circular* (Standards Source 1926.304)
Portable power-driven circular saws shall be equipped with guards above and below the base plate or shoe. The lower guard shall cover

the saw to the depth of the teeth, except for the minimum arc required to allow all proper retraction and contact with the work, and shall automatically return to the covering position when the blade is removed from the work.

45 *Saws, radial* (Standards Source 1926.304)

a Radial saws shall have an upper guard which completely encloses the upper half of the saw blade. The sides of the lower exposed portion of the blade shall be guarded by a device that will automatically adjust to the thickness of, and remain in contact with, the material being cut.

b Radial saws used for ripping shall have non-kickback fingers or dogs.

c Radial saws shall be installed so that the cutting head will return to the starting position when released by the operator.

46 *Saws, swing or sliding cut-off* (Standards Source 1926.304)

a All swing or sliding cut-off saws shall be provided with a hood that will completely enclose the upper half of the saw.

b Limit stops shall be provided to prevent swing or sliding type cut-off saws from extending beyond the front or back edges of the table.

c Each swing or sliding cut-off saw shall be provided with an effective device to return the saw automatically to the back of the table when released at any point of its travel.

d Inverted sawing of sliding cut-off saws shall be provided with a hood that will cover the part of the saw that protrudes above the top of the table or material being cut.

47 *Saws, table* (Standards Source 1926.304)

a Circular table saws shall have a hood over the portion of the saw above the table, so mounted that the hood will automatically adjust itself to the thickness of, and remain in contact with, the material being cut.

b Circular table saws shall have a spreader aligned with the blade, spaced no more than ½ inch behind the largest blade mounted in the saw. This provision does not apply when grooving, dadoing, or rabbiting.

c Circular table saws used for ripping shall have non-kickback fingers or dogs.

d Feed rolls and blades of self-feed circular saws shall be protected by a hood or guard to prevent the hands of the operator from coming in contact with the inrunning rolls at any time.

48 *Scaffolds, general* (Standards Source 1926.451)

a Scaffolds shall be erected on sound, rigid footing, capable of carrying the maximum intended load without settling or displacement.

b Scaffolds and their components shall be capable of supporting, without failure, at least 4 times the maximum intended load.

c Guardrails and toeboards shall be installed on all open sides and ends of platforms more than 10 feet above the ground or floor, except needle beam scaffolds and floats. Scaffolds 4 feet to 10 feet in height, having a maximum dimension in either direction of less than 45 inches, shall have standard guardrails installed on all open sides and ends of the platform.

d All planking shall be Scaffold Grade, or equivalent, as recognized by approved grading rules for the species of wood used.

49 *Scaffolds, swinging* (Standards Source 1926.451)

a On suspension scaffolds designed for a working load of 500 pounds, no more than two men shall be permitted to work at one time.

b On suspension scaffolds with a working load of 750 pounds, no more than three men shall be permitted to work at one time.

c Each employee shall be protected by an approved safety life belt attached to a lifeline. The lifeline shall be securely attached to substantial members of the structure (not the scaffold), or to securely rigged lines, which will securely suspend the worker in case of a fall.

50 *Scaffolds, tubular welded frame* (Standards Source 1926.451)

Scaffolds shall be properly braced by cross bracing or diagonal braces, or both, for securing vertical members together laterally, and the cross braces shall be of such length as will automatically square and align vertical members so that the erected scaffolds are always plumb, square and rigid. All brace connections shall be made secure.

51 *Stairs* (Standards Source 1926.500)

a Every flight of stairs having four or more risers shall be equipped with standard stair railings or standard handrails as specified below.

b On stairways less than 44 inches wide having one side open, there shall be at least one stair railing located on the open side.

c On stairways less than 44 inches wide having both sides open, there shall be one stair railing on each side.

d On stairways more than 44 inches wide, but less than 88 inches wide, there shall be one handrail on each enclosed side and one stair railing on each open side.

e On all structures 20 feet or over in height, stairways, ladders, or ramps shall be provided.

f Riser height and tread width shall be uniform throughout any flight of stairs.

52 *Steel erection* (Standards Source 1926.750)

a Permanent floors shall be installed so that there are not more than eight stories between the erection floor and the uppermost permanent floor, except when structural integrity is maintained by the design.

b During skeleton steel erection, a tightly planked temporary floor shall be maintained within two stories or 30 feet, whichever is less, below and directly under that portion of each tier of beams on which any work is being performed.

c During skeleton steel erection, where the requirements of the preceding paragraph cannot be met, and where scaffolds are not used, safety nets shall be installed and maintained whenever the potential fall distance exceeds two stories or 25 feet.

d A safety railing of ½-inch wire rope, or equivalent, approximately 42 inches high, shall be installed around the perimeter of all temporarily floored buildings during structural steel assembly.

e When placing structural members, the load shall not be released from the hoisting line until the member is secured by at least two bolts, or the equivalent, at each connection, drawn up wrench tight.

53 *Storage* (Standards Source 1926.250, 1926.151)

a All materials stored in tiers shall be secured to prevent sliding, falling, or collapse.

b Aisles and passageways shall be kept clear and in good repair.

c Storage of materials shall not obstruct exits.

d Materials shall be stored with due regard to their fire characteristics.

e Weeds and grass in outside storage areas shall be kept under control.

54 *Toeboards, floor and wall openings, and stairways* (Standards Source 1926.500)

a Railings protecting floor openings, platforms, scaffolds, etc. shall be equipped with toeboards wherever, beneath the open side, persons can pass, there is moving machinery, or there is equipment with which falling material could cause damage.

b A standard toeboard shall be at least 4 inches in height, and may be of any substantial material either solid or open, with openings not to exceed one inch in greatest dimension.

55 *Toilets* (Standards Source 1926.51)

a Toilets shall be provided according to the following:

(1) 20 or fewer persons—one facility

(2) 20 or more persons—one toilet seat and one urinal per 40 persons

(3) 200 or more persons—one toilet seat and one urinal per 50 persons

b This requirement does not apply to mobile crews having transportation readily available to nearby toilet facilities.

56 *Wall openings* (Standards Source 1926.500)

a Wall openings from which there is a drop of more than 4 feet, and where the bottom of the opening is less than 3 feet above the working surface, shall be guarded.

   b When the height and placement of the opening in relation to the working surface are such that a standard rail or intermediate rail will effectively reduce the danger of falling, one or both shall be provided.

   c If the bottom of a wall opening is less than 4 inches above the working surface, it too shall be protected by a standard toeboard or an enclosing screen.

57 *Washing facilities* (Standards Source 1926.51)

   a The employer shall provide adequate washing facilities for employees engaged in operations involving harmful substances.

   b Washing facilities shall be in near proximity to the work site and shall be so equipped as to enable employees to remove all harmful substances.

58 *Welding, cutting, and heating* (Standards Sources 1926.350, 1926.351, 1926.352, 1926.353)

   a Employers shall instruct employees in the safe use of welding equipment.

   b Proper precautions for fire prevention shall be taken in areas where welding or other "hot work" is being done. No welding, cutting, or heating shall be done where the application of flammable paints, or the presence of other flammable compounds, or heavy dust concentrations creates a fire hazard.

   c Arc welding and cutting operations shall be shielded by noncombustible or flameproof shields to protect employees from direct arc rays.

   d When electrode holders are to be left unattended, the electrodes shall be removed and the holder shall be placed or protected so that they cannot make electrical contact with employees or conducting objects.

   e All arc welding and cutting cables shall be completely insulated and be capable of handling the maximum current requirements for the job. There shall be no repairs or splices within 10 feet of the electrode holders, except where splices are insulated equal to the insulation of the cable. Defective cable shall be repaired or replaced.

   f Fuel gas and oxygen hose shall be easily distinguishable and shall not be interchangeable. Hoses shall be inspected at the beginning of each shift and shall be repaired or replaced if defective.

   g Proper eye protective equipment to prevent exposure of personnel shall be provided.

59 *Wire rope, chains, ropes, etc.* (Standards Source 1926.251)

   a Wire ropes, chains, ropes, and other rigging equipment shall be inspected prior to use and as necessary during use to assure their safety. Defective gear shall be removed from service.

b Job or shop hooks and links, or makeshift fasteners, formed from bolts, rods, etc., or other such attachments, shall not be used.

c When U-bolts are used for eye splices, the U-bolts shall be applied so that the "U" section is in contact with the dead end of the rope.

60 *Woodworking machinery* (Standards Source 1926.304)

a All fixed power-driven woodworking tools shall be provided with a disconnect switch that can be either locked or tagged in the off position.

b All woodworking tools and machinery shall meet the applicable requirements of ANSI 01.1–1961, "Safety Code for Woodworking Machinery."

WORD POWER

accepted engineering requirements (or practices) Those requirements or practices which are compatible with standards required by a registered architect, a registered professional engineer, or other duly licensed or recognized authority.

angle of repose The greatest angle above the horizontal plane at which a material can be without sliding.

bank A mass of soil rising above a digging level.

excavation Any manmade cavity or depression in the earth's surface, including its sides, walls, or faces, formed by earth removal and producing unsupported earth conditions by reason of the excavation.

sides, walls, or faces The vertical or inclined earth surfaces formed as a result of excavation work.

slope The angle with the horizontal at which a particular earth material will stand indefinitely without movement.

trench A narrow excavation made below the surface of the ground. In general, the depth is greater than the width, but the width of a trench is not greater than 15 feet.

QUESTIONS AND EXERCISES

1 According to 1926.302, abrasive wheels shall be inspected and ring-tested before mounting and subsequent use. What is a ring-test?

2 Do the Construction Industry Safety and Health Standards contain any fire protection requirements for compressed gas cylinders?

3 In 1926.51, the standard requires that an adequate supply of drinking water be provided in all places of employment. What is meant by potable water?

4 Is the Job Safety and Health poster required to be posted on construction sites?

5 Do the standards require that a physician-approved first aid kit be accessible and available at the work site? If so, what is the standards number?

6 What are the requirements for ladders made on the job?

7 Will observance of the standards make a hazardous job-site completely safe? Explain.

8 What are the principal disadvantages of a consensus standard?

9 From what source did OSHA derive its electrical standards?

10 What should a contractor do before beginning any excavation or trenching at a new job site?

11 List at least three organizations from which some of the OSHA Safety and Health Standards for the Construction Industry were derived.

12 *Situation:* You are a crane operator on a construction site. Every morning when you report for work, you are to make a visual inspection of your crane for safety purposes. What critical parts would you inspect?

13 What is the minimum height for the use of safety nets for work being performed above ground level in the absence of ladders, scaffolds, or safety belts?

14 What is the purpose of non-kickback "fingers" or "dogs" on a radial saw?

15 What is the standards requirement for fire extinguishers in storage locations for liquefied petroleum gas? In your opinion, is Part 1926.153(e) stringent enough for this purpose?

16 What is the required distance between the rest and the grinding wheel on a pedestal grinder? Why?

17 Must the employer (contractor) furnish toilet provisions for work crews on a construction project?

18 What are the minimum provisions for guarding wall openings?

19 What are the minimum provisions for splicing or repairing arc welding and cutting cables?

BIBLIOGRAPHY

Construction Industry, OSHA Safety and Health Standards Digest, U. S. Department of Labor, Occupational Safety and Health Administration, OSHA 2202, 1983.

Excavating and Trenching Operations, rev. U. S. Department of Labor, Occupational Safety and Health Administration. OSHA 2226, 1982, pp. 1, 19.

Safety Requirements for Portable Wood Ladders, American National Standards Institute, A14.1 1968, New York, N.Y.

13

INDUSTRIAL NOISE

Human beings have been lulled since time beyond reckoning by the sound of wind in the trees and murmuring brooks. They created music in response to some inner need for beauty of expression. As society advanced, it thrilled to the hunting horn, and later, dreamed of faraway places whenever a distant train whistle penetrated the night.

Unusual silence or unusual sounds also served a useful purpose in human history. Both cautioned men and women to beware of unwanted intrusions in their environment. But gradually, a new world of whining air conditioners, whirring machines, pneumatic tools, and the neighbor's television set have intruded on our peaceful world in a rising tide of unwanted sounds—noise, now considered an inevitable by-product of our modernizing society. It is unlikely that one needs more than a little reflection on one's own experience to be convinced of the problem.

Modern technology has created loud noise sources faster than it has learned to cope with them. Although building construction, acoustics, and better machine design have eliminated some of the problems, industrial noise is unfortunately still considered a necessary evil by many people.

Whether at home, at work, traveling, or even asleep, we are constantly exposed to some form of noise. Noise is defined as—at the least—an *unwanted* and *valueless sound*. The effects of noise can be far-reaching; noise can affect not only our hearing capacity, but also other

237

bodily functions. Noise can act on the body very much as other stresses do.

Noise is a problem encompassing many fields of endeavor, and will not be solved by any one scientific, engineering, or social discipline. Manufacturers and designers must turn to medicine, physics, architecture, psychology, communications, and law, in order to analyze the effects of noise on humans and remove the threat to the workplace trap.

Some research in architectural acoustics and noise control has been sponsored by HUD/FHA; the Department of Health, Education and Welfare; the Department of Defense; and the Federal Aviation Agency, largely on a contract basis.

Almost all states have made hearing loss compensable in their workers' compensation laws, whether from a traumatic occurrence or from long-term exposure to noise in the workplace. Great concern has been expressed by management, labor, and insurance carriers concerning the cost of compensation for the many hearing-loss cases attributable to the workplace.

EFFECTS OF NOISE

Some 20 million Americans have some form of hearing loss. It has been estimated that almost 16 million American workers are exposed daily to noise levels that can cause hearing damage. Regarding these figures there is always the question, "What is a safe level of noise?" Unfortunately, the matter of criteria remains unsettled. The known adverse effects of noise are:

Hearing loss
Structural damage to the ear drums
Speech interference
Annoyance
Distraction
Fatigue

Noise also may affect man indirectly through its effects on domestic animals, wildlife, and building structures.

Some of the major industries where significant noise-hearing loss hazards exist or are suspected are:

Iron and steel-making
Metal products manufacturing
Automotive vehicle production
Lumbering and wood products
Heavy construction
Ship building and repairing

Mechanized farming
Textile industry

Specific armed forces occupations which can be added to this list include flight line and carrier deck operations, engine test cell and weapons firing, armor operations and assorted repair and maintenance work.

In each of these industries and occupations, noise is unwanted, but is an unavoidable product of the technology involved. Each group performs different functions and, therefore, has different requirements with respect to maximum noise levels. As one can see, the effects of noise can be far-reaching; it not only affects hearing, but also can create physiological disturbances as well. Research on noise has been conducted over a long period of time. As a result, the technology has reached a stage where, with but few exceptions, we can cope with almost any indoor or outdoor noise problem provided that we are willing to go to sufficient lengths to do it.

NOISE

A rock group practicing for a concert may be considered pleasant by one group of individuals, but is considered a nuisance by others: to them it is unwanted, it is noise. Thus in a general sense, it may be considered as a random disturbance. Noise can occur at any location and may come from sources near and far. This composite of all-encompassing noise associated with a given environment is called *ambient noise*.

SOUND

When someone plays a flute it sets up a disturbance in the air called *sound*. Someone listening nearby may hear the sound of the flute, which has done two things: it has created a physical disturbance in the air, and it has created a physical sensation in the person's (the receiver) ear. Sound is created by *vibration*. When the flute was blown, it vibrated and caused *sound waves* to travel.

Sound can be generated basically in two ways. First, air flowing over a sharp edge is made turbulent and creates sound waves (as in blowing the flute). Second, a vibrating surface causes air to travel through liquids, gases, and solids in the form of *pressure waves*. Most sounds created in a factory environment are caused by vibrating surfaces of machinery or metal parts. The vibrating surfaces cause waves in the air, just as a paddle moved back and forward in water creates waves.

Sound travels like ripples on a pond, only instead of traveling in two dimensions (as in a circle), it travels in three (as in a sphere). The medium (air) does not travel. The energy is passed along from one particle to the next in the medium. The particles vibrate, but they do not travel.

Sources of Sound

The *source* of sound created by the rock group may represent not one, but many sources of vibrating energy, i.e., it may include all of the musical instruments being played together by the group, or the source may be just one instrument playing a solo passage. Other examples of sound sources include the rustle of leaves, a shotgun blast, the sound of water cascading over a waterfall, or the jingle of coins in a purse. A sound source may not radiate equally in all directions. The effects of propagation of sound depend very much on the ever-present characteristics, both in nature and man-made, of the surrounding atmosphere. Some of the more common factors that create these changes are deflection of sound off walls or buildings, smoke or dust particles in the air, reflection and/or absorption at ground level, and the fact that sound waves diffuse and spread out as they radiate away from their source.

The Receiver

The *receiver* of the noise may be a plant, animal, a single person, a group of people, or an entire community. The distance of travel between the source of noise and the receiver(s) is the *path* of the noise. The paths from the source to the receiver may be numerous, so they radiate different levels of sound, meaning different levels of sound waves. To the receiver, noise can be annoying or it can be harmful. As we know, there are pleasant sounds and unpleasant sounds, just as there are pleasant sights and unpleasant sights. But for purely physical reasons, unpleasant noise generates the strongest argument because it is most likely to spread from the source or person who produces it to the person or persons who are the receivers of it.

Human hearing never shuts off, even when a person is asleep, or unconscious. It is well known that patients are subconsciously aware of operating room conversations, and their recovery can be seriously affected by remarks made while they were unconscious.

Controlling the path of noise from the source to the receiver is the method used in 90 percent of all noise-control problems. When the noise path is in air, a barrier must be used to prevent the spread of energy through the air. The most effective barrier is a complete enclosure; that is, either completely enclose the source of the noise, or completely enclose the receiver.

THE HUMAN EAR

For purposes of discussion, the human ear (see Figure 13-1) is usually divided into three parts: the external, middle, and inner ear.

The *external ear* is bounded by the auricle externally and the external layer of the drumhead internally. It is about 25 millimeters long and has an S-shaped curve directed inward, upward, and then backward.

The external ear serves as a conducting channel for sound. The auricle assists in localization partly because of its shape and position but more because of the flexibility of the motions of the head. When testing the functioning of the ear, it is important to remember that the size and shape of the canal and auricle will influence the intensity of the stimulus at the eardrum. These influences are the result of cavity volume and standing waves, particularly when pure tones are used as stimuli.

The *middle ear* is bounded externally by the inner layer of the drumhead and internally by the mucous membrane covering the bony wall of the inner ear. Within this cavity are contained the three tiny ear bones: the "hammer, anvil, and stirrup" or—medically speaking—the malleus, incus, and stapes. There are also two small muscles: the stapedius, which

FIGURE 13-1
Parts of the ear. (From L. T. Edwards and G. R. L. Gaughran, Concise Anatomy, 3d ed., McGraw-Hill, New York, 1971, p. 314.)

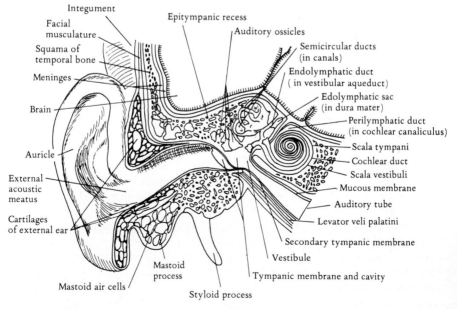

is attached to the stirrup (or stapes), and the tensor tympani, which is inserted at the hammer (or malleus). The middle ear also contains three openings: the oval and round windows and the opening to the eustachian tube. The oval window is fitted with the flattened bony plate of the stirrup (stapes). It is held in place by a fibrous membrane which surrounds the bony plate and acts as a hingelike structure to allow the stapes to move as a trapdoor. This movement originating at the drumhead transmits vibrations to the inner ear. The round window is a tiny opening located just below and in front of the oval window. It is covered by a thin membrane which resembles the drumhead.

The middle ear has three main functions: (1) Efficient transmission of sound vibrations from one medium—air—to another—fluid. This is accomplished by a reduction in area from drumhead to oval window of about 30 to 1 and by the shape and attachments of the three small bones. (2) Protection for the inner ear. (3) Equalization of pressure. The eustachian tube connects the middle ear with the posterior nasal cavity and allows air to get into the middle ear in order to maintain equal pressure on both sides of the drumhead.

The *inner ear* consists of two main parts: the semicircular canals, which contain the balance mechanism; and the cochlea, in which the hearing mechanism is housed. The cochlea is a long spiral-shaped bony canal having 2½ turns. It contains a system of membranous canals and the end organ of the nerve of hearing. One canal travels up the spiral from the oval window and then turns to double back and end at the round window. The second canal is essentially a closed sac situated in cross-sectional relation to the double canal, like a wedge of a pie which has been cut into thirds.

It is this closed sac which contains the highly specialized end organ of hearing, the "organ of corti." All these membranous canals contain fluid, making a hydrodynamic system of the inner ear. Because of this, and because fluid is not compressible, the two windows allow fluid movement when the stirrup vibrates in response to movements of the drumhead as a result of air vibrations in the external canal.

The ear, with its nerve tracts and physiological processes, is an extremely complex mechanism. The external and middle ears are relatively well understood; but the inner ear—and especially the organ of corti—is still pretty much a mystery, although recent advances in electronic instrumentation have made it possible to obtain a great deal more information about the function of the inner ear.

TYPES OF NOISE

There are different types of noise, and they further require different types of measuring techniques (discussed below). Basically, OSHA regulations define three types of noise: (1) steady-level noise, (2) mixed or varying noise, and (3) impact noise.

Steady-Level Noise

This type of noise, as the name suggests, is produced over a long period of time at a constant level. It is one of the simplest of the three types of noise that can be measured with a sound-level meter. Daily exposure can be computed by comparing measured sound level and duration with a table of proposed sound levels and permitted exposures given by OSHA (see Table 12-1).

Mixed Noise

A mixed, or varying-level, noise is usually made up of several layers of steady-level noise. Where machines operate intermittently or where employees move around different areas of a plant, mixed noise will typically be encountered. Measurement of mixed noise is more complicated than that of steady-level noise.

Impact Noise

Punch presses, power-actuated tools, forge hammers, and other such machines generate impact noise. This type of equipment produces an overwhelming noise of short duration, with sound waves continuously striking against the eardrum. As an example of impact noise, consider a punch press operating at 21 cycles per minute at a certain noise level. The number of cycles or impacts per day can be computed thus:

$$\text{Cycles/min} \times \text{min/hr} \times \text{hr/day} = 21 \times 60 \times 8 = 10{,}080$$

There are, then, 10,080 impacts per day. A proposed revision to the noise regulations of OSH Act would allow a maximum of 10,000 impacts per day at the same noise level. The operation of this punch press would be in violation of the proposed regulation. Either the number of cycles must be reduced to fewer than 10,000 per day or the noise level must be reduced.

MEASURING NOISE

The loudness or intensity of noise is measured in decibels. As the distance from the sound increases, the sound level decreases. To avoid working with large numbers in evaluating the intensity of sound, a logarithmic scale is used with the decibel as the unit of measurement. On this scale, one decibel is the lowest sound that one can hear, and 120 decibels is considered as the threshold of pain. There is no exact point at which sound becomes noise.

Sound-Level Meter

The sound-level meter is used to measure sound-pressure level. The meter consists of a microphone which detects sound, converts it into an electrical signal, and amplifies the signal to a level that can activate the meter, which indicates the *sound-pressure level*. Plant noises come from many different directions because sound waves from running equipment and machinery reflect off walls, ceilings, and floors. Measuring exposure under these conditions requires that the microphone and the attached sound-level meter be well coordinated. For special applications, other types of microphones may have to be utilized. To ensure accurate measurements of sound waves, the sound-level meter should be calibrated before and after each day's measurement.

Standard sound-level meters incorporate three weighting circuits (A, B, and C). Readings are indicated by the letter designating the scale used, such as 70 dB(A), meaning a reading of 70 decibels on the A scale. A small change in decibel value can represent a tremendous change in the intensity of noise. (See Table 13-1.)

Decibel Calculations

Because decibels are logarithmic units, they cannot be added or subtracted arithmetically. If the intensity of a sound is doubled, there will be a corresponding increase of only 3 dB, not double the number. If one machine created a noise exposure of 90 dB, a second identical machine placed adjacent to the first would result in a noise exposure of 93 dB, not 180 dB.

TABLE 13-1
COMPARATIVE NOISE LEVELS

Source of sound	Sound-level reading, dB(A)
Rustle of leaves outdoors	20
Quiet home	32
Ordinary conversation	60
Loud street noise; automobile traffic	80–85
Rivet presses at underbody line	92 (98 at impact)
Piercing machine at underbody line on cycle without pan in it	98
Inside powerhouse office	74–85
New compressor fans in powerhouse	91–104
Press shop in general	89–105
Jet airport	120
Shotgun blast (to the shooter)	140
Rifle blast; close-up jet engine	155

Decibel Addition The calculation in decibel addition is fundamental to safety. If one knows the sound levels of two separate sources, then what is the total noise when they are operating simultaneously? The following calculations can be used:

1 When two decibel levels are equal or within one dB of each other, their sum is 3 dB higher than the higher individual level. For example, 89 dB + 89 dB = 92 dB; or, 72 dB + 73 dB = 76 dB.

2 When two decibel levels are 2 or 3 dB apart, their sum is 2 dB higher than the higher individual level. For example, 87 dB + 89 dB = 91 dB; or 76 dB + 79 dB = 81 dB.

3 When two decibel levels are 4 to 9 dB apart, their sum is 1 dB higher than the higher individual level. For example, 82 dB + 86 dB = 87 dB; or 32 dB + 40 dB = 41 dB.

4 When two decibel levels are 10 or more dB apart, their sum is the same as the higher individual level.

When adding several decibel levels, begin with two lower levels to find their combined level, and add their sum to the next highest level. Continue until all levels are incorporated.

HEARING LOSS

Hearing loss is an impairment that interferes with the understanding of speech. Hearing loss is measured as a function of *frequency* or the number of vibrations in cycles per second, of a sound wave, called Hertz (Hz). Normal hearing detection ranges from 16 to 20,000 Hz. The understanding of speech ranges from 500 to 2,000 Hz, and loss of hearing generally occurs at 4,000 Hz. There are tremendous variations among individuals who have equal exposures to a given noise. Some may sustain no hearing losses, while others may sustain severe hearing losses. Hearing loss can probably occur at noise levels lower than those permitted by OSHA standards (see Table 12-1). Hearing loss usually begins at higher tones.

TYPES OF HEARING LOSS

There are two types of hearing loss, *temporary* and *permanent*. Temporary hearing loss is the loss of hearing resulting from a short-term exposure to loud noise—with normal hearing returning after a short period of rest. This condition generally occurs among workers who have experienced a day's exposure to general loud noise in the workplace and feel a slight amount of hearing loss. After a period of "rest" (from the end of the shift until the next day's return to work), normal hearing returns.

Temporary hearing loss is noncompensable under workers' compensation laws. Permanent hearing loss is the loss of hearing due to aging, disease, injury, or long-term exposure to loud noise.

There are two types of permanent hearing loss, *presbycusis* and *sociocusis*. Presbycusis is the normal loss of hearing due to aging. Many workers spend most of their working lives in noisy places, and, while still working, reach an age at which some loss of hearing can be expected because of the normal process of aging. Sociocusis is the loss of hearing due to almost constant exposure to loud noise. It generally involves deterioration of the tiny nerve cells within the inner ear. The ear's greatest sensitivity lies in the frequency range from 3,000 to 5,000 Hz. Loss of hearing can occur at about 4,000 Hz. Rest periods and job rotation can at least prolong the time required to develop a permanent hearing loss, or sociocusis.

NOISE EXPOSURE

Assessment of Exposure

Noise exposure cannot be assessed by noise measurements alone. The effects of continuous exposure to a given noise level differ from the effects of an exposure interrupted by periods of reduced noise. Therefore, one must know how the exposure to noise is distributed in time throughout a representative (typical) workday. And before an assessment of noise exposure is complete, one must determine the expected total exposure time during a work life.

One of the most difficult problems associated with a study of noise exposure is devising a method of describing and categorizing in a meaningful way the multitudinous types of noise exposure encountered. In order to describe noise exposure adequately, some numerical method of classification must be evolved which will combine a measurement of *quantity* of sound energy with a measurement of its *time distribution*. The problem of assessing noise exposure is very complex, and many data must be gathered and analyzed before it can be determined what a practical and useful assessment of noise exposure is.

The relationship of noise exposure to hearing loss is very intricate and not very well understood. But it is known that four major elements of noise exposure are important in producing hearing loss:

1 Overall noise level
2 Frequency composition, or spectrum, of the noise
3 Duration and time distribution of the noise exposure during a typical workday
4 Total duration of noise exposure during an expected work life

Effects of Noise Exposure

Although we would like to be able to predict and control the effects of any noise exposure on any individual or individuals, the chance of being able to do so is very small, not so much because of the difficulty of measuring noise exposure, but because of the wide variation in human responses to noise. While the measurement of intermittent and irregular noise exposure does pose some difficult problems, once techniques of measurement have been developed, they probably can be measured with reasonable accuracy. But individual responses to noise exposure are highly variable, and measurement of them encompasses a wide range of values. There are many causes of this variability, including individual differences in susceptibility to noise-induced hearing loss, differences in total time spent in industrial work, frequent changes of type of occupation, nonoccupational disease, and nonoccupational changes in hearing (e.g., changes that accompany advancing age). Nevertheless, some generalizations are possible.

The damage that noise can do depends on its exact nature and how long the listener is exposed to it. When a person is exposed to too much noise for too long, his or her hearing dulls. The effect of excessive noise on hearing is nerve deafness. This occurs when the cilia, tiny hair-capped cells that act as sensors within the inner ear, become damaged. If the sound is not too loud or too prolonged, the fatigued cells may recover in a few hours. If the hair cells are repeatedly overstimulated, however, they begin to deteriorate. A condition known as "sociocusis" sets in. Sociocusis is irreparable hearing damage; unlike people with many other types of deafness, those with sociocusis can get no help from hearing aids. It is also incurable. Estimates vary widely, but somewhere between 6 and 16 million American workers are exposed to noise levels severe enough to cause permanent hearing damage.

Noise-induced hearing loss is, then, a result of irreversible damage to nerves. No known therapy will reverse or "heal" this damage. In essence, we are dealing here with an injury which may be inflicted immediately upon contact with a known hazard, but which does not become apparent for some time—perhaps until after a few years of continued exposure. For this reason, it is of utmost importance that employers perform acoustical measurements in order to evaluate exposure. Measuring sound was discussed above (page 244, Table 13–1).

COMPLYING WITH OSHA NOISE STANDARDS

With the Occupational Safety and Health Act of 1970 came federal regulation of noise at the workplace. The act adopted noise regulations from the Walsh-Healy Act and extended them to cover all employees.

The OSHA noise regulations are brief but far-reaching. Basically, there are two sections. The first section sets the maximum levels of industrial noise to which an employee may be exposed. The second section explains what action the employer must take if these noise levels are exceeded.

The key part of the OSHA standards is Table G-16, permissible noise exposures (see Table 12-1). It sets the amount of time that an employee may be exposed to various levels of sound, as measured in decibels, or dB(A). The basic standard permits an employee to work 8 hours a day on a job where the sound intensity is 85 dB(A). This exposure is considered to be the upper limit of a daily dose that will not produce disabling loss of hearing in more than 20 percent of a population exposed through a working lifetime of 35 years. Reduced exposure time is specified for greater intensities. Table 13-2 shows the intensities of various representative sounds.

But, for the present, the OSHA standards should be regarded as a maximum guideline.

In many industries the sound intensity exceeds the maximum levels set by the standard. In these cases, the law is explicit about what must be done. Briefly, the program of action involves three basic steps:

1 Reducing the noise at its source through engineering controls, or reducing exposure by administrative controls. (*Note:* Administrative controls shall not be promulgated until all possible engineering controls have been tried.)
2 Providing hearing protection.
3 Carrying out a program of hearing conservation.

The specific parts of this standard can best be understood by examining each part separately and by looking at companies with programs illustrating the regulations.

Reducing Noise at Its Source

The standard states: "When employees are subjected to sounds exceeding those listed in Table 12-1, feasible administrative or engineering controls shall be utilized."

Engineering Controls   "Engineering controls" are really the heart of the regulation. They are more complex than administrative controls. The act requires that an employer's first approach to a noise problem must be to reduce the sound at its source through engineering design and innovations in equipment. Among possible steps recommended by OSHA are:

TABLE 13-2
REPRESENTATIVE SOUND LEVELS

Sound level, db(A)	Operation or equipment	
150	Jet engine test cell	
130	Pneumatic press (close range)	
	Pneumatic rock drill	
	Riveting steel tank	
125	Pneumatic chipper	
	Pneumatic riveter	
120	_____ Threshold of pain	
	Turbine generator	
112	Punch press	**Danger zone**
	Sandblasting	
110	Drills, shovels, operating trucks	
	Drop hammer	
105	Circular saw	
	Wire braiders, stranding machine	
	Pin routers	
	Riveting machines	
100	Can-manufacturing plant	
	Portable grinders	
	Ram turret lathes	
	Automatic screw machine	
90	Welding equipment	
	Weaving mill	
	Milling machine	
	Pneumatic diesel compressor	
	Engine lathes	**Risk zone**
	Portable sanders	
Hearing damage if continued exposure above this level		
80	Tabulating machines, electric typewriters	
75	Stenographic room	
70	Electronics assembly plant	
60	Conversation	**Safe zone**

Make sure the machine is in good repair and properly oiled, and that all worn or unbalanced parts are replaced.

Mount the machine on rubber or plastic to reduce vibration and noise.

Put silencers or mufflers on the noisy components.

Substitute a quiet process for a noisy one.

Confine the sound of the machine within an acoustical enclosure.

Isolate the operator within an acoustical booth.

Administrative Controls The term "administrative controls" means reducing the amount of time an employee is subjected to excessive noise. This can be done, for example, by dividing noisy jobs among two or more employees; another example would be performing very noisy operations at night or on shifts where few employees would be exposed.

Whenever noise cannot be reduced to permissible levels through engineering controls, administrative controls should be developed so that exposure to noise levels above 90 dB(A) does not exceed the limits shown in Table 12-1. Some possible administrative controls are:

Arrange an employee's work schedule so that the employee will work the major portion of the day at or very close to the 90-dB(A) limit.

Remove employees from an area with a high noise level to another area once they have reached the limit of duration given in Table 12-1.

Use noisy machines or operations less than full time—for only a portion of each day, for example, rather than all day for part of the week.

Measures such as these can be instituted if noise exposure is considered as part of production planning; if employees of a shop or plant are deployable enough to be able to work within the required noise levels without violating contractual agreements between union and management; and if management and unions will work cooperatively toward meeting requirements by administrative controls.

Providing Hearing Protection

Administrative and engineering controls are not feasible in every situation, and in any case can rarely be implemented immediately. The standard states: "If [administrative and engineering] controls fail to reduce sound levels to those of Table G-16 [Table 12-1], personal protective equipment shall be provided and used."

Only when engineering and administrative controls fail to reduce noise levels or duration of exposure to them is the use of personal protective equipment required by OSH Act. The use of this equipment can be considered only as an interim measure while engineering and administrative controls are being promulgated or perfected.

While there is still much disagreement as to the maximum intensity of sound to which the ear can be subjected without hearing damage, OSHA standards must be used for compliance purposes.

It should be noted that the law requires both the provision and the use of protective equipment. The responsibility for having employees use the equipment rests entirely with the employer. Persuading employees to do this is not always an easy task. OSHA recommends that each employer conduct an educational training program on the importance of using protective hearing equipment. This "educational" phase of an employer's program is especially important and deserves careful planning, for much of the success of the overall program will depend on how well employees have been "sold." If employees need protective devices but will not use them, the Department of Labor considers this a violation of the act: "lack of enforcement."

There are many types of devices that provide good hearing protection on the market today. The use of cotton stuffed in the ears is only of little value, if any, and is not regarded by OSH Act as compliance. Glass wool or impregnated cotton, also called "Swedish wool," is acceptable because the reduction of noise level—the "attenuation"—which can be achieved is good.

Commercially available (and properly fitted) earplugs can generally reduce noise levels by 25 to 30 dB(A). Ear plugs must generally be fitted to the individual. In addition, plugs—like any other type of protector inserted into the ear—must be fitted by a physician or by a trained person working under the direction of a physician. Frequent checks must be made to ensure that the plugs are being inserted properly.

Earmuffs, of the cup or muff type, provide an acoustical barrier over the entire ear. There are many models on the market, and the only fitting required is an adjustment of the headband. In the long term, muff-type hearing protectors are the least expensive type of protection. They require no "fitting" by trained personnel; jaw movements of the wearer will not "break the seal"; and the supervisor can tell from a distance whether they are being worn. Most important, they provide the highest degree of attenuation.

Regardless of the type of ear protector decided upon, its attenuation (as stated by the manufacturer) must be sufficient to reduce the noise level in the employee's ear to the level and for the duration prescribed in Table 12-1.

Hearing-Conservation Programs

The law requires that: "In all cases where the sound levels exceed the values shown herein [Table 12-1], a continuing, effective hearing conservation program shall be administered." Here, "continuing" means that the program will go on as long as noise levels exceed those permitted by law. "Effective" means that employees exposed to excessive noise will not suffer continuing deterioration of hearing; incipient loss of hearing will be detected and necessary steps taken to prevent further deterioration. "Hearing conservation program" refers to audiometry, periodic checks on the hearing ability of individual employees; and to noise surveys, periodic checks of the noise level in the areas in which employees are working.

Planning and Objectives Industrial hearing-conservation programs which fail are generally the result of poor planning, piecemeal planning, or no planning at all. Another reason for failure is a lack of education for employees and union officials on the merits and benefits of a hearing-conservation program. When a hearing-conservation program is planned for a company, a plant, or an area within a plant, serious efforts must be taken to develop it properly. The plan or program should determine objectives, establish procedures to reach those objectives, and determine criteria for measuring progress.

There are three basic reasons for initiating a hearing-conservation program:

1 The government requires it. Hearing-conservation programs are necessary to comply with the Walsh-Healy Act, the O'Hara-McNamara Regulations, and the recent amendments to the Occupational Safety and Health Act.

2 It is needed to protect the interests of a company by defining and limiting its liability under state workers' compensation laws.

3 It is needed to protect the health of the employee. An employer has a moral and legal obligation to provide a safe and healthy workplace.

Noise Surveys Table 13-2 gives some idea of what sound levels are generated by various types of operations and equipment. It is also a good cross-reference for many types of operations that are performed in industry today. Some of the decibel readings shown for various equipment and processes may be surprising—especially for everyday types of operations in various plants or jobsites. The "danger," "risk," and "safe" zones should be noted carefully.

Audiometry Audiometric tests should be made of all employees working infrequently or regularly in environments where the noise levels have been found to be above 85 dB(A). Audiometric testing determines the hearing level in each ear by means of an audiometer. Tests are given by a competent audiometric technician or a nurse certified in audiometry. Audiometric testing is the foundation of a hearing-conservation program.

The audiometer used to make tests should meet ANSI specifications for limited-range, pure-tone audiometers, must have a certificate of calibration before being placed in use, and must be recalibrated each year thereafter.

Another piece of equipment important for a hearing test is the test booth—an acoustically separate booth or room containing the audiometer and the person being tested. The booth or room should be both visually and acoustically separate from the surrounding environment. Its floor, walls, and ceiling should be acoustically treated to absorb sounds within and sounds from outside.

WORD POWER

acoustics   The science of sound, including its production, transmission, and effects.

anacusis   The total loss of hearing; the inability to perceive sound.

attenuate   To reduce the intensity or strength of sound as expressed in decibels.

audiologist   A person trained in the specialized problems of hearing and deafness.

audiometry   The science of measuring hearing ability.

deafness   The loss of the ability to hear, without designation of degree or cause of loss.

hearing   The subjective response to sound when translated from the physical stimuli into meaningful signals.

sonics   The technology of sound processing and analysis. Sonics includes the use of sound in any noncommunicative process.

sound absorption   A material that is constructed specifically for the purpose of absorbing sound readily.

sound field   A region containing sound waves.

tone   A sound wave capable of exciting an auditory sensation; a sound sensation having pitch.

QUESTIONS AND EXERCISES

1 At what tones does hearing loss generally begin?

2 Define the difference between administrative controls and engineering controls.

3 Are there any medical procedures or treatment that will "cure" or "heal" noise-induced hearing loss?

4 When is audiometric testing required of employees by the employer?

5 What is the definition of noise?

6 Where do you place workplace noise in your list of workplace hazards? Explain.

7 Define: "steady-level noise," "mixed noise," "impact noise." Give an example of each.

8 What is the difference between presbycusis and sociocusis?

9 Other than the steps outlined in this chapter, how many engineering methods can you give for reducing noise at its source?

10 Name five major industries where the greatest hearing loss hazards are known to exist.

11 Other than those mentioned in this chapter, how many other industries can you name where hearing loss hazards may exist?

12 Russian scientific literature relating to hearing conservation describes their standards as lower than ours in the United States. Despite the ideological and

political differences between our country and Russia, do you find this hard to believe? (For class discussion.)

13 What is the difference between background noise and ambient noise?

14 Debate exposure to noise from the side of labor and the side of management. (For class discussion.)

15 OSHA regulations require that "in all cases where the sound levels exceed the values shown in OSHA Table G-16, a continuing, effective hearing conservation program shall be administered." Design such a program for a steel mill employing 1,000 workers. Note: 50 percent of the employees work in areas showing decibel levels above 85 dBs, 30 percent of the employees work in areas showing exposure levels at between 75 and 86 dBs, and the balance of the employees are office workers.

16 Other than hearing loss, name the adverse effects of noise.

17 A worker is exposed to noise from two different machines located nearby. The sound pressure level for the two machines are 80 dB(A) and 84 dB(A), respectively. What is the total sound pressure when both machines are running?

18 Another worker is exposed to noise from three different machines near the work area. The sound pressure levels for the three machines are 86 dB(A), 88 dB(A), and 90 dB(A), respectively. What is the total sound pressure when all three machines are in operation?

19 What is the permissible noise exposure level under OSHA standards for (a) an 8-hour day, (b) 6 hours, (c) 1 hour?

20 Relative to noise exposure, which is the preferred requirement under OSHA standards: engineering controls or administrative controls?

21 Which type of hearing protection device provides the highest degree of attenuation: ear plugs or earmuffs? Why?

22 Explain the term "feasible engineering controls."

23 What are "sound waves?" What are they caused by?

24 What causes temporary hearing loss? What causes permanent hearing loss?

25 What is the value of job rotation in relationship to hearing loss in the workplace?

26 What are the ranges in Hz for normal hearing detection?

27 At what sound-pressure level is the threshold of pain considered to exist?

28 Does it seem possible that high noise levels may account for some accidents? Give some examples.

29 If noise reduction of existing equipment by engineering methods proves impractical, what are some other alternatives that management can provide?

30 Does the use of cotton or of swimmers' ear plugs provide any protection against excessive noise?

31 How might noise interfere with human performance, especially in tasks requiring communications?

32 People who live in residential communities are exposed to noise from many different sources. Name some stationary sources and name some moving sources of community noise.

33 Why would hearing loss present problems to employers in workers' compensation cases? Explain.

34 Should workers' compensation claims related to hearing loss be predicated solely on wage loss, or on physical loss regardless of loss of wages?

35 If you were asked to design a model noise ordinance for your community, what major consideration would you stress? Explain in detail.

BIBLIOGRAPHY

Accident Prevention Manual for Industrial Operations, 5th ed., National Safety Council, Chicago, Ill., 1964, pp. 43-9, 43-10.

Glorig, Aram: "The Doctor Studies Industry's Noise," National Safety Council reprint, National Safety Council, Chicago, Ill., 1979.

Noise, The Environmental Problem: A Guide to OSHA Standards, U. S. Department of Labor, Occupational Safety and Health Administration, OSHA 2067, 1979.

"Noise—Sound Without Value," *Committee Report on Environmental Quality of the Federal Council for Science and Technology,* Washington, D.C., September 1968, p. 1.

Olisfski, J. B., and E. R. Harford: *Industrial Noise and Hearing Conservation,* National Safety Council, Chicago, Ill., 1975. (The "word power" was taken from appendix F.)

"Proposed OSHA Noise Regulations, Guidelines for Compliance," *National Safety News,* vol. 114, no. 2, August 1976, p. 79.

14

FUNDAMENTALS OF MACHINE AND POWER TOOL GUARDING

Ideally, machine guarding should begin on the drawing board, instead of coming into being as an add-in component. Although machines are becoming more sophisticated with automation and computerized control systems, the fundamental purpose of machine guarding is to protect industry's most important tools—the hands of its employees.

Since the passage of the OSH Act, violations involving machine guarding have been among the top citations issued by the Department of Labor. The OSH Act regulations clearly state that machines and equipment *shall* be guarded to eliminate personnel hazards created by moving parts of machinery and equipment. Since the Industrial Revolution, these hazards have been responsible for countless injuries and fatalities in all types of industries.

Advanced technology in the machine tool industry has created newer methods for manufacturing products out of metal. Automation and robotization are already accomplished facts for many companies, and their use will spread to many more companies in the next 10, 20, or 30 years, increasing hazards for employees' hands and other parts of the body.

With respect specifically to machines, the hazards to employees who operate them and maintain them arise from lack of proper point-of-operation guarding or power-transmission guarding. No one disagrees with the intent and purposes of point-of-operation guarding; but people do disagree about how to accomplish it. The need for power-transmission

FIGURE 14-1
Automation and robotization require guarding, for the protection of both operators and other employees. Chain guards are often used to protect workers who are walking between moving automated machinery.

guarding is seldom questioned; it is accepted today because of two factors: (1) machine design and (2) third-party liability.

The records of the Occupational Safety and Health Administration show that, as was noted above, machine-guarding standards are among the most violated of all standards. This should, of course, be changed: machine guarding is visible evidence of management's interest in the safety of its employees. It is also of benefit to management because unguarded machinery is a principal source of accidents for which workmen's compensation must be paid.

PRINCIPLES OF GUARDING

Safeguarding

When pinch points or hazard points cannot be eliminated, guarding must be considered. In many cases, all that is needed is to place a barrier in front of the hazard point. Today machine and equipment manufacturers are doing this, especially for power-transmission components. Usually, a safeguard built into a machine or equipment is more effective and stronger than one added on later in the shop. Therefore, built-in safeguarding should be specified whenever new equipment is purchased.

FIGURE 14-2
This power press lacks point-of-operation guarding.

Older equipment may lack guarding, or the existing guards may be inadequate or damaged. Often a guard has been taken off and not replaced. Whatever the original reason for removing a guard, the guard must be replaced and used.

Safeguarding the point of operation is sometimes the most difficult guarding job of all, and it may require much engineering know-how. The point of operation is frequently the point of greatest hazard because this is where human being and machine interact most closely. A fixed barrier is the most effective type of guard because a worker who cannot get into the point of operation cannot be exposed to the hazard. Even machines that are operated automatically (and require only servicing of stock or monitoring) should have barrier guards to prevent people (even workers who service them) from inadvertently entering the hazard area or hazard point.

Some jobs preclude the use of fixed barrier guards because of the size or configuration of a part or the frequency of job changes. This does not eliminate the need for guarding, however.

It is a cardinal rule that guarding one hazard should not create another hazard. The created hazard can often be hidden, and may cause an accident because people are either forgetful or concentrating on something else. Forgetful persons may inadvertently enter a hazard area because they forget it is there; normally attentive persons may inadvertently enter a hazard area because they are thinking about other things. Forgetfulness and being distracted are only two examples of the many mental, emotional, psychological, and physical limitations and situations that must be considered in designing a guarding system. Considering such matters is called "human-factors engineering"; it is vital in effectively guarding machines and equipment.

Mechanical Motions

Movement of machinery (see Figures 14-3 and 14-4) consists of a few simple motions: (1) rotary, (2) reciprocating, (3) a combination of both. Both rotary and reciprocating motions can crush and shear; when this is clearly understood, all the danger points of any machine can be identified.

Rotary Motion A shaft is a good example of rotary motion. It is found in machines of all types and transmits power from one point to another by pulleys, belts, chains, gears, or cams. No matter what the characteristics of a shaft are, it is dangerous when revolving unless it is covered. The danger of a rotating shaft increases if pulleys are mounted on it or if there are collars, couplings, or projecting keys or set screws.

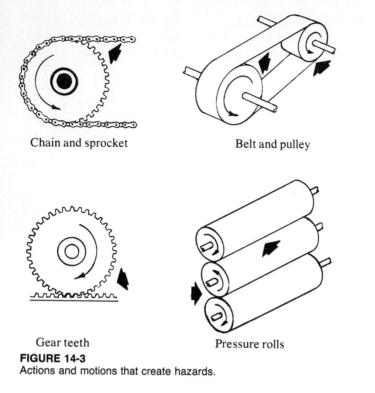

Chain and sprocket	Belt and pulley
Gear teeth	Pressure rolls

FIGURE 14-3
Actions and motions that create hazards.

Reciprocating Motion Reciprocating motion is an alternate backward-and-forward movement. An example would be an engine that uses pistons in cylinders moving rectilinearly.

TYPES OF GUARDS

Basically, there are four kinds of guards, involving relatively simple machinery found most frequently in industry. They are: (1) enclosure, (2) interlocking, (3) automatic, and (4) remote-control, placement, feeding, and ejecting.

Enclosure Guards

The fixed enclosure guard is probably most preferable because it prevents access to exposed moving parts by completely enclosing them. The guard can be limited to the size of the stock-feed opening, so that it does not allow the worker's hand to enter the danger zone. The adjustable enclosed guard can be made to form a barrier which can be fitted around different sizes and shapes of a die in a machine.

Interlocking Guards

An interlocking guard or barrier should be considered the best alternative when a fixed guard or enclosure is not possible. The basic function of an interlocking guard is that it shuts off the power to the machinery when the guard is open, preventing the machine from starting.

FIGURE 14-4
Examples of typical rotating, reciprocating, and transverse mechanisms.

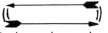
Transverse motion

Rotating motion

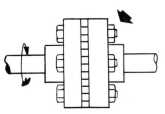
Reciprocating motion

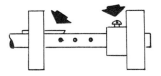
Rotating shaft and pulleys
with projecting key and set screw

Rotating coupling with
projecting bolt heads

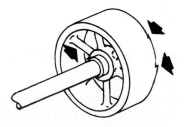
Rotating pulley with spokes and
projecting burrs on face of pulley

Automatic Guards

An automatic guard is used when an enclosure guard or an interlocking guard is not practical for an operation or creates a hindrance to engineering. Repeating its cycle as long as the machine is in motion, this type of guard actually pulls the operator's hand, arm, or body from the danger zone as a ram, a plunger, or some other tool closes on the work being done.

Remote-Control, Placement, Feeding, and Ejecting Guards

A two-handed trip system may be used to operate a machine. Such a device requires the simultaneous action of both the operator's hands on electric switches, air controls, or mechanical levers. If two-handed trip systems are used on machines requiring more than one operator, then each operator should have a separate set of two-handed controls; the machine will not function until both or all operators have depressed their switches or buttons at the same time.

Stocks can be fed automatically into a machine by chutes, hoppers, conveyors, revolving dies, and dial feeds.

Special jigs or feeding devices made of metal may be used to handle stock and keep hands away from the danger zone or point of operation. Mechanical or air-operated ejecting devices may be used to remove parts, again keeping the hands out of the danger zone or pinch point.

SAFETY ENGINEERING

It can be said that guards are only temporary expedients, awaiting the development of more fundamental means of eliminating hazards. Engineering revisions—safety engineering—mean improving or redesigning machinery, equipment, and processes so that hazards are not merely covered up but eliminated and, at the same time, efficiency and production are increased. This phase of safety engineering is often neglected; but it could well be made a major activity and might give unusually large returns on the time and effort invested in it.

GUARDING REQUIREMENTS FOR PRIME MOVERS

Flywheels

OSHA standards require that "one or more methods of guarding shall be provided to protect the operator, and other employees in the machine area, from hazards of flying or rotating parts." Any part of a flywheel 7 feet or less from the floor or platform must be guarded with an enclosure of sheet metal, perforated metal, or expanded metal, or woven wire. Wherever flywheels are above working areas, guards must be installed that have sufficient strength to hold the weight of the flywheel in case a shaft or wheel mounting should fail.

The principal cause of flywheel explosions is excessive speed. The safe rim velocity at which a flywheel may be run depends upon the materials from which it is made, its design, and the conditions of its use. Among the common causes of overspeeding are the breaking of an engine part, sudden loss of load, and incorrect adjustments or failure of the governor. Other causes of flywheel failure are defects in the material, damage to the flywheel, alteration of its original design, and unusual stresses due to sudden starting, sudden stopping, backfiring, excessive heat, loose bearings, or misalignment of the wheel.

Governors

Centrifugal governors, belts, and pulleys located where a person may come into contact with them should be guarded by a solid metal plate at least 4 inches greater in diameter than the pulley, and placed 1 to 2 inches from the pulley. To increase the degree of protection, governor belts and pulleys should be completely enclosed.

Cranks, Connecting Rods, Crossheads, Tail Rods, and Other Parts of Prime Movers

These require the same protection as flywheels—that is, standard guard rails or enclosures—especially if they are adjacent to aisles or working areas.

Friction Drives

The drive point of all friction drives exposed to contact should be enclosed. Arm or spoke friction drives and web friction drives with holes in the web should be entirely enclosed. Exposed projecting bolts on friction drives should be enclosed.

Couplings

Shaft couplings should be constructed so that bolts, nuts, setscrews, and other irregularities do not project. If the bolts are not countersunk or set flush and if they extend beyond the smooth surface of the coupling, they should be covered with a sleeve.

Jaw-clutch couplings should be provided with cylindrical sleeves which at least cover the jaws. The shifting parts of the friction clutch couplings and jaw clutch should be attached to the drive shaft when this is feasible.

Collars

Revolving collars, including split collars, should be cylindrical, without projecting screws or bolts, or should be enclosed with some type of a stationary guard.

Keys and Setscrews

All types of projecting screws and setscrews at collars and couplings should be set flush with, or countersunk beneath, the surface of the metal parts in which they are inserted, or they should be protected with cylindrical safety sleeves.

Starting and Stopping Devices—Clutches

Clutches, cutoff couplings, or clutch pulleys which have projecting parts or are 7 feet or less from the floor or working area should be enclosed by stationary guards. Although some clutches within a machine may be considered as "guarded by location," a complete enclosure should be provided if any possibility of a contact exists.

Remote-Control Apparatus

This category includes all devices used to start or stop machines where it is impossible for the operator to see the machine that is starting. The main precaution with regard to remote-control equipment is to make certain that the controls are guarded or constructed so as to eliminate the danger of accidental operation. A switch with a recessed starting button or protruding "stop" button should be utilized. The installation should conform to the requirements of the National Electrical Safety Code.

Belt Shifters

Belt shifters are of two general kinds: (1) fork and lever or (2) chain mechanisms. Cone pulley belts are usually shifted by means of levers or chain mechanisms. Belts should never be shifted by hand. Regardless of their type of action, all belt shifters should be provided with a positive locking device which will prevent the belt from creeping out of position or from being struck and shifted unintentionally. Such devices include recesses for operating handles, locking pins, ratchet controls, and cam devices; any of these can easily be built on any type of wood or metal belt shifter.

Horizontal Overhead Belts

Horizontal belts more than 7 feet above aisles or working areas should be guarded over their entire length. The guard should follow the line of the pulley to the ceiling or extend to the nearest wall.

Vertical and Inclined Belts

Vertical and inclined belts should be guarded to a height of 7 feet. On inclined belts, this distance is measured perpendicularly from the top of the guard to the floor or work platform. If the lower pulley is less than 7 feet above the floor, it should be guarded up to 7 feet above the floor.

Cone Pulley Belts

Cone pulley belts should be equipped with a belt shifter constructed so that it will guard the pinch point of the pulley belt. If the frame of the belt shifter does not guard the pinch point, the pinch point should be further protected by means of a vertical guard placed in front of the pulley and extended at least to the top of the largest step of the cone.

Shafting

Shafts sometimes have a smooth appearance, and therefore look harmless. Nevertheless, shafts still require guarding. Burrs, projections, and couplings on an otherwise smooth shaft have been known to catch on loose clothing or on an apron, a glove, or a rag, dragging a person against and around the moving part and causing serious injury and even death.

Horizontal Shafting Any portion of a horizontal shaft which is 7 feet or less from the floor or working platform, except for runways used exclusively for oiling or adjusting, should be protected by a trough on the sides and bottom or sides and top, as the location requires, or by a stationary casing completely enclosing the shaft.

Shafting under benches, fixtures, conveyors, and machines should be completely enclosed by stationary casing or by troughs on the sides and top or sides and bottom. If the shafting is located near the floor, the sides should come within 6 inches of the floor. In every case, the trough should extend at least 2 inches above or below the shafting, as the case may be.

Vertical and Inclined Shafting   Vertical or inclined shafting within 7 feet of the floor or working platform, except for runways used for oiling, should be enclosed in a stationary casing.

Pulleys

All exposed pulleys within 7 feet of the floor or working platform should be guarded at least to that height. Guards should be constructed with expandable metal, perforated or solid sheet steel, or wire mesh. Standing railings may also be used as guards as long as they are 42 inches in height with a midrail between the top rail and the floor. Posts, if required, should not be more than 8 feet apart and must be placed not less than 15 inches or more than 20 inches from the pulley.

Gears

Power-driven gears should be protected completely, if possible. If complete protection is not possible, they should be protected on all sides by a sheet-metal guard, an expanded-metal guard, or a perforated-metal guard. Openings in such guards should not exceed 2 inches. All gears should be examined, and where there is the slightest danger of injury, guards must be provided without question. In today's high-speed and automated machines, gears of all types are likely to be found in the most remote parts of the equipment and should be guarded in spite of their remoteness.

Sprockets and Chains

Sprocket wheels or chains (see Figure 14-5) less than 7 feet from the floor or working platform or station, or located so that the breaking point of a chain would cause injury to anyone nearby, should be guarded in the same manner as belts and pulleys. If a lubricator is designed on, or installed into, the equipment, it should have a small opening with a hinged or sliding cover; or, if possible, oil lines should be extended outside the guard for lubricating.

REQUIREMENTS FOR MACHINE CONTROLS

OSHA standards require that a mechanical or electrical control be provided on every machine, making it possible for the operator to cut off power to the machine without leaving the operating position. It is also required that the operator will not be endangered if the machine should restart after a power interruption or failure.

FIGURE 14-5
Sprockets and chains should be guarded.

Power and controls must be located within easy reach of the operator so that the operator is not exposed to pinch points or point-of-operation hazards.

The controls of the machine must be inoperative while adjustments or repairs to the equipment are being made.

GUARDING REQUIREMENTS FOR MACHINES

OSHA republished in the Federal Register of July 1, 1985, an updated set of standards for machinery and machine guarding. The rules and regulations of the standards in Subpart O spell out clearly, in more detail than can be given here, how operators can be protected from unguarded moving mechanical parts.

All employers should become familiar with the standards that apply to the machines used in their place of business, plant or factory. Employees, supervisors, and students will find the standards of great personal benefit in their trades and skills. Remembering that all mechanical motion or action is to some degree hazardous and understanding the basic techniques of guarding simple machinery is of great help in dealing with problems created by more complex machinery.

Under the OSH Act the employer is responsible for scheduling and recording inspections of machine guards and point-of-operation protection devices at frequent and regular intervals. Safe operating conditions depend on detecting hazards and potential hazards and taking immediate action to remedy them. Management is encouraged to develop and keep its own machine-guarding checklists. Students should do the same, for at some point in their careers they will be trained to inspect the same machinery they have learned to safely guard and operate.

Woodworking Machines

Woodworking machines operate at fast speeds. A woodworking machine must be constructed so that it is free from all but minimal vibration when the largest-size tool it will take is mounted and run idle at full speed.

Arbors and mandrels must be constructed so that they have firm and secure bearing and are free from play.

Wooden bandsaw wheels other than those commercially manufactured must not be used; and saw frames or tables must be constructed with lugs cast on the frame or with equivalent means to limit the size of the saw blade that can be mounted.

Circular-saw gages must be constructed to slide in grooves or tracks that are accurately machined. This will ensure exact alignment with the saw for all positions of the guide.

Hinged saw tables must be constructed so that the table can be firmly secured in any position in true alignment with the saw. Belts, pulleys, gears, shafts, and other moving parts must be guarded at all times.

All electrical machines and equipment must be grounded, and the ground wire must be provided with a separate polarized plug and receptacle. Three-wire grounding of cords is used for protection. The grounding lug must never be removed.

Every power-driven woodworking machine should be provided with a disconnect switch that can be locked in the "off" position.

Circular saws must be provided with a hood that covers the saw at all times to the depth of the teeth; and the hood should automatically adjust itself to, and remain in contact with, the material being cut. A spreader and anti-kickback device should be provided; and any exposed part of the saw behind or beneath the table should be guarded.

FIGURE 14-6
This overhead moving conveyor is completely guarded to prevent injury from falling parts.

In addition to the hood enclosing the blade of a radial saw, an adjustable stop should be provided to limit forward travel, and the head should automatically return to the starting position. When a saw is used for ripping, a spreader and anti-kickback device must be used to prevent material from squeezing the blade or being thrown back at the operator.

Revolving double-arbor saws must be fully guarded in accordance with all the requirements for circular crosscut saws or circular ripsaws.

No saw, cutter head, or tool collar should be placed or mounted on a machine arbor unless the tool has been accurately machined to fit the arbor.

Combs (featherbeds) or suitable jigs for holding work against the shaper or cutter must be provided when a standard guard cannot be used in such operations as dadoing, grooving, molding, and rabbeting.

All portions of a bandsaw blade, except the working portion between the guide rolls and the table, must be fully enclosed; and the machine must be provided with a tension-control device to indicate the proper tension for the saws used on the machine.

Hand-fed planers and jointers must be equipped with cylindrical, not square, heads; and the opening in the table must be kept as small as possible. The cutting heads of such machines, both horizontal and vertical, must be covered with a guard.

The chains and sprockets of all double-end tenoning machines must be entirely closed, except the portion of the chain used for conveying the stock. All cutting heads and saws, if used, must be covered by metal guards.

Whether rotating or not, the cutting heads of wood-turning lathes must be covered as completely as possible by hoods or shields hinged to the machine so that they can be lifted back to allow for adjustments.

Lathes, used for turning long pieces of wood stock and held only by the two centers, must be equipped with long curved guards extending over the tops of the lathes to prevent any stock that is loose from being thrown.

Sanding Machines

The feed rolls of self-fed sanding machines must be protected with a firmly secured semicylindrical guard to prevent the operator's hands from coming into contact with the inrunning rolls at any point.

Drum, disk, and belt sanding machines must be enclosed by exhaust-duct hoods that cover all but the work portions of those machines.

Belt sanders should have guards at each inrunning nip point on the power-transmission and feed-roll parts.

Veneer Cutters and Wringers

Veneer slicer knives must be guarded at both front and rear, and veneer clippers must have an automatic feed or a guard which makes it impossible to place fingers under the knife while feeding or removing stock.

Power-driven guillotine veneer cutters must have starting devices requiring simultaneous use of both hands or an automatic guard that will pull the hands away from the danger zone at every descent of the blade.

Abrasive Wheels

All abrasive wheels must be covered with guards strong enough to withstand the shock of a bursting wheel, and the exposed or work-contact portion of the wheel should not be more than one-fourth of the entire wheel. The operator should use only the exposed periphery or circumference of the wheel.

Work rests that support the material being ground must be strongly constructed and adjustable to the wearing down of the wheel. The work rest should have a maximum wheel clearance of ⅛ inch to prevent the work from jamming between the wheel and the rest itself.

When an operator stands in front of the wheel opening, the guard must be constructed so that the tongue guard can be adjusted to the decreasing diameter or wearing down of the wheel. The distance between the tongue guard and the wheel should never be more than ¼ inch.

Bench grinders must be permanently and securely mounted, and the table or bench to which they are attached should be free from vibration. The operator should wear a face shield or goggles at all times.

Mills and Calenders

The business end of a milling machine consists of two adjacent metal rolls, set horizontally, which revolve in opposite directions; the "bite" or nip point, between the inrunning rolls presents a danger to the operator. Most milling-machine accidents occur when operators are unloading or making adjustments to the machine.

The top of the operating mill rolls should not be less than 50 inches above the level on which the operator stands.

A safety control must be provided in the front and back of the machine and must be located so that the operator can easily reach it. The control must be one or a combination of the following:

Pressure-sensitive body bars that shut off the operation of the machine when the operator's body comes into contact with them.

A safety tripod installed within 2 inches of the front and rear rolls (whether pushed or pulled, the tripod immediately stops the action of the machine).

A safety trip-wire cable or wire-center cord installed within 2 inches of the front and back rolls (the trip wire or cord must operate readily, whether pushed or pulled).

Fixed guards should be installed at both the front and the back of the machine to protect the operator.

Mill dividers, support bars, spray pipes, feed conveyors, and strip knives must be located so that they do not interfere with the operation of the safety devices.

Calenders must be provided with a tripod, cable, or wire-center cord across each pair of inrunning rolls and extending the length and face of the rolls. These safety tripping devices must be located on both sides of the calender and near each end of the face of the roll. These lines should be not more than 12 inches from the faces of the rolls and not less than 2 inches from the floor or operator's platform.

All tripping and emergency switches must require manual resetting; and all mills and calenders should have stop actions that are fast enough to prevent or limit injury.

When the operator is not within sight or hearing of other workers, an alarm device should be provided to summon help in the event of an accident.

Table-Top Grinders

Wheel Guard   Abrasive wheels should be used only on machines that have safety guards. The guard should cover the spindle, end nut, and flange projection. The exposed area of the grinding wheel and sides for the guards should not exceed more than one-fourth the entire wheel.

Tool Rests   Tool rests should be adjustable to compensate for wheel wear. Work or tool rests must be kept in adjustment to the wheel, with a maximum clearance of ⅛ inch.

Tongue Guards   Tongue guards should be adjusted to the constantly decreasing diameter of the wheel, so that the distance between the tongue guard and the grinding wheel is never more than ¼ inch.

Flanges   All abrasive wheels must be mounted between flanges. Flanges should not be less than one-third of the entire diameter of the wheel.

Blotters   Compressible washers (blotters) should always be used between the flanges and abrasive wheel surfaces. Use of blotters distributes the flange pressure more uniformly.

Driving Flanges   Driving flanges must be securely attached to the spindle, and the bearing surface must run true.

Arbor Size   Grinding wheels should fit freely on the spindle and remain free during all grinding situations.

Bushings   Whenever a bushing is used in the wheel hole, it should not exceed the width of the wheel and should not come into contact with the flanges.

PURPOSE OF MACHINE GUARDING

1 To prevent direct contact with the moving parts of a machine or any other type of power-operated equipment.

2 To allow safe maintenance or lubrication of the machine or other power-driven equipment.

3 To prevent the movement of flying metal objects and/or the splashing of machine oils onto the operator.

4 To prevent electrical contact between the machine operator and any exposed electrical wiring, circuits or any other exposed electrical components of the machine or equipment.

5 To prevent accidents caused by some human failure, such as: distraction, curiosity, mental or physical fatigue, or deliberate chance-taking.

BASIC DESIGN CHARACTERISTICS OF A MACHINE GUARD

1 The guard must be designed for the machine, the type of operation, and the intended hazard.

2 It must prevent access to the hazard zone while the machine or the equipment is in operation.

3 The guard must be made impossible to remove while the machine or equipment is in operation.

4 The guard in itself should not constitute a hazard to the operator.

5 The guard should be designed to require a minimum of maintenance.

LOCKING OUT REQUIREMENTS

Machine accidents are the fourth most important cause of disabling injuries, and account for 15 percent of all permanent partial disabilities. Both manufacturers and plant maintenance departments should, of course, do their best to guard danger points, but there will always be some operations that cannot be completely guarded and times when an operation must be locked out. When machinery needs adjustment or maintenance, for example, it should be shut off and locked out; maintenance personnel can be seriously injured if they do not shut down and lock out the machines they are working on.

Lockout rules are important. On the basis of the type of equipment used, each plant or department should devise its own set of procedures. Locking out may seem like a nuisance sometimes, especially on small jobs, but many terrible accidents can occur if lockout rules are not in effect.

Definitions

"Locked Out"   For many years the term "locked out" has meant that a machine is put into a safe condition by a person or persons about to make adjustments or perform certain maintenance functions.

Thus compressed air, hydraulic pressure, and electricity, both alone and in combination, are put in a state or condition in which:

1 Every power source that can energize machinery is locked off.

2 The potential energy of the machine is at its lowest practical level and even the opening of pipes, tubing, or hose cannot produce a movement of the machine.

3 No electrical power can perform in any manner or for any function for which it was designed.

4 All accumulators and air-surge tanks are reduced to the lowest atmospheric pressure by venting off to the atmosphere or to a tank or open drain.

"Power"   Power is recognized as any type of energy that can operate equipment. Common types of power are electricity, air under pressure, oil, water under pressure, and steam.

"Locks"   Locks are recognized as padlocks. Special devices may be constructed or purchased so that padlocks can lock out means of connecting and disconnecting power, such as valves.

"Disconnecting Means"   This is a device which cuts off a source of power, such as an electrical disconnect or valve.

"Residual Pressure"   Shutting off the air to a machine leaves air under pressure in the machine: this is residual pressure. Unless the valve allows release of this pressure, a portion of the pipe should be disconnected to allow its release. Hydraulic systems are not usually opened in this manner, and the supervisor of hydraulic equipment should be consulted about the methods to be used.

How Power Is Locked Out

When the electrical disconnect switch is attached or adjacent to the equipment, the "motor stop" button is depressed, the disconnect handle placed in the "off" position, and the lock applied.

Air pressure on machinery is to be shut off and locked out only if the air pressure could result in an unexpected movement of the machine or its components.

Hydraulic pressure is locked out by opening the electrical disconnect switch to the hydraulic-pump motor and locking it out.

Multiple Locking

When more than one person works on a piece of equipment, multiple safety padlocks should be used, with each worker having his or her own lock. These padlocks are for the personal protection of the employee and should be used to lock electrical disconnect or control valves. *A safety tag should not be used as substitute for a safety padlock.*

Safety-lock adaptors are to be applied to electrical, air, mechanical, or hydraulic equipment in the deenergized state. An adaptor is designed so that subsequent employees working on equipment may utilize a common adaptor for locking it out. The last employee working on the equipment removes the adaptor and his or her padlock.

TOOLS

Today in industry there is hardly one job or job classification in which a worker does not use some type of hand or portable power tool in the course of a day's work. Accidents involving hand tools account for a very high proportion of injuries to the hand and other parts of the body.

Injuries from hand and portable power tools have four main causes:

1 Using the wrong tool or improvising a tool
2 Using a defective tool
3 Using a tool incorrectly
4 Failing to clean tools properly or failing to clean tools before putting them away

Because of the widespread use of hand and power tools, it is important that the prevention of accidents involving them be made a part of the safety program in any plant. All employees should be made aware of the fact that the tools they work with can be a source of disabling injuries if improperly used.

Even the simplest hammer or chisel can disable an employee who treats it or uses it carelessly. Hand tools are involved in 5 to 10 percent of all cases of compensable work injuries; and, as was noted above, the cause of a hand-tool injury can often be traced to some improper use of the tool. (This is also a cause of many power-tool injuries.) Unbelievable as it may sound, few workers are trained to select the proper tool for a specific job and to use it correctly. It is no wonder that simple hand tools account for a large percentage of on-the-job accidents. Surely, they are also the cause of many off-the-job accidents—the exact number could probably never be calculated.

Powered portable tools are time-savers and are used just as extensively at home as on the job. Their efficiency makes them useful in situations where fixed-type power tools and unpowered portable tools can be dangerous unless one follows the safe practices required for their use. The most common injuries involving powered hand tools are cuts, eye injuries, flash burns, and electric shock.

Manual Tools

A "manual" or "hand" tool is an unpowered tool. The potential of hand tools for producing injuries is greatly underestimated—they produce a larger percentage of injuries than one may suspect. One danger in using simple hand tools is the fact that nearly everyone has used them and is familiar with them, so that many people take too casual an attitude toward them. Another danger is that many people have not been trained, or taken the time to learn, how to use the proper tool for a specific job.

Striking and Struck Tools   There are some general rules that apply to almost all hammers (the hammer is the most commonly and widely used striking tool) and other striking tools:

• Never use a striking tool with a loose or damaged handle, a mushroomed head, or a dull cutting edge.
• Strike blows squarely. A glancing blow increases the chances of striking a finger or hand or chipping the head of the tool.
• Never strike with the side of the hammer.
• Never strike one hammer with another.
• When striking chisels, punches, wedges, etc., the hammer face should be larger than the head of the struck tool.
• Axes and hatchets are meant to strike wood. They should never be struck against metal, stone, or concrete.

Struck tools include rock and star drills, cold chisels, hot chisels, wood chisels, brick chisels, punches, drift pins, and wedges. Common rules for the safe use of struck tools include the following:

• Use the proper tool for the job. Never use cold chisels on stone or concrete; hot chisels on cold metal, stone, or concrete; wood chisels on metal; etc.
• Never use a chisel with a mushroomed head or dull cutting edge. Dull edges can be sharpened.
• Never use a drift pin as a punch or strike-on if the struck end is chipped or mushroomed.

• Never use a star drill with a dull cutting edge or damaged head, and never use it on anything but masonry.

• Never use brick chisels and sets in bad condition, and never use them on metal.

• Use only wedges in good condition.

Torsion Tools Any tool applying torque is potentially dangerous; and the more torque involved, the more serious the potential injury.

The most commonly used torsion tools are wrenches. There are many basic rules for the safe use of wrenches, one of which is never to use a "cheater" to increase the leverage of any type of wrench. Another basic rule is to discard any type of wrench with a broken or battered point; spread, nicked, or battered jaws; or a bent handle.

To prevent slipping of a wrench when in use, use only a wrench whose opening fits the nut. This is a basic rule for open-end, box, and socket wrenches.

In freeing a "frozen" nut or bolt, the application of penetrating oil is most important. Then, where practicable, use a striking face-box wrench or a heavy-duty box or socket wrench.

Pipe Wrenches Both straight and chain-tong pipe wrenches should be inspected before use to make sure they have sharp jaws and should be kept clean to prevent slipping.

Pipe Tongs These should be placed on the pipe only after the pipe has been lined up and is ready to be made up. If necessary, a small block of wood should be placed near the end of the travel of the tong handle and parallel to the pipe to prevent injury to the hands or feet if the tongs slip.

Machine Wrenches Machine wrenches are often misused as hammers. As a result, they soon become distorted and unsafe to use. A wrench of the proper size for the job should of course be selected.

Adjustable Wrenches These should be placed on the nut with the open jaws facing the user. For this reason, and for safety, wrenches, if possible, should be pulled, not pushed.

Pliers Pliers should never be used for cutting hardened wire, as a hammer, or as a pry. Avoid dropping them on hard surfaces.

Screwdrivers Screwdrivers are one of the most commonly used—and misused—tools in both industry and the home. They are misused in so many ways that it would be hard to list all the abuses. Some of the most common, however, are using screwdrivers as punches, wedges, pinch bars, and even pry bars. For electrical work, the blade and handle of a screwdriver should be insulated except at the tip, and the blade or rivet should not extend through the handle.

Note: Nonsparking Tools Nonsparking tools, made of nonferrous materials, are advisable for use where flammable gases, volatile liquids, and other explosive substances are stored or used. The heavy sparks from steel tools are capable of igniting many substances such as lint and fine metal dust, and chemicals such as carbon disulfide, xylene, ethyl ether, naphtha, and gasoline.

Nonferrous tools will reduce the hazard but may not eliminate it entirely. These tools need inspection before each use to ensure than they have not been attacked by foreign particles which could produce friction sparks.

Portable Power Tools

Portable power tools are divided into four groups, according to the source of power: (1) electric, (2) pneumatic, (3) gasoline, and (4) explosive (power-actuated).

In almost all accidents involving powered hand tools, the injured person had failed to follow a safety rule for the specific tool that was being used.

Electric shock is probably the most common hazard of electrically powered tools. Grounding of portable electric tools will prevent current in a defective or short-circuited tool from passing through the operator's body. Additional safety factors include insulated platforms, rubber mats, and rubber gloves if the tools must be used in rather wet locations such as wet floors. Low-voltage equipment should be used in such locations. Portable transformers will reduce the shock hazard in wet locations by 6 to 32 volts.

Electrical Standards for Powered Tools   Grounding is mandatory for portable powered hand tools. The ground must be provided by a separate ground wire and polarized plug and receptacle, except when the portable tools are protected by an approved system of double insulation or its equivalent. When such a system is employed, the equipment should be plainly marked.

Furthermore, portable powered tools should be equipped with an "on-off" control and may have a lock-on control if it can be turned off by a single motion of the same finger or fingers that turned it on.

Hand-held powered tools such as circular saws with a blade diameter greater than 2 inches, chain saws, and percussion tools without a hand-manipulated on-off switch should be equipped with a "deadman" switch—that is, a switch of the constant-pressure type.

Using Power Tools Safely   These days it is fairly easy to find a good-quality, safe tool for just about any job. But each tool must be used for the job for which it was designed. One of the major problems with hand power tools is that if the right tool for the job is not handy, a substitute may be used—this is an unsafe act and may cause an injury. Industry is constantly required to search for new tools with which to do better and more efficient jobs. Often this means designing new tools that must be not only better for particular jobs but also safer for employees.

"Jury-rigged" or reworked tools are often dangerous. Employees should not tamper with the design and shape of a tool, because reworking the metal may destroy its temper, creating an unsafe tool. Of course, some tools, such as chisels, are designed to be reworked and reused; but others should be discarded and replaced with new ones when they become damaged or worn.

Employees should keep their tools free of oil, grit, and dirt to maintain them in the best condition. Also, clean tools are more easily inspected for wear and damage, so that cleanliness discourages the use of possible unsafe tools.

In essence, hand-tool safety can be summarized in one simple rule: "Use the proper tool for the job." Knowing when and how to use a tool is the best way to prevent hand-tool injuries.

ZERO MECHANICAL STATE (ZMS)

Zero Mechanical State implies the mechanical state of a machine in which all power sources that can produce mechanical movement have been locked off in order to afford maximum protection against any unexpected mechanical movement. This concept includes not only the locking out of electrical energy, but also requires that all kinetic and potential energy be isolated, blocked, supported, retained, or controlled so that such energy will not unexpectedly be released.

Depending on the type of equipment involved, in addition to electrical power sources, consideration must also include the following power sources:

Compressed air
Pressurized fluids
Potential energy from suspended equipment
Potential energy stored in springs
Any other unexpected mechanical movement

All such energy sources must be brought to Zero Machine State before any maintenance or setup work can be performed safely on any equipment. American National Standard Z241.1-1975 defines and explains ZMS in detail. Table 14-1 illustrates a lockout and tagging program designed around the ZMS standard.

All of the machine-guarding standards are designed to protect employees by making some type of changes in the machine. Rarely do the stan-

TABLE 14-1
Lockout and Tagging Program

Superintendent	1. Develops lockout and tagging procedure.
Safety Department	1. Reviews lockout and tagging procedure.
	2. Audits lockout and tagging procedure.
Superintendent or Supervisor	1. Issues lockout procedure.
	2. Issues lockout devices.
	3. Issues Danger Tags.
Supervisor	1. Instructs personnel in lockout procedure.
	2. Instructs personnel in zero energy state procedure.
Operator or Mechanic	1. Turns off equipment.
	2. Turns off power supply.
	3. Installs personal lock on power shut-off.
	4. Writes out and attaches Danger Tags.
	5. Reduces system energy to zero state.
	6. Checks that equipment will not operate.
	7. Performs adjustments or repairs.
Other Operator or Mechanic	1. Installs personal lock on power shut-off.
	2. Performs adjustments or repairs.
	3. Checks that work is completed.
	4. Removes personal lock.
Operator or Mechanic	8. Verifies that all work is completed.
	9. Removes Danger Tag.
	10. Removes personal lock.
	11. Tests equipment for proper operation.

dards specify safe work practices or methods of motivating the worker to practice safety. It is readily understandable that an employee performing a repetitive operation near or at a hazardous machine component may eventually be injured, because the employee cannot be expected to be safety-conscious at all times; therefore, the employee must be separated from the hazard with a guard. However, it is also necessary that employers should constantly instill in employees positive attitudes toward safety; therefore, employers need standards that are employee-oriented. Furthermore, standards must be developed that are readily available and easy to use, so that the employer can reasonably be urged to acquire and use them. It makes no difference how good a standard is if it is not used. In summation, the overall conclusion is that machines must be guarded if machine-inflicted injuries are to be controlled, reduced, or eliminated.

WORD POWER

fencing   Guarding by means of a fence or rail enclosure that restricts access to a hazardous area.

guard   A barrier that prevents the entry of a part of the body into the hazard zone.

pinch point   Any point, other than the point of operation, for which it is possible for a part of the body to be caught between moving parts.

point of operation   The area in or on a machine where the work is being performed.

safety device   A machine control which stops or interrupts the operation of the machine if a part of the body is near the hazard zone.

squeeze point   Any point created by two solid objects, at least one of which is in motion.

run-in-points   Are created when two objects are in motion, with lessening separation until they finally touch.

QUESTIONS AND EXERCISES

1 List four types of actions and motions that create hazards in revolving machine parts.

2 What is the minimum requirement, in distance, for guarding an overhead exposed moving machine part?

3 How far from the machine operator must electrical operating controls be located?

4 What is the required distance between the tool rest and the grinding wheel on a pedestal grinder?

5 What are the four causes of injury from hand and portable power tools?

6 Portable power tools are divided into four groups, according to their source of power. Name the four groups.

7 Give a mechanical example of reciprocating motion.

8 What is the principal cause of flywheel explosions?

9 What are some secondary causes of flywheel explosions?

10 What is the most common hazard inherent in the use of electrically powered tools?

11 What is the difference between the pinch point and the point of operation?

12 What is the difference between a guard and a safety device?

13 Give an example of fencing used for guarding purposes.

14 Write up at least 5 general safety rules relating to hand-tool safety.

15 What does the term "Zero Mechanical State" imply?

16 What is the purpose of "locking off" or "locking out" a machine?

17 Write up 10 general basic safety rules for machine operators or production employees in the metal fabrication industry.

18 Does the justification for guarding hazardous equipment rest solely on humanitarian concerns or on the avoidance of legal penalties? Explain.

19 Explain the responsibilities for machine guarding by the following departments: (1) engineering, (b) maintenance, (c) production, (d) purchasing, and (e) safety.

20 What safety responsibilities does the worker have regarding machine guarding?

21 What are some of the possible economic results of unguarded or improperly guarded machinery to a manufacturing organization?

22 Discuss the use of safety inspection committees to further the awareness of the hazards surrounding mechanical and electrical machines and equipment.

23 At what voltage would you recommend guarding exposed electrical equipment?

24 Do you consider the "grounding" of electrically powered hand tools as a method of guarding the user of the tool? Explain.

25 Discuss the axiom: "Choose the right tool for the right job." How does it apply to safety?

26 What is the proper course of action to take for defective hand tools?

27 What are the types of defects the supervisor should be looking for in a "spot" check of hand tools?

28 OSHA standards stipulate that unguarded overhead equipment need not be guarded at or above 7 feet above the floor or the working platform. But what about an overhead conveyor carrying parts over a well-traveled aisle? Is an overhead guard legally required in this situation? Explain.

29 A drive shaft located at the rear of a machine and well out of the operator's reach revolves at 30 revolutions per minute. The shaft is smooth and without projecting keyway projections or attachments. Should the shaft be guarded? Explain your reasoning.

30 Today safety features that machine manufacturers install in machines are becoming definite sales assets to their products. Can you explain why?

31 Occasionally, supervisors and employees alike resist the use of machine guarding and other safety devices because they feel such devices will interfere with production. If you were a safety engineer, how would you counteract such an attitude? What would you say to the supervisor and the worker in defense of the use of machine guards and safety devices?

BIBLIOGRAPHY

American National Standard Safety Requirement for Sand Preparation, Molding and Coremaking in the Sand Foundry Industry, American Foundrymen's Association, Des Plaines, Ill., ANSI Z241.1, 1981.

Code of Federal Regulations, part 29: General Industry Standards. U.S. Department of Labor, Occupational Safety and Health Administration, July 1985.

Essentials of Machine Guarding, Safe Working Practices Series, U. S. Department of Labor, Occupational Safety and Health Administration, OSHA 2227, July 1975.

"Machine Guarding," *National Safety News,* vol. 112, no. 6, December 1975, p. 51.

Mechanical Power Transmission Apparatus, Safe Practices Pamphlet no. 110, National Safety Council, Chicago, Ill.

National Safety News, vol. 110, no. 4, October 1974, pp. 90–92.

"Take the Extra Step for Guards for Machines and Tools," *National Safety News,* vol. 113, no. 3, March 1976, p. 86.

CHAPTER 15

PERSONAL PROTECTIVE EQUIPMENT

Although personal protective equipment has been designed to provide workers protection from nearly every known workplace hazard, management should consider engineering and/or environmental controls in order to reduce exposures before resorting to the use of such equipment. Once the need for personal protective equipment has been determined, it is the employer's responsibility to select the approved type, in order to give maximum protection to employees. Only protective equipment and devices approved by responsible agencies and in accordance with OSHA standards should be purchased for use.

SELECTION OF EQUIPMENT

When purchasing safety equipment, management (purchasing and/or safety departments) should select only the best equipment available for the intended hazard protection. The specifications of certain standards or regulations (OSHA, NIOSH, ANSI) must be met. Some equipment standards, especially those relating to hazardous operations and rescue equipment, require that the equipment be certified, approved, listed, and tested for its intended use.

Research and development is constantly being carried out by safety equipment manufacturers in the face of new government regulations and the safety requirements by industry. Although engineering solutions for workplace hazards look promising, the need for the use of personal protective equipment will never diminish.

Once the need for personal protective equipment has been considered, the next step is to select the proper type. Two important criteria to consider in the selection process are the degree of protection offered and the ease of use.

PROPER USE OF EQUIPMENT

Once safety equipment is provided, the employer must insist that workers use it and use it properly. It is no longer a matter of choice; it is now a legal requirement that the employer must adhere to. The law not only requires that management provide safety equipment for its employees, but that management must also train them to use it and to enforce its use. Both supervisors and workers must be instructed how to use and to recognize the capabilities and limitations of the equipment. Furthermore, they must be taught how to fit and inspect it properly. Also, safety equipment must be inspected before and after use. The manufacturer's recommendation for fitting, testing, inspection, and maintenance must also be followed. Training should also include instruction in the nature of the hazard the equipment is to be used against.

HEAD PROTECTION

Helmets

There are many occupations that require the use of head protection to protect workers from impact and penetration from falling and flying objects and from limited electric shock and burn. Particularly hazardous operations requiring the use of head protection (hard hats) include: construction work, mining, shipbuilding, logging, general maintenance, and the foundry and steel industries.

Safety helmets (hard hats) are classified into two types: (1) full-brimmed and (2) brimless with peak. These two types are further classified as follows:

Class A: limited voltage resistance for general service
Class B: high voltage resistance
Class C: no voltage protection (metallic helmets)
Class D: limited protection for firefighting

All helmets must be identified on the inside of the shell with the manufacturer's name, American Standard designation 289.1, and the class (A, B, C, or D).

Headpieces designed for use around electrical hazards must meet performance voltage requirements (20,000 volts). All helmets must be designed to withstand a maximum average force of up to 850 pounds.

Metal helmets do not afford the high impact resistance of plastic ones; but because of their lighter weight, they are preferred by some workers. Metal helmets should never be used where electrical hazards or corrosive substances are present.

EYE AND FACE PROTECTION

Protective eye and face protection is required where there is a reasonable probability of injury that can be prevented by such equipment. Industrial operations expose the face and eyes to a variety of hazards, including flying objects, splashes of corrosive liquids or molten metal, dust, and radiation. Injuries to the eyes and face can not only disable, but also disfigure workers. Flying objects—metal or stone fragments and chips, nails, and abrasive grit from grinding operations—cause most eye injuries. Corrosive and harmful substances, hot splashing metal, and light and heat rays can also cause damage to the eyes.

General Requirements

Federal regulations stipulate that the design, construction, testing, and the use of eye and face protection must meet the requirements of the ANSI Standard for Occupational and Educational Eye and Face Protection, 287.1-1968. Federal regulations further require that employers make available the type of eye and face protection designed for the work to be performed, and enforce the wearing of such personal protection equipment.

Workers whose vision requires the use of corrective lenses in spectacles, and who are required by this standard to wear eye protection, must wear goggles or spectacles of one of the following types:

1 Spectacles whose protective lenses provide optical correction.

2 Goggles that can be worn over corrective spectacles without disturbing the adjustment of the spectacles.

3 Goggles that can incorporate corrective lenses by mounting them behind the corrective lenses.

Eye Protection

Eye protection devices, which are available in many types and styles, must first be considered as optical instruments to be properly selected and used. Today they can be stylish and can fit comfortably. Employers and supervisors should be familiar with the various types of eye protection and should know what type is the best for a particular job. Any equipment selected must meet all standard design and performance requirements. In order for protective eye wear to fit comfortably, it must be fitted properly and preferably by an ophthalmologist.

Cover Goggles Cover goggles are frequently worn over ordinary spectacles (glasses) to protect not only the wearer's eyes but also corrective lenses, because lenses that are not heat-treated or chemically treated can be easily shattered or broken. Cover goggles also protect the spectacles against pitting, scratching, or breaking. They are generally made of plastic material (both frames and lenses) and are designed with shielded vents on the sides.

Tempered Lenses Tempered lenses are widely used in ordinary eye corrective glasses for street wear, but do not meet the requirements for safety lenses. The requirements for safety lenses are 3 mm. in thickness, whereas the thickness for ordinary lenses are 2 mm.

Plastic Lenses Based on impact tests, plastic lenses used in industrial safety eyewear have proven to be very successful. For example, plastics are resistant to hot chemicals. Hot metal, which can ordinarily shatter glass, will not do so with plastic material. When plastic lenses are coated, they are almost as abrasion-resistant as are ordinary safety lenses. Without a coating, they scratch easily.

Protective Spectacles (Safety Glasses) Spectacles without attached side shields may be worn where it is highly unlikely that foreign particles will fly toward the side of the face. However, safety glasses with attached side shields are highly recommended for all industrial operations. ANSI, 287.1-1968 provides all of the specifications for eye and face protection.

Protective Goggles There are various types of goggles to protect the eyes:

Chemical Goggles These have soft vinyl or rubber frames; they protect the eyes against splashes of corrosive materials and exposure to fine dust or mist. The lenses can be of heat-treated glass or acid-resistant plastic. For protection against chemical splashes, there are goggles equipped with baffled ventilators on the sides. Goggles must also be ventilated when used as protection against vapor or gas. Many types are made to fit over plano or prescription glasses. ("Plano glasses" is a term used in industry to designate nonprescription safety glasses.)

Leather Mask Dust Goggles These should be worn by employees who work around or with noncorrosive dust, as in cement manufacturing, cinder deposits in steel mills, and flour mills. Such goggles should have heat-treated or filter lenses. Wire-screened ventilators around the eye cup provide for circulation of air.

Welders' Goggles Welders' goggles with filter lenses are available for operations such as oxyacetylene welding, cutting, lead burning, and brazing. OSH Act Regulation 29 CFR 1910.252(e) provides a guide for the selection of the proper shade numbers.

Chippers' Goggles These have contour rigid plastic eye cups and come in two styles—one for people who do not wear spectacles, and one to fit over corrective spectacles. Chipping goggles should be used where maximum protection from flying particles is needed.

Laser Goggles Although no one type of glass offers protection from all laser wavelengths, glass for protection against any single laser wavelength can be had on special order from reputable eyewear manufacturers. Laser goggles designed for protection from specific laser wavelengths should not be used with different wavelengths.

Face Protection

Many types of personal protective equipment shield the face (and also the head and neck) against light impact, splashes of chemicals or hot metals, heat radiation, and other hazards of similar nature.

Face Shields Face shields of clear plastic protect the eyes and face of a person who is sawing or buffing metal, sanding or grinding, or handling various chemicals, industrial detergents, and soaps. The shield should be of slow-burning plastic and must be replaced when warped, dirty, scratched, or brittle with age. The headgear and shield should be adjustable to the size and contour of the worker's head and should be easy to clean when necessary.

Metal screen face shields deflect heat but permit good visibility. They are used around blast furnaces, soaking pits, open-hearth furnaces, and other sources of radiant heat.

Babbitting Helmets These are used to protect the head and face against splashes of hot metal, rather than against head radiation. The helmet consists of a window made of extremely fine wire screen, a tilting support, an adjustable headgear, and a crown protector.

Welding Helmets Welding helmets, shields, and goggles protect the eyes and face from splashes of molten metal and radiation produced by arc welding. The shell of a welding helmet must resist sparks, molten metal, and flying particles. It should be a poor heat conductor and a nonconductor of electricity. Helmets should have the proper filter glass to keep ultraviolet and visible rays from harming the eyes. Furthermore, they should be constantly inspected for cracks or pinholes to prevent eye flash.

Impact goggles worn under the helmet protect the welder from flying particles when the helmet is raised. Eye protection of the spectacle type with side shields is recommended as minimum protection from flash from adjacent work or from popping scale of a fresh weld.

Acid-Proof Hoods Acid-proof hoods that cover the head, face, and neck are used by persons exposed to possible splashes from corrosive chemicals. This type of hood has a window of glass or plastic that is securely joined to the hood to prevent acid from seeping through. The hoods are made of various materials—such as rubber, neoprene, plastic film, and impregnated fabrics—which are resistant to different chemicals.

Air-Supplied Hoods These should be worn around toxic fumes, dusts, gases, or mists. Because an unventilated hood quickly becomes warm and humid, an air-line attachment makes the hood more comfortable for the user.

HAND PROTECTION

The hands and fingers are parts of an ingenious bodily tool which actually performs the work of pulleys, gears, and hoists (for pulling and lifting). Fingers and hands are a major location of disabling injuries because they are exposed to all kinds of cuts, scratches, bruises, and burns. Although fingers are hard to protect (because they are used in practically all work), they can be shielded from many common injuries by proper protective equipment. The following types of equipment are available.

Kelvar-treated gloves protect against burns and discomfort when the hands are exposed to sustained conductive heat.

Metal mesh gloves are used by those who work constantly with knives and other sharp tools (as are common in the meat industry) for protection against cuts and blows from the tools themselves and from sharp or rough objects.

Rubber gloves are worn by electricians. They should be tested regularly for dielectric strength.

Rubber, neoprene, or vinyl gloves are used when handling chemicals and corrosives. Neoprene and vinyl are particularly useful for handling petroleum products.

Leather gloves resist sparks, moderate heat, chips, and rough objects and provide some cushion against blows. They are generally used for heavy work. Chrome-tanned leather or horsehide gloves are usually used by welders.

Chrome-tanned cowhide leather gloves with a steel-stapled leather patch for palms and fingers are used in foundries and steel mills.

Cotton or fabric gloves are suitable for protection against dirt, slivers, chafing, or abrasions; they are not heavy enough for use with very rough, sharp, or heavy materials.

Coated fabric gloves protect against moderately concentrated chemicals. They are recommended for use in canneries, packing houses, food handling, and similar industries.

Heated industrial gloves are made for work in cold environments. They can also be part of a heated clothing system if necessary.

Hand leathers or hand pads are often more satisfactory than gloves for protecting against heat, abrasion, and splinters; wristlets or arm protectors (see the next section) are available in the same material as gloves.

ARM PROTECTION

Wristlets, or arm protectors, are also available to protect arms and forearms. They are made in many of the same materials as gloves. Protective sleeves and rubber sleeves for persons working on telephone poles or electrical wire poles perform the same functions for the arms as their counterparts do for hands and fingers.

SKIN PROTECTION

Protective Creams

Protective creams may be used to supplement protective devices and clothing or may be used where protective clothing is impossible. But they are not by any means sufficient by themselves. The following factors should be considered in the use of protective creams.

First, choose the correct cream for the exposure. Information on the protective capabilities of a particular cream can be obtained from the manufacturer. The major types are:

1 Those which fill the pores or cover the skin with water-repellent material to form a barrier.

2 Those which are solvent-resistant and specific for certain types of solvents.

3 Those which counteract specific irritants.

4 Those protecting against photosensitizing agents and sunlight.

Second, creams and ointments should be nonirritating.

Third, they should be easily applied and removed, and yet stay on while the worker is exposed.

Fourth, the skin should be thoroughly washed before and after the use of protective creams.

Skin Cleaners

Some skin cleaners may be irritating to the skin and cause dermatitis. Avoid strong abrasive soaps containing quartz, pumice, organic solvents, or cleaners with a high alkali content.

If a strong soap must be used, provide, for use after washing, skin lotions which replace natural oils and soften the skin.

FOOT AND LEG PROTECTION

In most instances of foot injury, the severity of the injury could have been significantly reduced if the injured person had been wearing appropriate foot protection. Foot protection is needed in almost all industrial operations, and employers should see to it that employees wear the proper protection whenever hazards exist.

Types of Foot Protection

Protectors for toes and feet are available in various types and in many styles, each of them for specific purposes. In rare instances, safety shoes with metal toe boxes are prohibited because there is a danger of contact with hot electrical equipment. Following are examples of safety shoes used for particular types of exposures which meet the criteria of ANSI Z41.1, Men's Safety-Toe Footwear.

Metal-free shoes, boots, and other footwear are available for use where there are severe electrical or fire and explosion hazards.

"Congress" or *gaiter-type shoes* are used to protect workers from splashes of molten metal or from welding sparks. They can be removed quickly to avoid serious burns and have no laces or eyelets to catch molten metal.

Reinforced shoes have inner soles of flexible metal and are designed for use where there are protruding nails but the likelihood of contact with energized electrical equipment is remote—for example, in the construction industry.

For wet work conditions (found, for example, in dairies and breweries) *leather shoes with wooden soles* or *wooden-soled sandals* are effective. Wooden soles provide good protection on jobs that require walking on hot surfaces (that are not hot enough to char wood), and they have been widely used by workers handling hot asphalt. These are sometimes called "pavers' sandals."

Safety shoes with metatarsal guards should always be worn during operations where heavy materials—such as pig iron, heavy castings, and timbers—are handled. They are recommended wherever there is a possibility that objects of any weight will fall and strike the foot above the toecap. Metal footguards are long enough to protect the foot back to the ankle and may be made of heavy-gage flanged and corrugated sheetmetal; they should be easy to adjust and remove.

Safety boots, shoes, and rubber overshoes are available in several styles and types to fit a variety of specific needs. They may be obtained with conductive soles that drain off static charges and with nonferrous construction to reduce the possibility of friction sparks in locations with a fire or explosion hazard. Initial and subsequent periodic tests should be made on conductive footwear to ensure that the maximum allowable resistance of 450,000 ohms is not exceeded.

Manufacturers of safety shoes are a prime example of an industry that is constantly stressing efficient protection and coming up with innovative ideas.

Leg Protection

Leggings, which encircle the leg from ankle to knee and have a flap at the bottom to protect the instep, protect the entire leg. The front part may be reinforced to give protection against impact. Such guards are used by persons who work around molten metal in foundries and steel mills. Leggings should permit rapid removal in case of an emergency. Hard fiber or metal guards are available to protect the shins against impact; these too should allow rapid removal.

Knee pads protect employees who work on their knees (as in cement tile setting, terrazzo cleaning, and other forms of floor work).

RESPIRATORY EQUIPMENT

The general requirements in this section have been drawn from OSH Act regulations. Additional information may be obtained from the referenced regulations or from American National Standards Z88.2, Practices for Respiratory Protection.

In the control of occupational diseases caused by breathing contaminated air, the primary objective is to prevent atmospheric contamination. This should be accomplished as far as possible by accepted engineering control measures (for example, enclosure or confinement of the operation, general and local ventilation, and substitution of less toxic materials). When effective engineering controls are not feasible, or while they are being instituted, appropriate respirators must be used.

Employer's Responsibility

Respirators must be provided by the employer when they are necessary for the protection of the health of the employee. The employer must provide respirators that are applicable and suitable for the purpose intended. Respiratory requirements are specified by OSH Act CFR 29, 1910.134.

The employer is required to establish a respiratory program that includes, but is not limited to: (1) written standard operating procedures governing the selection and use of respirators on the basis of hazards to which workers are exposed; (2) instruction and training of the wearer in the proper use of respirators and their limitations; (3) inspection, maintenance, cleaning, disinfecting, and storage of equipment; (4) appropriate surveillance of conditions in the work area, degree of exposure, and suitability of respiratory protections used; (5) regular inspection and evaluation to determine the continued effectiveness of the program.

Approval of Equipment

Approved respirators provide adequate respiratory protection against the particular hazards for which they are designed, in accordance with the standards.

The U.S. Department of Interior, Mining Enforcement Administration, and the National Institute of Occupational Safety and Health, are recognized as authorities for approving respirators.

Although respirators listed by the U.S. Department of Agriculture continue to be acceptable for protection against specified pesticides, the U.S. Department of Interior, Bureau of Mines, is the agency now responsible for testing and approving such respirators.

Selection of Respirators

It is essential to have some knowledge of various operational and environmental factors that will influence the selection of respiratory protection. Among the many factors to be considered are the following:

1 The nature of the hazardous operation or process
2 The type of air contaminant, including physical and chemical properties, physiological effects on the body, and concentration
3 The period of time for which respiratory protection must be provided

4 The location of the hazard area with respect to a source of uncontaminated respirable air

5 The functional and physical characteristics of respiratory protective devices

Hazards Presented by Respiratory Equipment

From the standpoint of fire hazard, neither pure oxygen nor air containing more than 21 percent oxygen is to be preferred to ordinary air for use in atmosphere (air) supplied to a respirator or self-contained breathing apparatus, for ventilating or for other purposes. To ignite, a flammable substance in the presence of oxygen requires only a small fraction of the energy it requires in the presence of air; and an explosion or fire can be much more violent in the presence of oxygen than in the presence of air.

Types of Respiratory Equipment

Respiratory protective devices can be classified as follows:

1 Air-purifying respirators
2 Atmosphere (air)-supplied respirators
3 Self-contained breathing devices

In turn, these are divided into a number of subclasses.

Air-Purifying Respirators Gas masks consist of a facepiece connected by a flexible tube to a cannister. Contaminated air is purified by chemicals in the cannister.

Because no one chemical has been found that will remove all gaseous contaminants, the cannister (purifying agent) must be carefully chosen to fit the specific need. A cannister designed for a specific gas or vapor (or for a single class of gas or vapor) will give longer protection than a cannister of the same size designed for protection against a multitude of gases or vapors.

Cannister Gas Masks with Full Facepiece These are for emergency protection in atmospheres immediately dangerous to life. They *do not* provide protection against oxygen deficiency. The effectiveness of the cannister type is limited to atmospheres containing at least 16 percent oxygen (by volume), and not more than 2 percent of those toxic gases for which it is designed (exceptions are ammonia, for which the limit is 3 percent; and phosphine, for which the limit is 0.5 percent). The period of protection that a gas mask provides depends upon the type of cannister, the concentration of the gas or vapor, and the physical activity of the user. When a respirator is used for protection against a gas or vapor that has little or no warning properties (like carbon monoxide), the cannister must have an indicator or timer that shows when it should be changed. Fresh cannisters should be used each time a person enters a toxic atmosphere or one that is suspected of being toxic.

Chemical Cartridge Respirators These consist of a half-mask facepiece connected directly to one or two small cannisters of chemicals. They are used only in nonemergency situations—that is, for atmospheres that are harmful only after prolonged or repeated exposures.

Particulate (Mechanical) Filter Respirators These can be designed to give satisfactory protection against any kind of particle. The major factors to be considered are the resistance to breathing offered by the filtering element, the adaptation of the facepiece to faces of various sizes, the use of earmuffs if necessary, the fineness of the particles to be filtered out, and the relative toxicity of the particles.

Combination Chemical and Mechanical Filter Respirators These utilize dust, mist, or fume filters with a chemical cartridge for dual or multiple exposure. The combination respirator is well suited for spray painting and welding.

Air-Supplied Respirators Several types of air-supplied breathing apparatus are available.

Hose Masks These are available with a blower (power-driven or hand-operated), without a blower, or connected to a source of respirable air under pressure. Of these three possibilities, only the hose mask with blower is to be used for work in atmospheres where the hazard is immediate. A hazard is considered immediate if, in the event of failure of the equipment, hasty escape from the dangerous (toxic, flammable, or oxygen-deficient) atmosphere would be impossible or could not be made without serious injury. When a hose mask with a blower is used, it is possible to breathe through the hose while making an escape.

Hose masks are used by workers entering tanks or pits where there may be dangerous concentrations of dust, mist, vapor, or gas, or insufficient oxygen. No one should enter, much less remain, in a tank (or similar space) which tests show has less than 19.5 percent oxygen at any time, unless approved respiratory protective equipment is worn (such as a hose mask or self-contained breathing apparatus). The atmosphere of an enclosed space should be tested to determine if there are any toxic or flammable contaminants present in dangerous or explosive concentrations. If so, or if oxygen is deficient, the space should be ventilated before anyone enters. The enclosure should be retested as long as someone must work in it, and reventilated if necessary.

Air-Line Respirators These can be used in atmospheres not immediately dangerous to life, especially where working conditions demand continuous use of a respirator. Air-line respirators are (as the name implies) connected to a compressed-air line. A trap and filter must be installed in the compressed-air line ahead of the masks to separate oil, water, grit, scale, or any other foreign matter from the airstream. When line pressures are more than 25 pounds per square inch gage (psig), a pressure regulator is required. A pressure-release valve, set to operate if the regulator fails, should be part of the equipment.

Self-Contained Breathing Devices When people must work in an atmosphere immediately hazardous to life at some distance from a source of fresh air, a self-contained breathing apparatus, which carries its own supply of oxygen, should be used. In an environment that contains a substance dangerously irritating or corrosive to the skin, the self-contained breathing device must be supplemented by impervious clothing (discussed below).

A self-contained breathing device affords complete respiratory protection in any toxic or oxygen-deficient atmosphere, regardless of the concentration of the contaminant. This type of equipment is frequently used in mine rescue work and in firefighting. Employees should be trained initially in the use of this equipment, and then given refresher training sessions at least every six months, to maintain efficiency.

Because of the extreme hazard, no one wearing a self-contained breathing apparatus should work alone in an irrespirable atmosphere; other persons similarly equipped should be standing by, ready to provide help.

Maintenance and Care of Respirators

A program for the maintenance and care of respirators should include the basic inspection for air leaks and defects; cleaning and disinfecting; and minor necessary repairs. Respirators should be inspected routinely before and after each use. Respirator regulators and warning devices should be checked for tightness of connections and the condition of the facepiece, headbands, valves, connecting tube, and cannisters. Records of inspections, findings, and necessary repairs should be kept.

Medical Surveillance

Workers should never be assigned to any operations requiring respiratory protection until a physician has determined that they are physically and psychologically capable of performing the work under these conditions. Many plants have preemployment physical examinations which aid in such evaluation, but additional checks may be necessary to determine a worker's suitability for a specific assignment.

If an employee is exposed to certain toxic materials, periodic examinations (such as urinalysis) may be necessary even though the worker wears the proper respiratory equipment.

SPECIAL CLOTHING

Manufacturers of safety equipment have been highly ingenious in developing many specialized types of clothing for protection against specific hazards of our technology. The development and use of many new types of fabrics, synthetic rubber, neoprene, and vinyl coatings have brought about many safeguards for the worker that in the past were hardly even considered possible.

Types of clothing include special headgear, face protection, aprons, gloves, and garments that completely enclose the entire body and contain their own air supply. Thermal undergarments popular for use in winter sports are also used by outdoor workers. Thermal knit cotton and polyester quilted materials are among the standouts in this field. Disposable clothing has also found its way onto the market for hospitals and the drug and electronics industries where contamination to the worker or the product may be a problem.

Materials for low-level nuclear radiation, machine-shop aprons, and materials for people engaged in work around hot sparks and molten metals provide vital protection not previously available. The synthetic rubbers, especially neoprene, have greatly diminished the burns and allergic reactions once so common among workers employed in handling chemicals, acids, and caustics.

QUESTIONS AND EXERCISES

1 Under what conditions does the OSH Act allow employers to use respiratory equipment over engineering controls?
2 What are the three principal types of respiratory protective equipment?
3 List the most important elements required in the employer's respiratory protection program.
4 What are the two basic classifications of safety helmets (hard hats)?
5 Why should metal hard hats never be worn where electrical or corrosive substances are present?
6 What is the required thickness for safety glass lenses?
7 What are some considerations that should be taken into account in the selection of personal protective equipment by the safety manager and/or the purchasing department?
8 Name three types of eye protection and elaborate on the purpose of each of them.

9 Make a list of general safety rules regarding the use of eye protection in a manufacturing environment.

10 What are the five requirements that must be included in a respiratory protection program by the employer under the OSH Act?

11 What is the purpose of conductive footwear?

12 What is the maximum allowable resistance for conductive footwear?

13 What are the general safety requirements for employees exposed to the hazards of welding, cutting, and brazing operations?

14 What are the general requirements for personal climbing equipment?

15 Name as many critical hazards as you can that can be eliminated by the use of personal protective safety equipment.

16 Identify the correct ANSI standard for each of the following: eye and face protection, foot protection, and respiratory equipment protection.

17 Can an air-purifying respirator be worn in an oxygen-deficient atmosphere? Why?

18 Can a cannister "gas mask" respirator with a full facepiece be used in an oxygen-deficient atmosphere? Why?

19 What is the difference between a mechanical filter-type respirator and a chemical cartridge-type respirator?

20 Name at least five different kinds or types of materials used in hand protection, and give the purpose for each of the materials used.

21 Assume you are the supervisor in a department which consists of the following manufacturing operations: welding, grinding, degreasing, press operations, and sawing. Prepare a list of environmental controls for these. Also prepare an additional list of personal protective equipment to be used as a supplement to the environmental controls.

22 What is dermatitis? How does it occur? List some personal hygiene methods that can be used to help control it.

23 What is the most common method used by industry for the prevention of foot injuries? Elaborate on a particular program.

24 Consider the possibility of allowing the union safety committee in a plant to make recommendations for the purchase of personal protective equipment. What might be the "pros" and "cons" for consideration of this idea?

25 Obtain a sales catalog from the manufacturer of personal protective equipment and review its contents. As you peruse the catalog, do you find the approval of the equipment listed? Does the catalog show the limitations of the various types of personal protective equipment in it?

BIBLIOGRAPHY

Code of Federal Regulations: General Industry Standards, U. S. Department of Labor, Occupational Safety and Health Administration, 29 CFR, 1910.

"Personal Protection," *National Safety News,* vol. 115, no. 3, March 1977, pp. 135–146.

"Take the Extra Step for Personal Protection," *National Safety News,* vol. 113, no. 3, March 1976, pp. 133–146.

16

FIRE PROTECTION, PREVENTION, AND CONTROL

Fire protection, prevention, and control in industry are sometimes thought of as activities separate from accident prevention. There may be some justification for this if we think of fire only in terms of damage to property. But in most accidental fires involving property damage, the danger of personal injury also exists. Since this is the case, prevention and control of the hazards from fire should be part of every plant's safety program. Fires are unpredictable, and even the best efforts do not stop all fires. But a good fire prevention program and knowledge of what to do when a fire starts will do much to keep fire loss down.

An effective fire-loss control program must have as its objective the prevention of loss of life and personal injury, and the loss of property. Emphasis must be placed on the following:

1 The layout of the plant and its structures
2 The availability of fire-fighting equipment
3 Keeping the accumulation of flammable and combustible materials to a minimum

THE CHEMISTRY OF FIRE

Fire

Fire is a chemical reaction between a flammable or combustible substance and oxygen. To produce fire, three factors must be present: fuel, heat, and oxygen. If any one of the three is missing, a fire will not start;

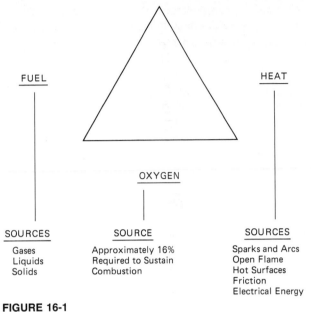

FIGURE 16-1
The fire triangle.

if any one of them is removed once a fire has started, the fire will be extinguished. This relationship is called the fire triangle (see Figure 16-1).

Fuel

A fuel is any substance or material that will combine with oxygen in the presence of heat, and burn as a result. Most ordinary fuels are compounds of carbon and hydrogen, in varying amounts, that can be oxidized rapidly and burned. Some typical examples are paper, wood, dusts, oil, and grease. Even certain metals which are not normally considered fuels will burn in some form in an atmosphere of pure oxygen. Fires are classified according to the type of fuel involved (as discussed later).

Heat

Heat is absorbed by a substance when it is converted from a solid to a liquid, and from a liquid to a gas. Most substances will burn only after the solid or liquid fuel has been vaporized or decomposed by heat to produce gas. The temperature at which a substance gives off these vapors or gases is called the "flash point" of the substance. The substance itself will not continue to burn when the source of ignition is removed until it has reached the "ignition temperature," which is the lowest temperature at which a substance will burn continuously. In order to have a fire, there must be enough heat to raise the temperature of the fuel to its ignition point. Heat can also reignite a fire that has been extinguished, if the extinguishing agent has not cooled it sufficiently.

Oxygen

Oxygen is a colorless, tasteless, and odorless gas. Fire normally draws its oxygen from the air. The air that we breathe contains 21 percent oxygen and 78 percent nitrogen. Fire requires only 16 percent oxygen to burn. The nitrogen in the air serves only to dilute the oxygen and does not ordinarily enter into the reaction. In an atmosphere of pure oxygen, many substances not normally considered as combustible will burn rapidly. Atmospheres of pure oxygen or even oxygen-enriched air will produce fires of great intensity. Variations in oxygen concentrations affect the ease of ignition.

Although smothering extinguishes fires by separating or excluding the oxygen, some fires cannot be extinguished by smothering. For example, many forms of plastics and combustible metals and other fuels do not depend on external oxygen to burn.

STAGES OF A FIRE

All fires start at some point, and if the proper conditions exist, expand in some predictable manner. If we assume that these essentials are constantly available, we can say that fire grows in stages. There are four stages in the growth and expansion of fire (see Table 16-1).

TABLE 16-1
STAGES OF A FIRE

Incipient Stage
No visible smoke
No flame
Very little heat
Combustion begins to take place
Smoldering Stage
Combustion increases
Smoke becomes visible
As of yet, there is no visible flame
Flame Stage
The point of ignition occurs and flames begin to become visible.
Heat Stage
Large amounts of heat, flame, smoke, and toxic gases are produced.

CLASSIFICATION OF FIRES

To help simplify the application of fire-fighting equipment, fire protection engineers have grouped fires into broad classifications (see Table 16-2).

TABLE 16-2
CLASSIFICATION OF FIRES

Classification	Method of Extinguishing
Class A: Fires involving ordinary combustible materials such as wood, paper, cloth, rubber, and many plastics	Water, water-based combinations of chemicals, dry chemicals, carbon dioxide. Approximately 75 percent of all fires are Class A fires.
Class B: Fires involving flammable or combustible liquids, flammable gases, and grease	Foam, dry chemicals.
Class C: Fires involving energized electrical equipment or other materials located near electrically energized equipment	Electrically nonconductive chemicals, multipurpose chemical compounds, carbon dioxide agents.
Class D: Fires involving certain combustible metals, such as magnesium, titanium, zirconium, sodium, potassium, aluminum	A heat-absorbing extinguishing medium not reactive with the burning metals. Commercial examples include Purple K, Monnex, Super K.

FUNDAMENTALS OF FIRE PROTECTION, PREVENTION AND CONTROL

Protection from, and prevention and control of, fire is extremely technical and complex. The literature relating to fire protection is voluminous. The *Handbook of Fire Protection* of the National Fire Protection Association and the *Handbook of Industrial Loss Prevention* by Factory Mutual Engineering Corporation are large volumes, yet they admittedly cover only the bare essentials. Still, the subject can be condensed or summarized as five statements or objectives:

1 Prevent the outbreak of fire.
2 Provide for the early detection of fire.
3 Prevent the spread of fire.
4 Provide for prompt extinguishing of fire.
5 Provide for immediate evacuation of personnel.

Oxygen and Fire

Consideration of the fire triangle (Figure 16-1) will indicate some of the general principles of effectuating the program. Oxygen is the most difficult of the three factors to control; first, because it is in the atmosphere, and second, because many natural and chemical compounds, called "oxidizers," have high concentrations of oxygen in them. Some very common examples of oxidizers are fluorine, chlorine, and methylene chloride. Pure oxygen is a stronger oxidizer than air. Fluorine is the only element which is a stronger oxidizer than oxygen. Fuel and heat are more controllable because under certain conditions, they can be separated from each other in a fire.

Fuels and Fire

Fuel must mix with air in certain proportions in order to burn. Temperature characteristics of fuels are important, especially with liquid fuels and to some extent with solid fuels as well. The "ignition temperature" is the lowest temperature at which a fuel, either solid or liquid, will ignite without an external source of ignition. With solids, this is also called the "kindling temperature." Solid fuels will ignite from an outside source of ignition after exposure to heat for varying lengths of time, but the ignition source must provide a temperature no lower than the ignition temperature at the surface of the fuel.

Three other temperatures are associated with liquid fuels. The "flash point" is the lowest temperature at which a liquid fuel will give off enough vapor to form a momentarily ignitable mixture with air. The "fire point" is the lowest temperature at which vapor will be formed rapidly enough to sustain a fire. The "boiling point" is the lowest temperature, at some given pressure, at which a liquid begins to evaporate or boil.

Materials that burn are innumerable. The following list indicates some of the most well-known materials that will burn:

Solvents
Lubricants
Wood products
Paper products
Cloth and other fiber material
Rubber and plastics

Heat and Fire

When organic materials come into contact with heat, they may eventually ignite—depending on the material, temperature, and the source of ignition. The ability of heat to aid in the ignition process depends on the intimacy between the fuel source and oxygen. Flammable mixtures can also be ignited by means of hot surfaces, either immediately or after a period of time. Some of the sources of hot surfaces are as follows:

1 Overhead electrical wiring
2 Burning cigarettes
3 Metals being welded
4 Boilers, furnaces, fireplaces, and chimneys
5 Friction of moving machine parts
6 Overhead motors
7 Electric heaters, hot plates, and irons

Heat also poses a physical danger to human beings exposed to superheated combustibles to the extent that it may cause death or serious physical harm.

Fire Gases

Fire gases are the gaseous products of combustion. Gases formed in a fire depend on the chemical combustion of the burning material, the amount of oxygen available for combustion, and the temperature. The greatest danger in most fire gases is carbon monoxide. The majority of fatalities from fires are caused by suffocation or inhalation of smoke or fire gases, not burns. Investigation into the hazardous properties of fire gases have shown the following gases to be among the greatest causes of death:

Carbon monoxide
Carbon dioxide
Hydrogen sulfide
Sulfur dioxide
Ammonia
Hydrogen cyanide

Other fire gases include nitrogen dioxide, acrolin, phosgene, hydrogen chloride, smoke, and soot.

PREVENTING THE OUTBREAK OF FIRE

It is well to examine the hazards from which most fires originate. These are sometimes described as "causes," but actually they are part cause and part source. While there are many possible causes, or sources, of fires, an analysis of previously reported data indicates that most fires are produced by a relatively small number of them. The hazards from which most fires originate can be ranked somewhat as follows:

1 Smoking and matches
2 Heating and cooking
3 Electrical equipment and faulty wiring
4 Rubbish
5 Flammable and combustible liquids
6 Open flames and sparks

With minor variations, which are due to the problem of classifying when two or more causes are apparent, these hazards remain from year to year as the origin of a large percentage of fires. However, there will be significant differences if we consider a specific industry or environment alone, as opposed to all fires in buildings and dwellings. For example, in some industries we might expect a large number of fires of electrical origin because there is extensive use of electrical equipment. In some industries, smoking and matches are subject to stringent control measures; and in mercantile establishments and hotels, the hazard of smoking and matches could be expected to be above average. In restaurants, of course, the hazards of heating and cooking rank high.

Smoking and Matches The National Fire Protection Association lists smoking and matches as the largest single cause of fire. With reference to smoking and matches, "cause" has to do with "careless disposal of," "improper handling of," or "use in hazardous locations." In view of how widespread smoking is (hundreds of billions of cigarettes are sold annually in the United States), this is not surprising. Every cigarette smoked has to be lit first, and disposed of later; and disposal of matches and cigarette butts can entail very imposing fire hazards. In some situations, careless disposal of a match or butt can mean the end of a life or lives and the loss of buildings and plants. Added to this is the hazard of lighting any combustible in an area which might contain highly flammable or explosive substances. The gravity of smoking and matches as a cause of fire is very clear.

Some general rules that should be adopted to minimize the hazards of smoking and matches are:

Prohibit smoking in all areas where flames or heat might be a serious hazard.

Provide clearly marked smoking areas and display signs in areas where smoking is prohibited.

In plants where highly hazardous products are processed or handled, establish and enforce more stringent "no smoking" rules.

Provide and identify adequate and convenient receptacles for the disposal of cigarettes and other smoking materials.

While it might be desirable to prohibit all smoking in some industrial plants, such a practice is impractical to enforce and would lead to smoking in hidden places. This would transfer the fire hazard to seldom-visited areas. For this reason, it is better to regulate smoking than to prohibit it.

Heating and Cooking   There are many types of stoves, heaters, ranges, boilers, and furnaces designed for a wide variety of uses in industry. Certain general principles can be given for the selection, installation, and use of such equipment:

Equipment should be *selected* to fit the needs of the job.

All equipment should be *approved* by a nationally recognized agency or conform to the code of the legal agency having jurisdiction.

Installation should permit safe operation and comply with recognized standards for safe operations.

There should be a safe *clearance* between any heating device and any combustible materials.

Provisions should be made for safe *storage and handling* for all fuels.

Provisions should be made for the safe *disposal* of all wastes, both solid and liquid.

Provisions should be made for connections to adequate *chimneys, vents,* and *ducts.* Furthermore, adequate *air supply* for heater rooms should be assured.

Personnel should be trained in the proper and safe use of the equipment.

Equipment should be *maintained* in a safe condition.

Electrical Equipment   Fires which originate in electrical equipment or installations or as a result of the use of electricity account for a substantial part of all fires. An examination of figures in industry indicates that fires of electrical origin are a major hazard.

The National Fire Protection Association has compiled and summarized the reports of fires of electrical origin from the International Association of Electrical Inspectors, and with minor variations from year to year the causes remain about the same. Some of the statistics are shown in Tables 16-3 and 16-4.

Even a casual analysis of these figures will suggest certain remedies for industrial manufacturing plants:

Install approved equipment selected for the job.

TABLE 16-3
CAUSES OF FIRES OF ELECTRICAL ORIGIN

	Approximate percentage of total
1 Equipment worn out in service	28
2 Improper use of approved equipment	14
3 Accidental occurrence	12
4 Defective installation	11
5 Unknown or not reported	35

TABLE 16-4
ORIGINS OF ELECTRICAL FIRES

	Approximate percentage of total
1 Appliances	52
2 Wires, cords, cables	27
3 Terminal equipment	12
4 Other	9

Comply with legal code requirements for installation.

Provide for an adequate program of inspection and maintenance.

Prepare standard operating procedures for use of electrical equipment, train personnel to use it, and provide adequately trained supervision.

Plan for adequate electrical power for new processes and equipment.

Rubbish Accumulations of rubbish, waste, or trash, while they may not of themselves cause a fire, can serve as fuel for a fire. In fact, many industrial waste products by their very nature provide a fuel that requires only a small amount of heat to burn. Combining poor housekeeping with careless smoking, poor electrical equipment, welding sparks, or fumes and vapors will almost inevitably lead to fires.

Poor housekeeping contributes greatly to the danger of spontaneous heating resulting in ignition when the materials involved reach their respective ignition temperatures. Accumulations of such items as grease, oil, various chemical solutions, and oily rags are susceptible to spontaneous heating and ignition if proper containers are not provided and prompt disposal is lacking. But even some materials employed in housekeeping can add to the hazard because they may contain hazardous flammable solvents. The following controls are suggested:

Provide a program of good housekeeping designed specifically for the plant and its operations and processes.

Provide for adequate disposal of all combustible waste and rubbish.

Provide safe containers for all substances subject to spontaneous heating and arrange for prompt and regular disposal of their contents.

Where large amounts of packing paper, wastepaper, or combustible wastes cannot be removed from the building immediately, make provision for them to be stored under overhead sprinklers.

Prohibit storage in locations which are "out of the way" and where there is no fire protection equipment available.

Flammable Liquids   Flammable liquids are an important cause of fire even though they account for relatively few fires. They must be vaporized before they burn, however; and it is somewhat difficult to distinguish between liquids and gases because of the wide range of temperatures at which the liquids become gas. Flammable liquids furnish the fuel for fires, and many can be ignited at relatively low temperatures. Those which have low flash points and low ignition temperatures are especially dangerous.

The fact that flammable liquids are the source of a smaller number of fires than some other sources does not mean that they are less hazardous; it could mean simply that industry is aware of the hazard and commonly takes certain precautions to control it. The extreme violence of explosions which can be produced by vapors of flammable liquids dictates caution in dealing with them. A single gallon of gasoline, when vaporized, can create a disaster of major proportions. The following precautions should be taken by users of flammable liquids:

Whenever possible, the use of a highly flammable liquid should be avoided; a nonflammable or less flammable liquid should be used instead.

All flammable liquids should be kept in closed containers or safety cans except when in actual use.

Limit the supply of flammable liquid in the work area to the amount necessary for one work shift.

Provide a safe operating procedure for all processes.

Provide for bonding and grounding of equipment as well as for grounding of all electrical equipment involved in a process where flammable gases are used or a place where flammable gases are stored.

Use only approved electrical equipment installed in conformity with code requirements of the agency having legal jurisdiction.

Prohibit smoking, open flames, and spark-producing devices or equipment in the vicinity of flammable volatile liquids.

Provide for storage of flammable liquids at a safe distance from any heat source.

Provide adequate ventilation for all operations involving the use or storage of flammable liquids.

FIGURE 16-2
Flammable material in any form, open or containerized, improperly stored (as shown here) can cause fires and/or explosions.

Provide for proper venting of buildings and storage tanks.

Provide for safe disposal of flammable liquid wastes.

Provide sand or other noncombustible substances for use in cleaning up spills of flammable liquids.

Open Flames and Sparks   The fire hazard of open flames is so obvious that users would be expected to take proper precautions. The prevalence of fires from this origin, however, belies such an assumption. It is true that cutting and welding torches, salamanders, and other sources of open flame are used under conditions which make adequate control measures difficult, but this merely intensifies the necessity for precautions.

Sparks also cause many fires. The hazard from sparks from an outside source can be controlled by the use of incombustible roofs and spark arresters. Sparks from an outside source—static electricity, furnaces, heating equipment, grinding wheels, welding and cutting torches, or the use of steel tools—likewise require control.

The following are general rules for controlling hazards from open flames and sparks:

Provide housekeeping to avoid accumulations of fuels and combustibles in the vicinity of open flames.

Where open flames are unavoidable, establish strict controls and provide adequate portable or fixed fire extinguishers.

Provide for safe clearances between combustible substances and sources of heat.

Where clearance is not feasible, provide for covering or insulation of combustibles. This is important in welding or cutting.

Ground and bond all electrical equipment to prevent the danger of electric or static sparks.

Use only properly designed and constructed equipment.

Provide a safe procedure for the operation of all devices which might produce flames or sparks.

Provide periodic maintenance and inspection programs.

FIGURE 16-3
These liquid propane tanks are stored in a safe manner.

Provide for Early Detection of Fire

Except for explosions, most fires start out as small ones. At the beginning, then, extinguishing a fire seldom presents much of a problem; but once the fire begins to gain headway, it may develop into a conflagration of disastrous proportions. The prompt detection of fire and the signaling of an alarm are therefore of prime importance. The alarm, of course, warns of the need to evacuate the building and also summons those charged with and trained in firefighting. When a fire is detected by personnel, it is important that they have been trained to act effectively. The first impulse of many individuals on discovering a fire is to try to extinguish it; this has frequently led to long delays in sounding the alarm. All employees should be trained to sound the alarm as soon as the fire is discovered and, if possible, *then* take action to try to extinguish it. (Of course, each plant may have its own fire procedure.)

Guards   Employment of a fire guard during "off" times when plants are not operating is probably the oldest type of alarm system. If there is any qualification a guard must have, it is reliability. A 60-minute-per-hour watch system is needed, not just a casual tour of the plant by the guard. The use of a clock which the guard "punches" at specified locations offers some protection. However, a system in which the guard "punches in" at given locations and this fact is recorded is much superior, since any failure on the part of the guard will be indicated and can be investigated immediately.

Automatic Alarm Systems   The next most desirable system of fire detection is the use of automatic fire alarms. Types of automatic alarms include:

The *fixed-temperature type,* which is designed to operate when the temperature in the vicinity of the alarm reaches a predetermined level. For example, such an alarm can be set to operate when the temperature reaches 135°, 150°, or 200°.

The *rate-of-rise type,* which is designed to operate when there is a rapid rise of temperature.

A recently developed *nuclear detector* device, which is designed to be activated by the products of combustion rather than by a given heat or a rapid rise in temperature. Smoke is not required to activate it.

Each of these systems has its own advantages for particular installations. Which type to use for a certain location should be determined only by competent engineering personnel after careful analysis of needs. All automatic alarms share one advantage: they eliminate the possibility that

FIGURE 16-4
Firefighting equipment should never be blocked by materials, storage racks, or anything that might impede access to it.

a fire will not be discovered because no one is in the vicinity; that is, they eliminate the factor of human failure. The use of an automatic alarm which operates simultaneously with an automatic extinguisher is the next refinement.

Where an automatic alarm system is installed, certain general requirements are important:

A reliable signal must be transmitted.

The signal must reach all personnel, including those trained for fire protection, no matter where they are located.

The signal must be of such a nature that it will be instantly recognized.

The signal must be used for no other purpose than to sound the alarm for a fire.

A method for signaling outside help—either the local fire department or any other plant participating in a mutual-aid program—should be located immediately adjacent to the local alarm system.

Prevent the Spread of Fire

Once a fire is discovered, it is of prime importance to confine it to the smallest area possible—that is, to prevent its spread. This can be accomplished by details of construction and by safe practices, but neither is sufficient alone. An understanding of the means by which heat is transmitted will be of value in taking the necessary steps to prevent the spread of fire.

Transmission of Heat   Heat is transmitted by conduction, convection, and radiation. In addition, it can spread by contact of fuel with the fire itself.

It is well-known that a flammable or combustible material brought into contact with flame will catch fire if the contact is maintained long enough to raise the temperature of the substance to its ignition temperature. It is, therefore, important to store all combustible materials so that they are well removed from any source of heat. In many cases, combustibles may not be "stored" but rather are wastes or rubbish which have been allowed to accumulate in close proximity to flames or heat sources. One of the advantages of a good housekeeping program is that it prevents the accumulation of substances which could furnish fuel for a fire.

Conduction   Heat is transmitted through solids by conduction. Materials vary greatly in their ability to transmit heat. Metals are good conductors of heat. On the other hand, wood, glass, pottery, asbestos, and many similar substances are very poor conductors of heat; these are termed "insulators." It should be remembered, however, that there are no perfect insulators of heat. All materials will conduct heat to some extent; and, if the heat continues long enough, it will be transmitted by the solid. The hazard of heat transmission is illustrated by the fact that a fire on one side of a metal wall could start a fire on the other side if combustibles were close to the wall.

Convection   Convection currents occur in fluids—that is, liquids and gases. The self-circulating hot-water furnace is an excellent example of convection currents in water. The water in the furnace expands when heated and is lighter than the cold water in the radiators. The cold water, therefore, moves downward and pushes the hot water up into the radiators.

One often hears the statement that hot air rises and pulls in cold air. This is the reverse of the truth. In fact, when a gas is heated, it expands and is therefore lighter per unit of volume. The cold gas, being heavier per unit of volume, moves in and displaces (pushes upward) the hot gas.

Convection currents are important in fires. The hot gases which develop in a fire are pushed up through any vertical opening and can spread a fire from floor to floor. The hotter the fire on the lower floors, the greater the upward push of the hot gases.

Radiation In radiation, heat rays travel from one body to another in the same manner that light is carried through space by light rays. Heat rays travel in a straight line and are not absorbed to any great extent by the air or by a transparent substance like glass; but they are absorbed by any opaque object they encounter. Thus, heat radiated by a hot furnace would not be absorbed by the air through which it travels but would be absorbed by a wall.

All other things being equal, the amount of radiant heat reaching an exposed object from a heat source depends upon the temperature difference between the heat source and the object and varies inversely as the square of the distance between them. The amount of heat emanating from a heat source increases very rapidly as the temperature goes up. With very hot sources of heat, great quantities of radiant heat are produced and must be dealt with.

The distance of the exposed object from the heat source is also important. It is apparent that increasing the distance an exposed material is removed from a heat source will decrease the possibility of a fire. As a practical matter, however, where the area or intensity of the heat source is large in comparison with the distance—as in the case of a large building burning across an alley from another building—small variations in distance have little practical effect on the severity of the exposure.

Barriers to Limit the Spread of Fire It is not always possible to extinguish a fire promptly. Barriers are one means of control that will limit the area of a fire or at least retard its spread. The use of fire-resistant materials in construction and content will aid significantly in accomplishing this goal. The following list covers only some of the bare essentials:

Fire walls
Fire doors
Shutters or louvers
Fire stops
Baffles
Fire dampers
Fire windows
Parapets
Dikes
Enclosures of vertical openings

Provide for Prompt Extinguishing of Fire

Fires can be extinguished by eliminating the fuel, the oxygen, or the heat.

Removing Fuel Extinguishment by removing the fuel is possible in comparatively few fires. If an oil burner in a basement were causing a fire, and the fuel tank were outside, it might be possible to close a valve at the tank and thus extinguish the oil fire. Once a fuel is on fire, however, its removal usually poses a major problem.

Most fires are extinguished by (1) excluding the supply of oxygen or (2) reducing the temperature of the fuel below its ignition temperature.

Excluding Oxygen Excluding oxygen from a fire usually involves the use of some agents that will tend to "blanket" the fire and thus "smother" it. Four extinguishing agents are used for this purpose: (1) foam, (2) carbon dioxide, (3) dry chemical, (4) vaporizing liquids. In most cases these substances are heavier than air and settle around the fire in such a way as to shut out the oxygen.

There are many compounds—classified as "oxidizing agents"—which have a large supply of oxygen held rather loosely in combination. When these compounds come into contact with fuels under favorable conditions, combustion can result. Since the oxygen is being supplied from the compounds themselves, it would be impossible to extinguish such a fire by excluding oxygen. Once a fire involving these substances is started, the heat will promote the disintegration of the compounds, and large supplies of oxygen may be released to feed the fire. Extinguishment is possible only by cooling. Among the chemical groups that can intensify fire by their oxidizing action are nitrates, peroxides, chlorates, perchlorates, and permanganates. In particular, it should be noted that commercial hydrogen peroxide (90 percent) is highly dangerous since it liberates great volumes of oxygen easily. Nitrates also liberate oxygen easily.

Reducing Temperature Water is the most common agent used to reduce the temperature of fuels below their ignition temperature. Where there is much heat, it is necessary to use large amounts of water to reduce the heat and to keep the fuels below their ignition temperature after the fire has been extinguished.

Fire Extinguishers Fire extinguishers fall into two categories: permanent and portable.

Permanent Permanent or "built in" means of extinguishment, such as standpipe and hose, automatic sprinkler systems, and automatic extinguishing systems, are used extensively to lessen the danger from fire. The number and location of such devices are highly technical matters, and safety personnel should get competent technical assistance if installation of a built-in system is contemplated. After such a system is installed, its maintenance is equal in importance to other aspects of the fire

FIGURE 16-5
The first line of defense against fire is a good supply of fire extinguishers located in a strategic area of the plant and made easily accessible.

safety program. A system of regular inspection and maintenance is suggested.

Portable   Portable fire extinguishers have been called "first-aid" fire extinguishers. They contain a limited supply of an extinguishing medium. These appliances are designed for use on fires of specific classes. Each type of fire extinguisher is of value, but no one type is of equal value or effectiveness on all kinds of fires.

Familiarity with the correct use of portable fire extinguishers is essential for all personnel likely to be involved in the use of this equipment. Any fire-protection program that does not include thorough training of personnel can create the possibility of a costly or even a disastrous fire. Lack of knowledge of how to use first-aid fire extinguishers may lead to confusion, clumsy application, and ineffectiveness. The following *general rules* are suggested:

Reliable extinguishers, carrying the approval label of a nationally recognized testing laboratory, should be used.

The right type of extinguisher should be provided for each class of fire that may occur in the area.

There should be enough units to afford protection for the area.

Extinguishers should be located where they will be readily accessible for immediate use.

Extinguishers should be maintained in perfect operating condition, inspected frequently, checked against tempering, and recharged as required.

Personnel should know the location of extinguishers and be trained to use them effectively and promptly.

Provide for Immediate Evacuation of Personnel

Once a fire is discovered in a building, the first and most important step is the prompt evacuation of all personnel to a safe place. This matter takes precedence over every other consideration. There is no way to account for the actions of persons who are faced with fire, but behavior can be improved by training in orderly evacuation through fire drills.

Fortunately, it is not excessively difficult to provide and maintain exits that will empty the ordinary structure in ample time to prevent loss of life or injury, except in the case of explosions—but where the possibility of explosion exists, extraordinary precautions should be taken. The provisions of the Building Exits Code, developed by the National Fire Protection Association's Committee on Safety to Life, should be used as a guide in determining the adequacy of exits.

Following are suggestions for evacuation of personnel:

1 Any location where persons work or congregate for any purpose should be provided with at least two exits, located so as to render the cutting off of both at one time highly improbable.

2 Exits should be readily available, with an unmistakable path to safety, clearly marked and suitably lighted. Where work is done at night, an emergency lighting system is suggested.

3 Exit stairways should be enclosed in fire-resistant walls with openings protected by fire doors. All doors in exits should open in the direction of egress and should be operable from the inside without the use of a key. Panic hardware (that is, a brass bar across the width of a door which automatically pushes the door open when depressed) is suggested.

4 The use of outside fire escapes is not recommended unless windows or other openings in walls are designed for protection against flame and smoke.

5 The minimum clear stair width should not be less than 44 inches.

6 No storage or accumulation of materials in escape ways should be allowed.

7 Thorough training of all personnel for evacuation should be provided. Periodic drills are a must in such a program. Fire wardens and all supervisory personnel should be trained to supervise evacuation.

8 An unmistakable evacuation alarm system, used for no other purpose, should be provided.

9 Adequate provision should be made for the evacuation of physically handicapped persons at all times.

10 The location to which personnel are evacuated should provide assured safety.

ARSON

Insurance companies, with the cooperation of the fire and police departments, are doing all they can to take the profit out of arson. But despite these efforts, more and more insurance dollars are being paid out for fire losses resulting from arson; as a result, fire and casualty insurance premiums increase. Fire insurance, after all, is simply a means of spreading risk; insurance companies collect premiums from many people and compensate the few who have losses.

An increasing number of fires in the United States are being categorized as "incendiarism" or "cause unknown; incendiarism suspected." Unfortunately, incendiarism has become one of the fastest growing crimes in this country.

In 1986, arson was the largest single cause of property damage in the United States. That year recorded 111,000 cases of suspected arson. Of the 6,000 fatalities due to fire that year, 705 were reportedly caused by arson.

Incendiarism is a disturbing, difficult, social and economic problem that should be of concern to us all. The fact is, incendiarism is fast becoming a major threat to business and industry. Loss investigators find it difficult to determine the *who, why,* and *how* of incendiarism because the evidence is often consumed. Fire setters range from juveniles with matches to disgruntled employees, from pyromaniacs to professionals. The motives for man-made fires (arson) are:

1 Spite or revenge
2 Vanity
3 Fraud
4 Pyromania
5 Concealment of a crime
6 Civil disobedience

WORD POWER

combustible liquid A liquid having a flash point at or above 140 degrees Fahrenheit.

fire point or ignition point The lowest temperature at which a liquid fuel will give off enough vapor to burn continuously.

flammable gases Gases that will burn in normal concentrations of oxygen in the atmosphere.

flammable liquids Liquids having a flash point below 140 degrees Fahrenheit.

gases Formless fluids that will occupy whatever space is available to them.

spontaneous combustion A form of ignition caused by the internal development of heat.

QUESTIONS AND EXERCISES

1 What is the difference between the flash point and the ignition temperature?
2 How is it possible for heat to reignite a fire once it has been extinguished?
3 Name the four stages of a fire.
4 Is water considered as a good extinguisher for a Class C fire? Why?
5 Name three commercial examples of a Class D fire extinguisher.
6 Would sand be a good extinguisher for a Class D fire? Why?
7 What is combustion?
8 How many materials can you name that can ignite by spontaneous combustion?
9 What is the difference between a flammable and a combustible liquid?
10 There are three classifications of gases: flammable, nonflammable, and toxic. Define toxic gases.
11 The production of carbon monoxide gas is considered as one of the greatest dangers in a fire. How is carbon monoxide produced in a fire?
12 Hydrogen cyanide is produced in residential and aircraft fires during the combustion of materials containing nitrogen. Name at least five of these materials.
13 What are the three methods of heat transmission?
14 Name five barriers used to limit the spread of fire in a structure or building.
15 Are portable fire extinguishers necessary in a building equipped with an overhead sprinkler system? Why?
16 Name four extinguishing agents used for smothering or blanketing fire.
17 What characteristics of water make it such an excellent fire-extinguishing agent?
18 Name at least five hazards from which fires originate.
19 Give some general principles for the selection, installation, and use of heating and cooking equipment.
20 Name three common causes of fires of electrical origin.
21 Is it possible for rubbish to ignite spontaneously in the workplace?
22 Name some common precautions used by industry for the storage and use of flammable liquids.
23 Describe what is meant by a "fixed-temperature" type of automatic alarm system.

24 Describe how a nuclear detector device is activated. Is smoke required to activate it?

25 What are the general requirements for the installation of an automatic alarm system?

26 Name at least five barriers used by industry to limit the spread of fire in the event of a fire.

27 What are oxidizers? Name the five chemical groups containing oxidizers.

28 Name five general rules relating to the installation and use of portable fire extinguishers.

29 Procure a portable fire extinguisher and demonstrate its proper use for the class.

30 Name the classes of fires and their extinguishment agents.

BIBLIOGRAPHY

Accident Prevention Manual for Industrial Operations, 8th ed., National Safety Council, Chicago, Ill., 1980, p. 626.

Fire Protection Handbook, 15th ed., National Fire Protection Association, Boston, Mass., 1981, section 18-4.

Marshall, Gilbert: *Safety Engineering,* PWS-Kent Publishing Company, Boston, Mass., 1982, p. 199.

CHAPTER 17

SAFETY IN WELDING AND CUTTING WITH OXYACETYLENE EQUIPMENT

WELDING GASES

Welding gases are compounds of carbon and hydrogen—that is, they are hydrocarbons. When welding gases are burned, the product is heat energy, the same kind of energy released by burning coal, wood, or oil—which are also hydrocarbons. Heat energy is released from welding gases by burning these gases with oxygen; the burning creates a reaction between oxygen and carbon, and oxygen and hydrogen. Burning—combustion—is a chemical process involving the oxidation of substances like carbon and hydrogen in welding gases during burning. The temperature to which a substance must be raised in order to burn is called its "ignition point." The ignition point differs in different combustibles. A few examples of combustibles are hydrogen, carbon, carbon monoxide, acetylene, methane, ethane, ethylene, and propane.

The four important characteristics of a welding gas are heat content, flame temperature, combustion ratio, and combustion products.

Oxygen

Oxygen is an element which at atmospheric temperatures and pressures exists as a colorless, odorless, tasteless gas. About one-fifth of the atmosphere is oxygen: 20.99 percent by volume. Since almost 21 percent of air is oxygen, air is the natural source from which to obtain pure oxygen. To accomplish this, the air is cooled down to a temperature of almost −300°F at which point it becomes liquid in a process called "liquefaction."

The most common industrial use of oxygen is in welding. Oxygen is combined with acetylene and other fuel gases for use in such processes as welding, metal cutting, flame hardening, scarfing, and cleaning. When burned with oxygen, acetylene is an excellent welding process that produces a temperature of 5,800°F.

Acetylene

Acetylene is a compound formed by the combination of two elements: carbon and hydrogen. At atmospheric temperatures and pressures, acetylene is a colorless gas with a garliclike odor. It is manufactured from calcium carbide: water and calcium carbide are combined to produce acetylene, with calcium hydroxide as a by-product.

In gaseous form, acetylene is subject to explosive decomposition when ignited under pressure. Therefore, a limit of 15 pounds per square inch (psi) has been set as the maximum allowable pressure for the generation and handling of acetylene in general industrial situations.

Acetylene is stored in cylinders having a porous-mass packing material made so as to compensate for the decomposition characteristics of the gas. The porous mass is saturated with acetone in which the acetylene dissolves. The combination of the porous filler and acetone solvent allows the acetylene to be safely stored at a pressure of not more than 250 psi without danger of explosive decomposition.

Although acetylene has many commercial uses, it is most widely used in welding. Used with oxygen, it produces one of the hottest known flames.

Hydrogen

Hydrogen, combined with oxygen, is used for low-temperature and brazing operations. The oxyhydrogen flame produces a lower heat than the oxyacetylene flame and is preferable for low-melting-point metals. Lead burning or welding is another application of oxyhydrogen low-heat flame, which also burns clean. This feature of the oxyhydrogen flame, with the low melting point of lead, dictates the use of oxygen and hydrogen.

Hydrogen is also used in atomic hydrogen welding. In this process, the metal is welded by being heated with an electric arc maintained between two metal electrodes in a hydrogen atmosphere. By means of this arc between two electrodes, the temperature in the arc stream rises to approximately 11,000°F, thereby dissociating the molecular hydrogen into its atomic form (hence the term "atomic hydrogen"). This form of welding is applicable to practically all nonferrous, as well as ferrous, metals and alloys.

Hydrogen is used in preference to acetylene in underwater cutting because it can be safely compressed to the pressures necessary to overcome water pressures at the depths where salvage operations are being undertaken. Acetylene does not conduct itself as tamely under water as hydrogen does; acetylene will pocket under water and possibly explode.

This chapter deals with oxyacetylene equipment—that is, equipment using the gases oxygen and acetylene—because it is most common (90 percent of all welding is oxyacetylene welding).

Efficiency and safety in the use of the oxyacetylene process for welding, cutting, and heating go hand in hand with careful observance of suitable operating procedures, precautions, and safe practices. The oxyacetylene process is one of modern industry's most versatile tools, and sensible care will ensure its full economic usefulness. Welding or cutting is not particularly hazardous, but common-sense precautions and strict adherence to good safety practices must always prevail.

SETUP OF OXYACETYLENE EQUIPMENT

Equipment Needed

In addition to the gases required, the complete equipment for welding, heating, and cutting with oxygen and dissolved acetylene consists of the following items:

> Welding blowpipe (or torch)
> Cutting blowpipe (or torch), or cutting attachment
> Oxygen pressure-reducing regulator
> Acetylene pressure-reducing regulator
> Oxygen hose
> Safety glasses and gloves
> Torch lighter
> Blowpipe wrench
> Acetylene hose
> Hose connections
> Regulator wrench

In the welding industry, the words "blowpipe" and "torch" are synonymous.

For welding or heating only, a cutting blowpipe or attachment is not needed. Likewise, a welding blowpipe is not required if the outfit is for cutting only.

No acetylene regulator will be needed if acetylene is supplied from a low-pressure generator. With a medium-pressure generator, a regulator may or may not be required, depending on the kind of generator used and the nature of the work.

Connecting the Oxygen Supply to the Welding Torch

Oil and grease may ignite violently in the presence of oxygen under pressure. Therefore, every piece of equipment through which oxygen may pass must be kept entirely free from oil or grease. This includes cylinder outlets, regulators, manifolds, oxygen pipelines, hose lines, blowpipes, and all connections. Such oxyacetylene apparatus is designed so that it can be made tight without the aid of any pipefitting compounds or lubricants. Since such compounds and lubricants usually contain oil or grease, they should never be used.

Oxygen Cylinders Markings or labeling on a cylinder should be checked to be sure that the cylinder contains oxygen and not some other gas. A cylinder which is not clearly marked or labeled should not be used but should instead be returned to the supplier.

Oxygen should never be used from a high-pressure cylinder unless an oxygen regulator is attached to the cylinder valve, so as to obtain a safe, constant working pressure. The only exception to this rule occurs when suitable oxygen-cylinder manifolds or coupled units are used, in which case a pressure-reducing regulator is included in each assembly to care for the entire manifold supply. An oxygen pressure-reducing regulator should never be connected to a cylinder containing acetylene or other fuel gas, nor should a pressure-reducing regulator for acetylene or other fuel gas be connected to an oxygen cylinder.

If the cylinder is not on a suitable cylinder truck or otherwise firmly held, it should be securely tied to a workbench, wall post, or large-size acetylene cylinder so that it cannot be accidentally knocked over or pulled over. The valve-protection cap should not be removed until the cylinder is firmly secured. Do not lay a high-pressure cylinder down unless there is nothing to tie it to.

Oxygen cylinder valves should be "cracked." To do this, the welding operator (welder) should stand at the side or rear of the oxygen cylinder outlet, open the oxygen cylinder valve slightly for an instant, and then close it. This will clear the valve of dust or dirt which may have accumulated during shipment or storage. Otherwise, the dust or dirt might mar the seat of the regulator inlet nipple or be carried into the regulator and cause leakage or creeping.

Oxygen regulators and oxygen-cylinder valve outlets have right-hand

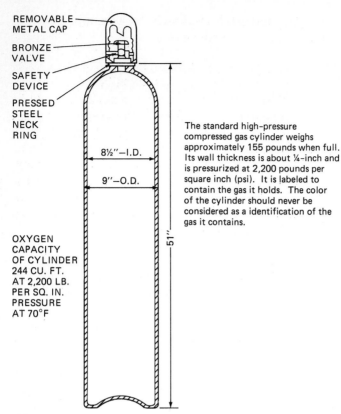

REMOVABLE
METAL CAP

BRONZE
VALVE

SAFETY
DEVICE

PRESSED
STEEL
NECK
RING

8½"–I.D.

9"–O.D.

51"

OXYGEN
CAPACITY
OF CYLINDER
244 CU. FT.
AT 2,200 LB.
PER SQ. IN.
PRESSURE
AT 70°F

The standard high-pressure
compressed gas cylinder weighs
approximately 155 pounds when full.
Its wall thickness is about ¼-inch and
is pressurized at 2,200 pounds per
square inch (psi). It is labeled to
contain the gas it holds. The color
of the cylinder should never be
considered as a identification of the
gas it contains.

FIGURE 17-1
High-pressure compressed gas cylinder.

thread connections. The union nut should be tightened with a regulator wrench. To prevent leakage, the nut should be pulled up tight.

It is important that the regulator pressure-adjusting screw be released—that is, that it be turned counterclockwise (to the left) until it is loose—before the cylinder valve is opened.

The cylinder valve on the high-pressure cylinders should never be opened suddenly, as the rush of high-pressure oxygen might strain the cylinder-pressure gage mechanism. The oxygen cylinder valve should first be opened just enough to allow the hand of the cylinder-pressure gage to move up slowly. When the gage hand stops moving, the cylinder valve should be fully opened. The pressure in a high-pressure oxygen cylinder is shown on the high-pressure or cylinder-contents gage whenever the cylinder valve is open. **Precaution:** A welder should always stand to one side of the gage and away from the gage faces and the front of the regulator when opening the cylinder valve.

Connecting the Hose   Oxygen hose (green) and acetylene hose (red) should never be interchanged or used for any other purpose.

It is recommended that factory-fitted hose assemblies be used. Factory fitting is convenient and reliable; furthermore, fitted hose is available which is made of an oilproof, flame-resistant synthetic that is lighter than rubber yet stronger and more flexible.

If the oxygen hose (green) does not have hose connections at both ends, put these connections on next. The most common types of connections are the push-type nipple and the screw-type nipple. For securing to a push-type nipple, first put the nipple and nut together and attach the nut to the regulator. Slide the clamp over the hose, push the nipple into the hose, and tighten the clamp in the desired position. Circle-type clamps are recommended because they make a gas-tight connection that will not loosen under pressure, and they can be put on and tightened easily by squeezing the sides of the clamp fold together with pincers. To install screw-type nipples, attach the nut to the regulator tightly enough to prevent turning the nipple while the hose is being screwed on.

Push a ferrule onto one end of the hose. Twist the hose onto the nipple attached to the pressure-reducing regulator. The end of the hose should be cut off square before the ferrule is pushed on. If the hose has been used before, the end should be cut off square beyond the mark of the old fitting. When double hose joined with web or its equivalent is being used, the two pieces must be parted for a short distance and a clamp used to prevent further separation. Care must be taken to cut ends to the proper length to avoid unnecessary straining or buckling of the separated ends. Attach connections as with single hose. As has been noted, the green hose is for oxygen; the red is for acetylene.

The ferrule should be pushed on up to its shoulder. Use water or a soap-and-water solution to make the connections slide together more easily if the ferrule is tight. It will also help if a little of the liquid is applied to the nipple and inside the end of the hose. The hose should be screwed onto the nipple tight until the back of the ferrule just clears the connections nut. Then the hose connection can be unscrewed from the regulator and the second hose connection put on in the same manner.

Attach one end of the hose to the oxygen pressure-reducing regulator. If the hose is new, blow it out to remove loose talc. Then attach the other end to the torch connection marked "oxygen." To blow out hose, attach one end to the outlet connection of the oxygen regulator and leave the other end free. Turn the pressure-adjusting screw of the regulator clockwise (to the right), permitting oxygen to blow through the hose. Keep turning the handle until a pressure of 5 pounds per square inch (psi) shows on the low-pressure gage. Then turn the pressure-adjusting screw of the regulator to the left until the flow of oxygen stops. Since new hose is dusted in the inside with fine talc, this procedure will blow out any talc that is loose.

After blowing out the hose, attach the free end to the torch connection marked "oxygen."

Testing Connections for Leaks To test connections for leaks, close the torch oxygen valve and turn the pressure-adjusting screw of the oxygen regulator clockwise (to the right) to give about normal working pressure. Test the following for leakage: oxygen cylinder valve stem; oxygen regulator inlet connection at the cylinder valve; all hose connections; torch oxygen valves. Use nothing but an approved leak-test solution or soapy water.

Connecting the Acetylene Supply to the Torch

Use only hose and connections made especially for oxyacetylene welding and cutting. Acetylene hose and oxygen hose, as has already been mentioned, should not be interchanged or used for any other purpose. As with oxygen hose, do not use pipefitting compounds, oil, or grease for making connections in acetylene hose. Never force connections that do not fit.

Acetylene Cylinders Check the cylinder to be sure it contains acetylene. If there is no marking or labeling on a cylinder, if the marking is not clear, or if there is any doubt about the type of gas in a cylinder, do not use it. Return it to the supplier and explain why it is being returned.

Always attach an acetylene pressure-reducing regulator when using acetylene from a cylinder. Acetylene should never be used from the cylinder unless an acetylene regulator is attached to the cylinder valve, so as to obtain a safe, constant working pressure. The only exception to this rule occurs when suitable acetylene cylinder manifolds or coupler units are used, in which case a regulator is included in each assembly. To attach the acetylene pressure-reducing regulator, fasten the acetylene cylinder in an upright position so that it cannot be knocked over. If the cylinder is not on a suitable truck or otherwise firmly held, it should be securely tied to a workbench, a wall, a post, or an oxygen cylinder so that it cannot be accidentally knocked or pulled over. Never connect an acetylene regulator containing oxygen, or an oxygen regulator, to a cylinder containing acetylene or other fuel gas.

Acetylene cylinders should always be stored and used with the valve up. They should not be allowed to lie on their sides.

"Crack" the acetylene cylinder valve. To do this, stand at the side of the acetylene outlet, open the acetylene cylinder valve one-quarter turn, and then close it immediately. This is intended to clear the valve of dust or dirt that may have accumulated during shipment or storage. Otherwise, the dust or dirt might mar the seat of the inlet nipple or be carried into the regulator and cause leakage or creeping. **Precaution:** Never crack an acetylene cylinder valve near other welding or cutting work or near sparks, flame, or any other possible source of ignition.

Connect the acetylene pressure-reducing regulator to the acetylene cylinder. Acetylene regulators and acetylene cylinder valve outlets have left-handed thread connections, so that an acetylene regulator cannot be accidentally connected to an oxygen cylinder. Tighten the union nut with a regulator wrench. To prevent leakage, be sure the nut is pulled up tight. If the thread on the cylinder connection does not match the thread on the regulator, first connect the proper adaptor to the cylinder valve and then connect the regulator to the adaptor.

Precaution: Should it ever be necessary to tighten the union nut after the regulator is connected, be sure to close the cylinder valve first. If the connection still leaks when reasonable force has been used in tightening the nut, close the cylinder valve, take off the regulator, and clean both the inside of the cylinder valve seat and the regulator inlet nipple seat. If excessive force is used in tightening the connections, the seats may be marred, and the threads in the nut may become so distorted that it will not fit any other cylinder. If this happens, the regulator must be returned to the manufacturer for repair.

Turn out the pressure-adjusting screw of the regulator until it is loose. The regulator pressure-adjusting screw must be released—that is, it must be turned counterclockwise (to the left) until it is loose—before the cylinder valve is opened. This avoids abuse and possible damage to the regulator and gages.

Open the acetylene cylinder valve slightly at first, and then 1½ turns but no more. The acetylene cylinder valve should never be opened suddenly, as the rush of acetylene might strain the cylinder-pressure gage mechanism. **Precaution:** Always stand to one side of and away from the gage faces and the front of the regulator when opening the cylinder valve.

Always leave the T-wrench in place on the valve while the cylinder is in use.

The pressure in the acetylene cylinder is shown on the high-pressure or cylinder-contents gage whenever the cylinder valve is open.

Connecting Hose   If the acetylene hose is new, blow it out to remove loose talc. A suitable way to do this is:

1 Attach an oxygen-hose connection nipple and nut to the oxygen regulator.

2 Twist one end of the acetylene hose partway onto the nipple.

3 Blow out the hose with oxygen at a pressure of about 5 psi.

4 Remove the hose from the oxygen nipple and blow through it from the mouth.

Precaution: Do not blow out the hose with acetylene. A combustible gas should not be released into the atmosphere except in special cases, and then only with extreme care and when there is no source of ignition. A gas should not be ignited at the end of any hose connection.

If the acetylene hose (red) does not have connections at both ends, put these on next. Do this the same as for oxygen hose, except, of course, that the acetylene hose connection nut, which has a left-handed thread, should be attached to the acetylene-regulator outlet.

Attach one end of the hose to the acetylene pressure-reducing regulator and the other end to the blowpipe connection marked "acetylene."

Testing Connections for Leaks   To test connections for leaks, close the torch acetylene valve and turn the pressure-adjusting screw of the acetylene regulator clockwise (to the right) to give a pressure of about 10 psi. Test the following points for leakage: acetylene cylinder valve stem; acetylene regulator inlet connection at the cylinder valve; all hose connections; torch acetylene valve. Use nothing but a leak-test solution or soapy water.

Adjusting Pressures

Attach the proper head, tip, or nozzle to the welding torch. Refer to the instruction leaflet furnished with the torch for information on what size head, tip, or nozzle and what oxygen and acetylene pressures to use for a particular job.

Some welding heads require only slight force in being tightened to the torch handle; others require the use of a wrench and firm tightening. Be sure to follow the recommendations of the torch manufacturer. Always be sure the seating surfaces between the head, tip, or nozzle and torch are uninjured (undamaged) and free of dirt before attaching.

With the torch oxygen valve *open,* turn in the pressure-adjusting screw on the oxygen regulator to the pressure desired. Then close the torch oxygen valve. When the torch oxygen valve is open and the regulator pressure-adjusting screw is turned in to the right, the delivery-pressure (low-pressure) gage indicates the actual operating pressure.

For cutting attachments or cutting torches, open both the torch oxygen valve and the cutting valve before adjusting the oxygen pressure. Should the cutting torch be connected to two oxygen regulators and hose lines, adjust each regulator separately with the corresponding torch valve open. Always close the torch valves after pressures are adjusted.

Always avoid using oxygen pressures higher than those recommended by the manufacturer.

With the torch acetylene valve *closed,* turn in the pressure-adjusting screw on the acetylene regulator to the pressure desired. Open the torch acetylene valve, light the flame, and readjust the regulator to the correct pressure. Then close the torch acetylene valve.

Precaution: Never release acetylene into the air near other welding work or near sparks, flame, or any other possible source of ignition, or into any space which is not adequately ventilated. To avoid such a release is why the acetylene pressure should first be adjusted with the torch acetylene valve closed.

When the torch acetylene valve is *closed* and the regulator pressure-adjusting screw is turned in to the right, the delivery-pressure (low-pressure) gage indicates the approximate operating pressure.

The next step is to open the torch acetylene valve two turns and light immediately at the tip. Adjust to the proper acetylene pressure by turning the pressure-adjusting screw either in or out, depending upon whether more or less pressure is needed. Then close the torch acetylene valve.

OPERATION OF OXYACETYLENE EQUIPMENT

General Instructions for Operating the Torch

Never allow oil or grease to come into contact with oxygen under pressure. As has been noted, no lubrication of the apparatus is necessary.

Never use oxygen as a substitute for compressed air, as a source of pressure, or for ventilation.

Before starting to weld or cut, look around to make certain that flame, sparks, hot slag, or hot metal will not be likely to start a fire.

Always use the proper size tip and the proper gas pressures. It is especially important to use the proper gas pressures for the head, tip, or nozzle selected. Tip sizes and gas pressure for the work involved are usually shown in the literature furnished by the manufacturer with the torch.

Be sure to keep a clear space between cylinders and the work. This is important so that cylinders and pressure-reducing regulators can always be reached quickly.

Never use matches for lighting torches. Hand burns may result from this practice. Use friction lighters, stationary pilot flames, or some other similar suitable source of ignition.

Do not relight flames or hot work in a pocket or small confined space. Always relight with a lighter. When a flame is relit from hot metal, the gases do not always ignite instantly and, in a small pocket, ignition may be violent if it is delayed for even a second.

Never use acetylene at pressures above 15 psi. Using acetylene at pressure in excess of 15 pounds per square inch gage (psig) or 30 psi absolute pressure is a hazardous practice. The limit of 30 psi absolute pressure is intended to prevent unsafe use of acetylene in pressurized chambers such as caissons, underground excavations, and tunnel construction.

Never release acetylene where it might be the cause of fire or explosion. For example, acetylene should never be released into the air near other welding or cutting work or near sparks or flame from any other source. If it is necessary to release acetylene, release it out in the open in a place where a mixture with air will not create a hazard.

Always see that hose is securely connected before using the equipment. When using equipment for the first time—and always after making or remaking connections at the torch and regulators—test for leakage.

Never connect an oxygen pressure-reducing regulator to a cylinder containing combustible gas. Also, never connect a regulator for a combustible-gas cylinder to a cylinder containing oxygen. Always remember that oxygen should never be mixed with acetylene or any other fuel gas except in a suitable torch. Do not experiment with pressure-reducing regulators or torches, and do not alter them in any way.

Disconnecting Pressure-Reducing Regulators

To disconnect a regulator:

1 Close the cylinder valve.
2 Open the torch valve to release all pressure from the hose and regulator.
3 Turn out the pressure-adjusting screw.
4 Close the torch valve.
5 Uncouple the regulator.

Closing the cylinder valve and then opening the torch valve relieves all pressure in the regulator and hose line. After the gage readings have reached zero, the pressure-adjusting valve should always be released, since this must be done before the valve is opened on the new cylinder. The acetylene and oxygen pressure should not be relieved simultaneously, and care should be taken that the release of the acetylene does not create a fire hazard. If the cylinder connection nut on the regulator is provided with a dust plug, this should be screwed into place, unless the regulator is to be connected to another cylinder immediately.

Backfire

Improper operation of the torch can cause the flame to go out with a loud snap or pop, which is called a "backfire." A flame that has backfired can be relighted at once. It can be relighted instantly if the metal being welded is hot enough to ignite the gases; otherwise, a lighter should be used. Causes of backfire are: touching the tip against the work, overheating the tip, or operating the torch at other than recommended gas pressures; a loose tip or head; or dirt on the seating surfaces in the torch.

Flashback

A "flashback" occurs when the flame burns back inside the torch, usually with a shrill hissing or squealing. Should the flame flash back inside the torch, close the torch oxygen valve at once, and then close the acetylene valve. Closing the torch oxygen valve, which controls the flame, stops the flashback at once. Then the acetylene valve should be closed and the torch allowed to cool off before being relit. Also, blow oxygen through the tip for a few seconds to clear out any soot that may have accumulated in the passages. In a cutting torch, oxygen should be blown through the preheating as well as the cutting orifices before relighting. Flashbacks may extend back into hoses or regulators. When flashbacks occur, it indicates that something is radically wrong, either with the torch or with the way it is being operated. Any such occurrence should be investigated to determine the cause before the torch is relit. Incorrect oxygen and acetylene pressures are often the cause. As has already been said, always avoid using gas pressures higher than those recommended by the manufacturer.

STORAGE, HANDLING, AND USE

General Storage, Handling, and Use of Cylinders

Avoid abusing cylinders. When cylinders leave the charging plant or warehouse, they are in proper condition and safe for the purpose intended. To keep them in this safe condition, it is necessary to prevent their abuse on the job.

Cylinders should be stored only in approved, safe places:

1 Store cylinders in definitely assigned places where they will not be knocked over.

2 Cylinders should be kept away from stoves, radiators, furnaces, and other hot places. They should be stored well away from highly combustible materials such as oil, grease, paper, and excelsior.

3 Inside buildings, cylinders of oxygen should not be stored in the same compartment with cylinders of acetylene or other fuel gas. Unless they are separated by a distance of at least 20 feet, there should be a fire-resistant partition of enough strength to contain a fire for ½ hour or longer. This partition should be between the oxygen cylinders and the acetylene or fuel-gas cylinders.

4 Where cylinders are stored in the open, they should be protected from accumulations of ice and snow, as well as from the direct rays of the sun in localities where extreme temperatures prevail.

When cylinders are being moved, keep them from being knocked over or from falling. When moving cylinders by crane or derrick, use a suitable cradle, boat, or platform. Never use slings or an electric magnet. Unless cylinders being moved are conveyed or handled on a suitable truck, regulators should be removed and valve-protection caps should be put in place hand-tight. Cylinder valves should always be closed before cylinders are moved.

Never use valve-protection caps on high pressure for lifting cylinders. Valve-protection caps are designed to protect valves from damage. Before raising oxygen cylinders from a horizontal to a vertical position, be sure that the cap is properly in place and hand-tight; then raise the cylinder by grasping the cap firmly in the hand. Valve-protection caps should never be used to lift cylinders from one vertical position to another. It is a good practice not to allow acetylene cylinders to lie in a horizontal position.

Oxygen Cylinders Always call oxygen by its proper name—"oxygen." Oxygen should never be called "air" and should never be confused with compressed air.

Warning: Never use oxygen for compressed air. A serious accident can easily result if oxygen is used as a substitute for compressed air. Oxygen should never be used in pneumatic tools, in oil preheating burners, to start internal combustion engines, to blow out pipelines, to "dust" clothing or work, for pressure tests of any kind, or for ventilation.

Oxygen, or air rich in oxygen, should never be allowed to saturate any part of the clothing, since a spark can quickly start a fire.

Keep oxygen cylinders and fittings away from oil or grease; as has already been noted, oil or grease may ignite violently in the presence of oxygen under pressure. Oily or greasy substances must be kept away from cylinders, cylinder valves, couplings, regulators, hose, and apparatus. Do not handle oxygen cylinders or apparatus with oily hands or gloves.

A jet of oxygen should never strike an oily surface or greasy clothes or enter a fuel-oil or storage tank that has previously contained a flammable substance.

Oxygen cylinders should not be stored near reserve stocks of acetylene or other fuel-gas cylinders or near any other substance likely to cause or accelerate fire. Oxygen will not burn, but it supports and accelerates combustion and thus will cause oil and other similar materials to burn with great intensity.

Never allow oxygen cylinders to be stored near furnaces, radiators, or any other source of heat. Gaseous oxygen cylinder valves are equipped with a safety device which acts as an excess-pressure release. Excessive heat will increase the temperature of the oxygen and cause a corresponding increase in the pressure within the cylinder. At abnormally high pressures, the safety release blows or bursts, and the oxygen escapes to the air.

Acetylene Cylinders Call acetylene by its proper name—"acetylene." Acetylene should not be called "gas." It is far different from city gas or furnace gas. Mixtures of acetylene between 2.6 percent and 80 percent acetylene are explosive if ignited. Under certain conditions, acetylene may ignite at a temperature as low as 650°F.

In storing cylinders, remember that acetylene is a fuel gas: since it will burn, it must be kept away from fire. Acetylene cylinders should not be stored near stoves, radiators, furnaces, or other sources of heat; furthermore, they should be stored well away from combustible materials.

Always stand acetylene cylinders with the valve end up. Never allow them to lie on their sides when in storage or while being used.

Rough handling, dropping, or knocking of acetylene cylinders should be avoided. Such treatment might damage the cylinder, valves, or fusible plugs and cause leakage.

Fusible plugs should never be tampered with. The fusible safety plugs with which all acetylene cylinders are provided act as safety releases when the cylinder is exposed to excessive temperatures. They melt at about the temperature of boiling water and release acetylene from the cylinder.

Never use a cylinder that is leaking. If acetylene leaks around the valve spindle when the valve is opened, close the valve and tighten the gland nut. This compresses the packing around the spindle. If this does not stop the leak, close the valve and attach a tag to the cylinder stating that the valve is unserviceable. Notify the supplier and have the supplier pick it up (or follow the supplier's instructions for returning it).

If acetylene leaks from the valve even when the valve is closed, or if rough handling should cause any of the fusible safety plugs to leak, move the cylinder to an open place well away from any possible source of ignition and plainly tag it as having an unserviceable valve or fusible plug. Open the valve slightly to let the acetylene escape slowly. Place a sign at the cylinder warning everyone against coming near it with a lighted cigarette or other source of ignition. When the cylinder is empty, close the valve and move the cylinder to a suitable location awaiting shipment. Again, notify the supplier, who will pick it up or give instructions about how it should be returned.

Use the special T-wrench for opening or closing cylinder valves. Always leave the T-wrench in position, ready for immediate use, so that the acetylene can be quickly turned off in case of emergency.

Do not open an acetylene cylinder valve more than 1½ turns. This permits an ample flow of acetylene. Always open the acetylene valve slightly at first, so as not to risk damaging the regulator mechanism or gages; and always stand away from and in front of the regulator and gage faces when opening the cylinder valves.

Handling and Use of Hose

Use only hose and connections made specially for oxyacetylene welding and cutting. Green is the recognized color for oxygen hose, and red for acetylene or other fuel gas. As has been said, oxygen hose and acetylene hose should not be interchanged. (The ¼-inch and larger sizes of hose are frequently designated "oxygen" and "acetylene.") It is essential to differentiate sharply at all times between "air" and "oxygen." The hose must be of good-quality rubber having a sufficient number of plies of fabric to withstand the service for which it is being used. All hose should comply with the requirements of "Specifications for Rubber Welding Hose" adopted jointly by the International Acetylene Association and the Rubber Manufacturers Association.

Pneumatic hose or hose fittings should not be used on welding and cutting equipment, because sooner or later a piece of hose oily or greasy from previous use with air tools would be used.

Metal-covered hose should never be used; it is not sufficiently flexible to permit proper handling of the torch. If some combination of accidental causes should build up a bursting pressure in metal-covered hose, the metal covering would create an additional hazard.

Only standardized hose connections of the correct size should be used for connecting hose to torches and regulators. It is very important that these connections be tight enough to withstand, without leakage, a pressure twice as great as may be obtained through the regulating valves. Use nothing but leak-test solution or soapy water for testing. Wire should never be used for binding hose to the hose nipple; suitable hose clamps or ferrules should be used.

Blow out new hose with oxygen before using. New hose is dusted on the inside with fine talc, and any loose talc should be blown out before the hose is used. Lengths of both oxygen and acetylene hose can be blown out with a pressure of about 5 psi of oxygen. When acetylene hose has been cleared by oxygen, blow through it from the mouth before attaching it to the acetylene regulator. **Precaution:** Do not blow out hose with acetylene.

Always protect hose from damage or interference. Protect it from being trampled on or run over; avoid tangles and kinks; and place the hose so that it will not be tripped over. Connections might be pulled off, or the cylinders and equipment might be pulled over, by a sudden strong tug on the hose. Do not allow hose to come into contact with oil or grease—these deteriorate the rubber and constitute a hazard with oxygen. Protect the hose from flying sparks, hot slag, or other hot objects and open flames.

WELDING HAZARDS

Although it involves potential hazards, including exposures to toxic agents, welding can be a safe occupation.

The generation of fumes and gases appears to be the most significant possible hazard in the welding process, along with fire and explosion. The amount and types of fumes and gases involved will depend on the welding process, the filler material, and the shielding gas, if any. The toxicity of the contaminants depends primarily upon their concentrations and upon the physiological responses of the human body.

Following is a list of substances that may be involved in the welding of common metals, with some of their effects upon the human body:

Cadmium The use of cadmium containing silver solder or brazing alloys should be avoided because toxic fumes can be produced which, in turn, can cause respiratory irritation, chest pains, and difficult breathing, if inhaled. Cadmium appears in the form of both fumes and dust.

Fluorides Welding electrodes whose coatings contain fluorides offer a definite hazard. Fluorides have a ceiling value of 2.5 milligrams (mg) per cubic meter. This is the approximate number of milligrams of particulate per cubic meter of air given in Table Z-1 in OSHA Subpart G, 1910.1000 of the Federal Register.

Iron oxide If fumes formed from the welding of carbon steel are inhaled, iron oxide fumes can be deposited in the lungs. This may result in a condition known as "siderosis"; but siderosis is not known to result in disability or to create a predisposition to tuberculosis or lung cancer. Iron oxide fumes have a ceiling value of 10 mg per cubic meter of air.

Lead Welding materials which contain lead or whose surfaces are covered by lead-based paints represent a hazard to a welder who inhales the fumes. In its most serious form, the end result could be lead poisoning.

Zinc Welding on brass, galvanized steel, or any metal coated with zinc-based paints may result in metal-fume fever if the exposure is excessive. Metal-fume fever is characterized by a chill which is not usually serious and whose signs and symptoms are usually gone in a day. Zinc oxide fumes have a ceiling value of 5 mg per cubic meter of air.

Ozone Ozone is an intensely irritating gas produced by the action of the electric arc during the welding process. Ozone concentrations are usually low in the breathing zone of the welder, so that a health hazard usually does not exist in most welding operations. Welders who do have excessive exposure to ozone may develop headaches, chest pains, and dry throat. Severe exposures to ozone can cause pulmonary edema. Ozone has a ceiling value of 0.1 ppm (parts per million) in vapor form and 0.2 mg per cubic meter of air.

Although the list above is obviously far from complete, it indicates that hazards can exist in a workplace where welding is being done.

If management is aware of any potential hazards associated with any of its welding processes, it should take steps to find and correct them. One method of protecting the welder is to engineer any required ventilation. OSHA standard 1910.252 states that local exhaust or general ventilation systems shall be provided and arranged to keep the amount of toxic fumes, gases, or dusts below the maximum concentrations as specified in OSHA Standard 1910.1000.

It is often difficult to accurately determine the degree of exposure to welding fumes and gases, since there are many factors that may have a bearing on the actual exposures. Whatever the need for proper protection, it can only be determined by properly obtained air samples that are representative of the actual exposure. Proper protection can be mechanical ventilation, local exhaust ventilation, or the use of personal protective equipment—that is, the respirator.

Precautions to lessen the danger from fumes and fire or explosion are discussed in the next section.

PRECAUTIONS IN WELDING AND CUTTING WORK

Confined Spaces

Special care and special precautions are necessary for welding or cutting in confined spaces. Special clothing should be worn—preferably fireproofed, but at least wool, which is relatively resistant to sparks and hot slag.

Be sure ventilation is proper and adequate; this can be accomplished by natural means or by an air fan or blower.

Never feed oxygen from a cylinder into a confined space; it is unsafe to do so. Remember, oxygen will not burn, but it supports and accelerates combustion and thus will cause oil, wood, clothing, and other similar materials to burn with great intensity. Clothing saturated with oxygen, or air rich in oxygen, may need only a spark to burst into flame. Always have a helper present outside the confined space to close the cylinder valve or give other help in case of emergency. Test all equipment, including hose, for leaks before taking it into a confined space, and bring it out with you when work is interrupted for any reason, even for a short time.

Fumes

Arrange for good ventilation when welding on brass, bronze, or galvanized iron. When welding is done on brass, bronze, or zinc-coated surfaces, the degree of protection necessary depends upon the location of the work. In fairly open locations, it is usually sufficient to drive the vapors away from the worker by using a stream of air or by suction, provided this does not drive a concentrated stream of vapors to an unprotected co-worker. In more confined spaces, masks providing air from an outside source should be used. A worker who feels sick during or after welding brass, bronze, or galvanized iron should drink milk freely.

For welding or cutting metals containing or coated with lead, cadmium, beryllium or mercury, a suitable air-line mask should be worn. No operator should be considered immune from the effects of these fumes. Straight filter-type masks are inadequate; nothing short of air-supply respirators is suitable. There must be sure protection against breathing the fumes which occur when lead, mercury, beryllium, cadmium, or their components are heated.

Dusty or Gassy Locations

Particular caution must be used when welding or cutting in dusty or gassy locations. Dusty and gassy atmospheres in certain mines, mills, industrial plants, and plants spraying lacquers and the like, require extra precautions to avoid explosions or fires from electric sparks, matches, and open fire. Welding or cutting in such places should therefore be done only when proper precautions have been taken, and only after the responsible supervisor has inspected the situation and has personally given instructions to proceed.

Underground Locations

When welding or cutting is done underground, as in mines or caissons, acetylene generators or hydraulic seals should not be used, since it is difficult or impossible to vent them properly. Approved flash arrestors should be used between acetylene cylinder valves and acetylene regulators. Testing for leakage should be performed very frequently. Great care should be taken to protect hoses and cylinders from damage, and mine timbers and props from sparks and hot slag. Atmospheric testing should be thoroughly conducted at frequent intervals before and during underground welding.

Preventing Fires

When welding or cutting has to be done near materials that will burn, special precautions should be taken to make certain that flame, sparks, hot slag, or hot metal do not reach combustible material and thus start a fire. It is especially important to take precautions in portable cutting operations. Since cutting produces a greater quantity of sparks and hot slag than welding does, locations where portable cutting equipment is used must be thoroughly safeguarded against fire.

Never use cutting or welding torches where sparks or an open flame of any kind would be a hazard. Flames are a hazard in rooms containing flammable gas, vapors, liquids, or dust, or any material that easily catches fire.

It is not safe to use cutting or welding equipment near rooms containing flammable materials unless there is absolutely no chance that sparks can pass through cracks or holes in walls or floors, through open or broken windows, or through open doorways. There is also the possibility that flammable vapors may escape from other rooms through doors or other openings, making it doubly necessary to keep sparks and flames away.

Cutting and welding work should be located where there will be no possibility of setting fires. This must, of course, always be done when the metal to be welded or cut is in a place where open flames are barred; but it may also be sensible in many other locations, even if open flames are permitted.

If flammable materials cannot be moved, sheet-metal guards, asbestos paper or curtains, or similar protection should be used to keep sparks close to the work being done.

Correct oxygen pressure should be used when cutting. An oxygen pressure greater than necessary will cause extra sparks and increase the slag flow, to say nothing of the increase in oxygen.

Extra oxygen and acetylene cylinders should be stored away from work areas. Only enough cylinders should be near the work to ensure an adequate supply of gases for the job at hand.

If a sprinkler system is available, welding or cutting should be done under full utilization of it. If the sprinkler system must be shut down for a short time, it should be done when welding or cutting work is not in progress.

MAINTENANCE OF OXYACETYLENE EQUIPMENT
Torches

Reasonable care should be taken in changing heads, attachments, nozzles, and tips. Although torches are ruggedly built to withstand more than ordinary use, care should be taken to prevent damage by careless usage.

If orifices in the tips or nozzles become clogged, they should be cleaned with a soft brass or copper wire. The orifices in the torch, tip, and nozzle are carefully sized for the best operation. Therefore, any tool which would enlarge or bell-mouth the orifices should not be used to clean them.

If a leakage develops around the torch valve stem, the packing nut should be tightened and, if necessary, repacked. Use only packing supplied or recommended by the manufacturer of the torch. Oil or oil-base packing should never be used.

If a valve does not shut off completely, clean the seat by removing the valve assembly and wiping the seating portion of the valve stem with a clean rag. The internal replaceable seat should be used or the valve should be reseated.

If the passages become clogged, blow them clear by blowing oxygen backward through the torch. To do this, remove both lengths of hose from the torch and put in the largest-size welding head, tip, or nozzle. Hold the oxygen hose over the end of the tip, set the oxygen regulator at about 20 psi pressure, and blow oxygen backward through the torch. First have only the acetylene valve open, and then only the oxygen valve. Note that this specialized purpose is one of the few exceptions to the rule that oxygen should not be used as a source of pressure or a cleaning jet. Oxygen is preferred for cleaning torch passages at the rare intervals when this is needed because ordinary compressed air contains considerable oil and moisture. **Precaution:** All hose should be disconnected from the torch before blowing out with oxygen.

If backfires or flashbacks occur, check over the torch for loose connections or improper seating.

Flashbacks and backfires can be dangerous: they can cause a flame to go into the acetylene tank if not immediately corrected. Closing the torch oxygen valve which controls the flame stops the backflash immediately. The acetylene valve should be closed next. The torch should be blown with oxygen through the tip for a few seconds to clear out any soot or other debris that may have caused a disruption in the flow of gas. Flashback means that something is radically wrong and that an investigation should be made at once.

Pressure-Reducing Regulators

Steady, efficient operation of the torch depends on the accuracy with which pressure-reducing regulators perform their functions. Regulators should be tested at specified intervals and repaired if necessary.

The following suggestions may be helpful:

1 If a regulator "creeps," have it repaired at once. Creeping of a regulator is indicated on the low-pressure (working-pressure) gage by a buildup of pressure after the torch valves are closed.

2 If the gage hand will not go back to the pin when pressure is released, the gage should be repaired.

3 If the safety-release disk ruptures, a new one should be installed immediately.

4 If a regulator inlet filter needs to be repaired, do so immediately. Never operate the regulator without a filter.

5 Gas-tight connections should always be maintained between the pressure-reducing regulator and the cylinder.

Precaution: Never tighten a leaky connection between the regulator and cylinder without first closing the cylinder valve.

Hoses

Hoses should be examined at least once a week (more often if working conditions require it) for leaks, worn places, and loose connections. This can be done by immersing the hose, under normal working pressure, in water. Hose leaks not only are a safety hazard but also waste oxygen and acetylene. Escaping acetylene from a hose is likely to build up, ignite, and cause an explosion and fire.

Should a flashback occur and burn the hose, discard the entire length of hose. A flashback renders the hose unsafe because it burns the inner wall and can cause further problems by clogging or otherwise interfering with proper operation of the torch. Never repair hose with tape.

QUESTIONS AND EXERCISES

1 If oil or grease is under pressure in the presence of oxygen, what could occur?

2 What color is oxygen hose? Acetylene hose?

3 An acetylene cylinder valve should never be opened suddenly. What is the reason for this?

4 Why is it safer to use a friction lighter than a match to light a torch?

5 A welder should never use acetylene at pressures above 15 psi. Why is this important?

6 Describe (a) a backfire and (b) a flashback.

7 For what reason should oxygen cylinders never be stored against acetylene or other fuel-gas cylinders?

8 What does OSHA say in the Federal Register about storing fuel cylinders inside buildings?

9 What is OSHA's interpretation of the separation of oxygen cylinders from fuel-gas cylinders?

10 Where can a guide be found for selecting the proper shades of welding lenses to be used in various types of welding?

11 Why is woolen clothing preferable to cotton clothing for a person who is doing welding or burning?

12 Ventilation is required where welders must work in a confined space. Name some other OSHA requirements regarding ventilation for general welding and cutting.

13 Does OSHA require any record keeping with regard to welding equipment? If so, what equipment is covered and how often should inspections be made?

14 What are the general characteristics of welding gases?

15 What is oxygen? Acetylene?

BIBLIOGRAPHY

Code of Federal Regulations, Part 29: General Industry Standards, U. S. Department of Labor–OSHA. July 1, 1985.

Federal Regulations, Occupational Safety and Health Standards, Subpart G: Occupational Health and Environmental Control, Department of Labor, OSHA 2206 (rev.), January 1976.

Precautions and Safe Practices in Welding and Cutting with Oxy-Acetylene Equipment, Union Carbide Corporation, Linde Company Division, New York, 1960.

Precautions and Safe Practices in Welding and Cutting with Oxy-Acetylene Equipment, L-Tec Welding and Cutting Systems, Florence, S.C., 1987.

CHAPTER 18

RADIATION SAFETY

On April 26, 1986, an explosion and fire at the Chernobyl nuclear reactor plant spread a huge cloud of invisible radiation from the plant—located 80 miles north of Kiev, capital of the Soviet Ukraine—over Europe. Eventually this cloud encircled the entire globe. At least 31 persons were reported dead, and more than 200 others suffered radiation sickness, in a catastrophe that led to the evacuation of 135,000 men, women, and children in the world's worst nuclear accident. Soviet scientists could reach no consensus on the number of deaths that eventually may be linked to radiation exposure from this disaster.

Every day we are being bombarded by radiation from natural sources as far away as the sun and as close as the person standing next to us in a supermarket check-out lane. Since the beginning of time, man has been exposed to bombardment by various types of natural radioactivity. In addition to cosmic radiation, the most penetrating radiation known, we are exposed to approximately fifty elements in nature that are radioactive. So far as we know at present, the vast variety of materials in the world is made up of 102 elements. Fourteen of these elements have been created by man in the laboratory.

The development of the X-ray machine, nuclear power, nuclear testing equipment, and other sources of nuclear technology do, however, demonstrate the fact that biological damage can result if the amount of radiation exposure becomes excessive. Nonetheless, radiation has been of immeasurable benefit to man. Unfortunately, the human body has no defense mechanism against radiation, nor can any of its five senses detect it.

Ionizing Radiation

Ionizing radiation is high energy radiation that causes damage to the tissues and organs of the body. These radiations have been identified as *alpha* and *beta particles, gamma rays* and *neutrons*. Each of these may cause injury to the tissues and organs of the body by creating enough energy to cause functional changes in body cells. The capability of a cell to withstand exposure to radiation is directly proportional to its reproductive processes. Damage to cells that can be reproduced is reversible, but if the cells cannot be reproduced or replaced, the radiation damage is irreversible. The most serious hazard of radiation is cancer.

Alpha particles are the least penetrating type of radiation. Although they travel at high speed with high energy, they are short-ranged in dense materials. Externally, there is no real danger, but, internally, they are a hazard if they attack a specific organ of the body. They can be shielded against by a thin sheet of paper.

Beta particles Like alpha particles, beta particles are small electrically charged particles emitted from radioactive materials. However, they have considerably more penetrating power than alpha particles. The range of beta particles is sufficient to penetrate the dead layers of skin. If they find their way into the body, they are particularly hazardous since practically all of their energy will be given up in the body tissue. Their method of entry into the body is by ingestion, inhalation, and penetration through the skin.

Gamma rays are very penetrating and can pass through thick layers of material. They are produced by radioactive materials, either in nature or manmade. They are not atomic particles, but are waves of energy, similar to light, that can travel millions of miles and are hazardous at considerable distances. They are extremely difficult to stop. They easily pass through most solid substances and thick sections of lead or concrete are required for shielding against them.

X-rays are produced in a vacuum tube. They can penetrate through solid materials and can ionize gases (change their molecular structure). X-rays are produced for the following purposes useful for man:

1 Medical conditions requiring a photograph for diagnostic purposes.
2 The treatment of certain types of cancer.
3 Examination of packages and baggage for illegal articles.
4 Measuring metal thicknesses; inspecting for cracked welds in metal; in industry, inspecting castings for flaws.

All work with X-ray equipment involves some danger and requires that considerable safety precautions be taken. First consideration should be given to proper shielding and continuous monitoring.

Neutron particles are very penetrating and require very heavy shielding. They have no known industrial application at the present time. They are used primarily in research and are also under consideration for military purposes.

Radioactive Decay or Half-Life

The atoms of radioactive elements in their process of change to more stable forms of the same element, or of a different element, do so in a random manner. However, the average rate of change (or decay, as it is called) of the atoms of any given element is always the same. The measure used in describing the rate of decay is called "half-life." This is the time required for the radioactivity in a sample of material to be reduced to half its original amount. Since the average rate of decay is constant, after two half-lives, one-fourth of the original amount or degree of radioactivity will remain; after three half-lives, one-eighth will remain and so on. At the end of each half-life, half the total number of atoms will still remain. The half-lives of radioactive materials that have been measured range from less than a millionth of a second for Polonium 212 to more than 4 billion years for Uranium 238.

Measuring Ionizing Radiation

Low-energy radiation will have its maximum effect on the surface of the body, while high-energy radiation will penetrate and produce effects deeper in the body. Radiation cannot be measured directly. It is measured by the ionization produced by the passage of the radiation through a medium. The units for measuring radiation refer either to the charge, to energy, or to biological effect. There are four basic units used to measure a stated quantity of radiation. They are *roentgen, rep, rad,* and *rem.*

The *roentgen* is a measure of a unit of exposure dose of ionization of air due to X or gamma radiation. The *rep* is a measure of radiation in human tissues. The *rad* is the unit of energy absorbed by radiation in any material. Although the roentgen is strictly applicable only to X or gamma radiation, the rad unit is applied regardless of the type of ionizing radiation or the type of absorbing medium. The *rem* provides the amount of biological injury, of a given type, that would result from the absorption of nuclear radiation. Therefore, the rem is a unit of biological dose. Thus, in general, for gamma radiation, the biological dose in rems is numerically equal to the absorbed dose in rads, and is also roughly equal to the exposure dose in roentgens.

Biological Effects of Radiation Exposure

Ionizing radiation in large doses can produce biological damage. The biological effects produced by one type of radiation can be produced by any other type of radiation. Whether it comes from the sun or other forms of radiation, radiation can damage the cell or tissue. The total amount of radiation absorbed in a tissue is a function of many variables, including the type of radiation, the energy of the radiation, and the substance being irradiated. In most biological applications, alpha and beta radiation will be completely absorbed by the tissue, while gamma rays are only partially absorbed. In general, for a given type of radiation—the greater the energy, the greater the penetration.

Ordinarily, when reference is being made to dosage or exposure, it is assumed to mean dosage delivered to the whole body. The qualification of whole-body radiation is important because large doses of radiation can be applied to local areas (as in therapy) with little danger, but are lethal if applied to the whole body. For example, a person could expose a finger to 1000 R and experience little effect except localized injury to the finger, with subsequent healing and scar formation; whereas, if the same dosage were delivered to the entire body, it would prove fatal.

The effects of radiation exposure are divided into *acute* and long-term effects. Short-term exposures to the body are said to be "acute," and can produce both immediate or delayed effects on the body. Exposures, the result of which appear years or decades later, are called "long-term." Acute exposures usually result from mishaps, whereas *chronic* exposures (long-term) are due principally to sustained and unrecognized conditions. During the period of time between the initial radiation exposure event and the time it is detected, there is a time lag. This period is referred to as the "latent period."

Pathological Effects of Radiation Exposure

Exposure to radiation can present two problems: one by radiation originating from a source outside the body and coming at the body from the outside; the other by exposure resulting from radioactive materials which have been taken into the body. Certain radioactive materials are not hazardous outside the body; whereas the same materials inside the body would cause radioactive poisoning.

One of the pathological effects of radiation exposure is *radiation sickness*. This is caused by massive overdoses of penetrating external radiation and results in nausea, vomiting, diarrhea, malaise, and possibly hemorrhaging. If serious enough, it can cause death.

Radiation injury results from localized overdoses of less penetrating external radiation. It can cause burns, loss of hair, and skin lesions. It is generally limited to the hands because contact is usually with the hands. Radiation sickness, injury, and poisoning are not contagious or infectious. Treating or helping a victim who has been exposed to radiation will not expose the emergency personnel to radiation, unless the victim is covered with a radioactive material like dust, which in itself could contaminate the rescuer.

The genetic effects of long-term radiation exposure can become significant when the reproductive organs have been exposed to radiation. Radiation exposure of other organs of the body has no genetic effects whatsoever. The problem with the genetic effects of radiation is not one of protecting a specific individual, but of protecting the entire population. This can be done only by keeping all exposure to radiation down to the lowest practical limits.

Radioactive Waste

The tragic consequences of hazardous waste mismanagement in the past are reflected in polluted ground water, lakes, and rivers, and broken health and lives of many innocent people. Another example of mismanaged waste is the concern for radioactive materials used in medicine and in industry. A typical example, and one of great concern, occurred in Brazil, South America, in the fall of 1987.

A six-year-old girl and her aunt died of radiation poisoning, the first fatalities in a nuclear accident that contaminated at least 246 people. The poisoning resulted from a lead capsule containing radioactive cesium 137 that was carelessly left at an abandoned radiation treatment clinic. The

TABLE 18-1
SUMMARY OF 24-HOUR SHORT-TERM
(ACUTE) EXPOSURE EFFECTS

25 rem	No detectable effect
50 rem	Slight repairable blood damage
100 rem	Nausea, fatigue
250 rem	1st death
500 rem	50% death
1,000 rem	1,000% death

Source: Ronan Engineering Company.

TABLE 18-2
RADIATION MAXIMUM PERMISSIBLE DOSES

Type	Dose
Worker (restricted area)	5 times number of years beyond age 18
a) Whole body, head and trunk, active blood-forming organs, gonads, or lens of eye	3 rem per calendar quarter
b) Skin of body	7.5 rem per year
c) Hands, forearms, feet, and ankles	75 rem/year OR 25 rem/calendar quarter
Worker (unrestricted area)	
a) Individual whole body	0.5 rem/year
b) Average (genetic)	5 rem/30 years OR 0.17 rem/year

Source: Ronan Engineering Company.

capsule was found by a junk dealer who showed it to his wife. She broke open the protective casing with a hammer and took out the glowing powder and showed it to relatives, friends, and neighbors who were fascinated by it. They passed it among themselves and, for reasons unknown, rubbed it on their bodies. The junk dealer's young niece ate some of it.

Forty-three of the victims were hospitalized, and 11 of the most seriously affected were transferred to the navy hospital in Rio de Janeiro. The victims suffered from open sores, hair loss, burns, and serious depletion of white blood cells. The junk dealer's wife died shortly after. Her niece, who had severe lesions of her throat and tongue, also died from the exposure.

The bodies of the two victims were buried in a common cemetery in specially made fiberglass coffins lined with lead in order to prevent the spread of radiation. The World Health Organization called it the worst radiation exposure in the western world and the most serious after the Chernobyl accident in the Soviet Union.

With the advent of the use of nuclear materials in power plants, medicine, industry, military, and research, the question is—what to do with the nuclear wastes? Another question—what is the connection between radiation exposure and cancer? The answer to the latter question becomes logarithmic because it takes so long between exposure and the detection of the disease. The fact that nuclear energy may provide the answer for our ever-growing future needs has created serious concern for safety in nuclear programs and the disposal of their waste. These concerns have become one of the most serious safety questions since the dawn of civilization.

Radioisotopes

As atomic energy is being harnessed and put to work for the benefit of mankind, radioisotopes, one of the by-products of this process, are being used in industry, research, medicine, quality control, and measurement. Radioisotopes are the radioactive form of chemicals. For example, radioiodine is a radioactive isotope used in the diagnosis and treatment of thyroid disorders. A radioisotope can also be defined as a form of an element that is unstable and that exhibits the property of radioactivity. Most radioisotopes have been made artificially in reactors and particle accelerators.

A common industrial application of radioisotopes is in gaging. Industry is utilizing radioisotope gages for four general applications: thickness gaging, density measurement, level measurement, and belt-weighing. Electronic signals from such gages can be used for automatic process control.

Radioisotope-equipped gages are packaged and shielded in such a manner as almost to eliminate any danger to the personnel operating them.

Radiography

Radiography is the use of X-rays and gamma rays for such purposes as detecting cracks in metal and also for measuring metal thickness. There are a number of manufacturers who make the instrumentation for such work. These instruments employ ultrasonic and various electronic ray-producing and measuring means. Metal thicknesses can be determined to within a percent of the actual thickness.

The inspection of pipeline field welds is made in the field as natural gas and petroleum pipelines are laid in the ground. Soon after the pipe lengths are welded together, a certain percentage of the welds is inspected radiologically. This method of inspection verifies the quality of the welding and provides for the correction of improper welding techniques, if recurring defects are found by radiography.

Film is placed on the weld surface located opposite the radiation source and is then exposed by the radiation that passes through the metal. Depending on the type of instrumentation used, the radiation source can be located inside the pipe; otherwise, the radiograph may be taken from outside the pipe. The danger of accidental or careless exposure to ionizing radiation is the major hazard peculiar to radiography. The radioactive element that is generally used in pipeline radiography is iridium 192 or cobalt 60.

Potential Health Hazards Associated with Ionizing Radiation

Industrial plants, medical facilities, and chemical laboratories are frequent users of radioactive materials, generally in small amounts, but with the possibility of danger to individuals due to mishandling or accidental exposure. The release of radioactive materials in such situations could result in the inhalation or absorption of the material through skin cuts or abrasions or any other point of entry in the body.

Thorium, an alph emitter, is a classic example. It is used widely in tungsten welding rods for industrial and construction welding. It is also used to improve strength and heat resistance in magnesium alloys in the aircraft and aerospace industry. Its principal hazard exists when it is machined, ground into dust, or during welding—when fumes and fire-dust particles are generated. Precautions must be taken against inhaling the fumes and ingesting them into the digestive system when handling and consuming foods and liquids.

Radioactive isotopes present the same problem. Very small amounts can contaminate the hands and enter the body through cuts and abrasions, with food and drink, and by smoking contaminated cigarettes. Generally, many isotopes are contained in a vacuum tube that must be carefully handled. Personal protective equipment such as masks, respirators, gloves, shoe covers, and coveralls are required to prevent accidental or careless exposure.

Exposure Control

The factors that affect exposure to radiation are *time* ar d *intensity*. *Time* relates to the length of the exposure. The less time spent in a radiation exposure, the lower the exposure to the total dose. Unnecessary exposure to radiation is avoided by good time management.

Intensity refers to the strength of the dose, the distance from the source, and the amount of shielding present. The intensity of radiation received depends on the type of radiation emitted by the source. Certain types constitute hazards only at short distances.

Observing a safe distance is necessary with gamma rays. Maintaining distance between the gamma source and the person reduces the chance for radiation injury. A good rule that applies to distance is: WHEN THE DISTANCE FROM THE RADIATION SOURCE IS DOUBLED, THE INTENSITY (AND THE SAFETY FACTOR) IS INCREASED BY A MARGIN OF FOUR. As a matter of principle, it is best to avoid all unnecessary exposure to ionizing radiation.

The establishment of safe levels of long-term radiation exposure requires knowledge of the relationship between radiation exposure and biological damage. Observations and studies involving human and animal life have resulted in the accumulation of significant amounts of data—data that are sometimes difficult to evaluate and are often controversial.

Radiation Safety Program

The area of areas in the plant containing radiation sources should be classified as restricted areas and posted with the appropriate caution signs. The standard symbol is shown in Figure 18-1. Other notations can be made to identify the source, specify the frequency, and so on. The use of warning signs is important because RF radiation, like other forms of radiation, is undectable.

A *radiation area* is one where personnel could receive, in one hour, a whole-body dose exceeding 5 mR (milliroentgen), or, in five consecutive days, a dose in excess of 100 mR.

A *high radiation area* is one in which an individual could receive, in one hour, a whole-body dose in excess of 100 mR.

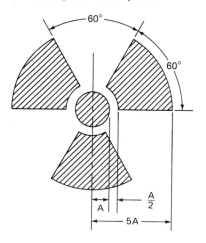

RADIATION SYMBOL
1. Cross-hatched area is to be magenta or purple.
2. Background is to be yellow.

FIGURE 18-1
Each room, enclosure, or operating area in which radioactive material is used or stored shall be conspicuously posted with a sign or signs bearing the radiation caution symbol. (*Source: OSHA 1910.96*)

Management must establish radiation safety training for all individuals, both management and employees, working in or frequenting any portion of a restricted area. Management must be kept informed of the storage, transfer, or use of radioactive materials. They must also be instructed in the health protection problems associated with exposure to radioactive materials or radiation.

Management must also establish radiation safety responsibilities for all levels of management, and employees whose functions are related to radiation. A radiation safety policy must be established for the following:

1 A radiation safety procedure for all employees working in the plant, and subsequent policies for employees working directly in radiation areas.

2 An evacuation and identification procedure (identify and isolate all suspected contaminated employees).

3 A record-keeping and reporting procedure for documenting radiation exposures of employees and reporting it to the Nuclear Regulatory Commission.

4 A procedure for the establishment of a decontamination crew, contamination cleanup, and the legal disposal of waste materials.

Precautionary measures against ionizing radiation would include the following:

1 Emergency procedures for radiation safety should be posted with the required warning signs at the access to all areas where ionizing radiation conditions exist.

2 Depending on the level of radiation present, employees should be provided with personal dosimeters and the proper protective clothing.

3 Edible foods, cosmetics, and tobacco products should not be used in areas where the possibility of radioactive contamination exists.

4 Access to areas in which radioactive materials are present should be restricted to personnel directly concerned with the operation.

5 Each entrance or access point to a high radiation area should be maintained locked, except during periods when access to the area is required, with positive control over each individual entry.

6 Emergency personnel should not be exposed to a 25 roentgen source for over one hour.

7 Cleanup and decontamination should be undertaken by personnel trained for such purposes.

8 No other persons should be permitted to enter a decontaminated area until it has been monitored, and the fact that no contamination continues to exist has been ascertained.

Radiological Emergencies

Preplanning for radiological emergencies is an important aspect of the Radiological Protection Officer (RPO), who must not only design an emergency plan but also familiarize all employees of the procedures for the areas in which they work. Radiological emergencies in the workplace can also include fires of unknown origin. In many smaller organizations, the safety specialist or safety engineer is also the RPO.

In dealing with the design of the Radiological Emergency Plan, the RPO must take into account the following considerations for the safety of the workers:

1 Evacuate the immediate area of all personnel.

2 Identify and isolate all persons who may have received high radiation exposures and been subsequently contaminated.

3 Regulate entry to the scene of the accident so as to minimize any subsequent exposures.

4 Decontaminate where personnel decontamination is immediately required. Priority to human safety should be given in all emergencies.

5 Report all radiological emergencies where required by local, state, and federal agencies.

6 Maintain complete records of the accident and follow-up procedures. These are not only required by law, but provide information for the possible adoption of subsequent remedial procedures.

Fire Control

The storage and use of radioactive materials, with rare exceptions, will affect neither the frequency of fires, nor the extinguishing media required. However, because of the radiation hazard during and after a fire, greater reliance should be placed upon automatic means of extinguishment. Special consideration should be taken into account regarding the disposal of sprinkler and hose-stream water to avoid the spread of contamination throughout the building and its equipment.

The initial procedure in a fire situation at a radiological emergency is to always assume that the smoke is contaminated and always approach the situation from upwind and stay back about 300 to 500 feet. All emergency personnel should utilize full structural protective clothing, gloves, and self-contained breathing apparatus. Monitoring devices should be read and personal dosimeters worn at all times during the situation.

An outer-limit "Support Zone" should be established, and if there is measurable radiation, a second limit should be established no closer than the 1 milliroentgen reading for decontamination, triage, and transporting.

If broken containers are found, an attempt to identify their contents should be made. Nothing should be removed from the radiation area, beyond the victims. Any equipment that is taken into the area, including self-contained breathing apparatus, must be left inside the 1 milliroentgen perimeter in a designated location. Radiological specialists, cleanup crews, and personnel versed in decontamination must be summoned to the scene. Emergency personnel must attempt to contain the spread of the substance, isolate the area, withdraw to the support zone, and leave the situation alone until qualified personnel arrive.

Nonionizing Radiation

Nonionizing radiation is radiation that lacks the energy to deform or change the structure of atoms, but can still cause bodily harm by raising the temperature of the body tissue it contacts with the energy it imparts in the form of heat waves. Three types of nonionizing radiation that can cause injury are ultraviolet, infrared, and microwaves.

Ultraviolet Radiation

Ultraviolet radiation is created by the heat waves given off by energy sources that radiate light. The chief source of ultraviolet radiation is the sun, which can cause painful sunburn. However, much of the ultraviolet radiation emitted by the sun is filtered out by an ozone layer in the atmosphere, thus reducing some of the exposure to the ultraviolet rays of the sun.

In industry, the chief hazard is from the arcs given off by welding equipment. The intensity of the light created by welding processes can become so great that it causes what welding personnel call "eye flash." The burn caused by eye flash creates a sensation of sand in the victim's eyes. Safety glasses worn for such protection are made of material that will not permit the ultraviolet rays to cause harm to the eyes.

TABLE 18-3
SHIELDING REQUIREMENTS FOR IONIZING RADIATION

Radiation	Shielding Material
Alpha Particles	Sheet of paper, or aluminum foil, outer layer of skin, or ordinary clothing
Beta Particles	1/25-inch of aluminum
X-Rays	Several inches of lead
Gamma Rays	Several inches of lead

Note: These figures are general requirements. For purposes of easy clarification, voltages are not given.

Other sources of ultraviolet radiation include tanning lamps, medical therapeutic lamps, and devices used in sanitizing and sterilizing.

Infrared Radiation

Infrared radiation is considered as electromagnetic or heat-producing, creating a thermal effect. The injurious effects of infrared radiation are skin burns, caused by the heat from the high temperatures it creates.

Industries where infrared radiation is most common are: foundries and open hearths in steel producing, glass making, the brick and clay industries, and heat processes required in paint and enamel drying. Continued exposure to infrared radiation and the heat it produces can result in excessive body perspiration, heat cramps, heat exhaustion, and heat stroke. The eyes are particularly sensitive to infrared radiation, especially when involved in laser work. Special safety glasses have been designed for protection of the eyes from the high intensities created by laser beams.

Clothing, gloves, cloth face masks, and leggings will protect the skin from the effects of infrared radiation.

Microwave Protection

Microwave absorption can penetrate the body deeply enough to create a temperature change because of the body's inability to dissipate the absorbed energy. Extended periods of exposure of the eyes to microwave radiation at high frequencies can produce cataracts. Since the effects of microwave radiation are thermal, the rising temperatures create burns.

In high-intensity fields, microwave radiation can cause inductive heating of metals and induced currents which can produce sparks. Rings, watches, keys, and other similar objects worn by persons can be heated until they cause a burn to the wearer.

Some of the commonest sources of microwave radiation are microwave ovens, industrial dryers and heaters, and radar and microwave communications systems. Medical diathermy equipment will also produce radiation in the microwave range. The Bureau of Radiological Health has set standards for the manufacture of all microwave ovens sold in the United States. The standard also requires that a warning device be provided on each piece of microwave equipment to indicate when it is radiating.

Responsibilities and Duties of the Radiological Protection Officer

One of the first responsibilities of the RPO is to review all plans and conditions for the proposed use of radioisotopes by the organization or plant.

He or she should require that the purchasing department permit review of all requisitions for radioisotopes prior to purchase, and ensure that a safe storage area exists prior to arrival of the materials. The RPO should also require or provide a radiation survey procedure for all incoming shipments of radioisotopes, their storage, and distribution. He or she must also survey and determine radiation exposure levels at the workplace, and, if necessary, provide personnel monitoring devices—such as film badges, pocket meters, or dosimeters—to employees.

"Survey" means an evaluation of the radiation hazards incident to the production, use of, release, disposal, or presence of radioactive materials under a specific set of conditions. Such evaluations include a physical survey of the location of the materials and equipment, and also the measurements of levels of radiation present.

The RPO should require the posting of warning signs in areas in which radioisotopes are used or stored and should also measure the intensities of these areas as frequently as possible (see Figure 18-1).

Planning to handle emergencies such as spills or personal contamination is an important aspect of the RPO's responsibilities, including the training of all employees to familiarize themselves with such emergencies.

The RPO must, according to federal regulations, maintain all records incident to the receipt, storage, transfer, and disposal of radioisotopes and their wastes. The RPO must also report exposure, spills, losses, or theft of radioactive materials to the Nuclear Regulatory Commission, or—if his or her state has its own program—to the state's commission. These are called Agreement States.

The Atomic Energy Act provides for inspection of facilities, premises, and records of all holders of radioactive material licenses. The act further provides for the inspection of licenses if a worker or worker's representative believes that a violation of the act has occurred in the plant regarding radiological working conditions.

Because of their knowledge of the product, environment, and the general aspects of the organization's radiological activities, it would behoove the RPOs to accompany NRC inspectors during inspection of the facilities. The RPOs would be acting in their capacity as management's representative, in the same manner that the safety professional would during an inspection by OSHA compliance officers.

WORD POWER

atom   The smallest constituent of an aggregate of protons, neutrons, and electrons whose number and arrangement determine the element.

electron   A minute atomic particle possessing the smallest negative charge found in nature.

film badge   A small piece of film sensitive to radiation and placed in a light-tight holder and carried by a person who works with radiation. When the film is developed, it is measured to determine the total radiation dose, in millirems (abbreviated mRem).

geiger counter   A gas-filled electrical device which detects the presence of radioactivity.

molecule   The smallest physical unit of an element or a compound. For example, a molecule of water consists of two hydrogen atoms combined with an oxygen atom. Hence, the well-known formula H_2O.

neutron   A neutral particle that has no electrical charge and has the same mass as a proton.

nuclear energy   The energy released in a nuclear reaction, such as fission or fusion.

nucleus   The center of an atom, made up of protons and neutrons. The total positive charge and most of the mass are contained in the nucleus.

personnel monitor   A device worn by a person to determine the radiation dose in mRem that that person has received.

proton   A positively charged subatomic particle; one of the constituents of every atomic particle.

survey meter   A portable device for making radiation surveys that measures field strength in mRem.

QUESTIONS AND EXERCISES

1 Distinguish between ionizing radiation and nonionizing radiation.

2 In what ways do ionizing radiation and nonionizing radiation cause harm to the body?

3 Name five sources of nonionizing radiation.

4 What sources in nature produce radiation?

5 What are the five forms of ionizing radiation?

6 What is the term for the length of time it takes for a radioactive source to decay 50 percent of its initial strength or power?

7 Which part of the body is the most sensitive to, and easily damaged by, laser beams?

8 Radiation is emitted from both natural and manmade sources. Name some manmade sources of radiation.

9 What are the four basic units for measuring a stated quantity of radiation?

10 What are the two effects of exposure to ionizing radiation?

11 During the period of time between the initial radiation exposure and when it is detected there is a time lag. What is this period of time called?

12 Distinguish between radiation sickness and radiation injury?

13 Is radiation sickness or radiation injury contagious or infectious? Explain.

14 What are the genetic effects of long-term radiation exposure? Does the problem affect the protection of a single individual or an entire population? Explain.

15 Does the problem of radioactive waste, as described in the South American case earlier in this chapter, create any great concern for us here in the United States? Why?

16 What are radioisotopes and what is the value to our lives of their use?

17 In what physical form are radioactive materials when their presence might become a health hazard through inhalation or ingestion?

18 What are the two factors that affect exposure to ionizing radiation?

19 What is the rule or safety factor that applies to distance for gamma exposure to gamma ray sources?

20 What is considered a high radiation area?

21 What are the considerations that the RPO must take into account in the design of a radiological safety program?

22 What are three types of nonionizing radiation that can cause bodily harm to an individual?

23 What is the form of energy imparted by nonionizing radiation sources?

24 Name some examples of ultraviolet radiation exposure in industry.

25 In which industries is infrared radiation exposure most common?

26 How does microwave radiation create problems to the body's system?

27 What are some of the most common sources of microwave radiation?

28 Other than what you've read in the text, can you name any other responsibilities and duties that the RPO should undertake in the interest of radiological safety.

29 What are some considerations that the RPO should take for reducing the hazard of radioactive spills?

30 Discuss some steps that can be taken to overcome possible public opinion against nuclear power for generating electricity.

BIBLIOGRAPHY

Hazardous Materials for First Responders, International Fire Service Training Association, Fire Protection Publications, Oklahoma State University, Stillwater, Okla., 1988, p. 310.

Radiological Health Handbook, U. S. Department of Health, Education, and Welfare, Washington, D.C., 1970.

Ronan Engineering Safety School Manual, Ronan Engineering Company, Florence, Ky., pp. 5, 9, 14.

"Notices, Instructions, and Reports to Workers," *10 Code of Federal Regulations CFR: part 19,* Federal Register, no. 22217, Nuclear Regulatory Commission, Washington, D.C., August 17, 1973.

CHAPTER 19

HAZARDOUS WASTE MANAGEMENT

"Oh, purge the land of toxic waste;
Go sweep it, scrape it, pump it;
But when the job at last is done;
Where do you plan to dump it?"

Richard F. Barrett,

The Wall Street Journal, August 18, 1983, p. 23. New York, N.Y.

The technology of industrial waste disposal, and the laws and regulations affecting such problems, have become another area of concern for managers and safety professionals alike. In order for an organization to comply with or implement a hazardous-waste program and because of the potential of hazardous exposure to employees and the public alike, it must turn to its safety and health experts for guidance.

The problem of waste disposal is complicated by the fact that nature's methods for processing wastes were once well within the means by which it could reduce or dilute them to safe limits. Eventually all of this changed. Following World War II, our nation's phenomenal growth was triggered by a surge in consumer demand for new products and goods as fast as they could be produced. This demand in turn created problems never before imaginable: how to manage the increasing amounts of waste that was generated by industry and consumer alike.

THE PROBLEM

Every year, billions of tons of solid waste are discarded in the United States. These wastes range in nature from household trash to complex materials in industrial wastes, sewer sludge, agricultural residue, mining refuse, and pathological wastes from institutions such as hospitals and laboratories.

Unfortunately, many dangerous materials that society has "thrown away" over recent decades have endured in the environment. The Environmental Protection Agency (EPA) has on file hundreds of documented cases of damage to life and the environment, resulting from indiscriminate or improper management of hazardous wastes. The vast majority of these cases involve pollution of the ground water—the source of drinking water for about half of the United States population—from improperly sited or operated landfills and surface impoundments such as open pits, ponds, and lagoons.

Hazardous waste can also contaminate rivers and lakes, pollute the air we breathe, poison us via the food chain, poison us by direct contact, or burn or explode.

In 1978 and 1979, over 200 families, residents along Love Canal near Niagara Falls, were evacuated because the soil in their yards was contaminated and their basements were flooded with water contaminated by chemicals in the ground. As buried drums holding the chemical waste corroded, their contents permeated the soil and eventually leaked into the basements of many of the homes. Many of these chemicals were later identified as suspected carcinogens that were buried in the area a quarter of a century earlier.

There are many similar examples that provide evidence of damage to life and the environment from the mismanagement of hazardous waste. In addition there are the abandoned and uncontrollable sites where wastes of all kinds have been indiscriminately dumped.

Yet another, and more common, occurrence are the freight train and truck spills of liquid toxic chemicals reported daily by the media. Many products that require hazardous raw materials are transported daily over the nation's highways, railroads, and pipelines.

The discoveries and revelations are documented daily by EPA:

"A tanker bloated with toxic chemicals is found abandoned at a truck stop."

"Radioactive waste is found buried in a dump near a refinery."

"Hydrochloric acid from a derailed railroad tank car spews a cloud of chlorine gas over a city in Pennsylvania forcing the evacuation of more than 2,000 people in a 2-mile square area."

"A pipeline carrying fuel oil from Indiana to southeastern Wisconsin ruptured, spilling thousands of gallons of heating oil into the Menomonee River in a suburban community near Milwaukee in June, 1988."

A century of abuse by industry and neglect by local and state enforcement agencies has created a threat of catastrophic proportions. There are no statistics available of the amount of hazardous waste that is moved daily throughout this nation by truck and train. Federal officials can only estimate at somewhere between "monumental" and "phenomenal."

Economic Impact

In recent years, with increasing frequency, society has been forced to properly dispose of waste that was previously disposed of haphazardly. The best known example of this is Love Canal, where 20,000 tons of waste were buried over a period of years. The diagnosis of a severe health hazard in the area due to wastes seeping into house basements and surfacing in backyards caused society to take remedial action. The price tag to the state and federal governments was expected to be almost $36 million for cleanup, relocating residents, health and environmental testing services, and other expenses associated with the disaster. Thus, society spent about $1,800 per ton in its effort to clean up waste improperly disposed of. Furthermore, the $1,800 per ton excluded human health and suffering, which most probably easily outweighed dollar costs.

Risk Assessment

Industrial managers have long become accustomed to the use of incidence and severity rates as one of the tools for assessing risks to workers at the workplace. These assessments were not only used for financial purposes, but were also used for training employees to use good judgment when evaluating potential risks, whether from a voluntary or an involuntary experience.

Members of the public have risks imposed on them daily without their consent or knowledge. The key elements for the public's acceptance or rejection of the risks of hazardous waste are not available to them.

Meaningful statistical analysis for purposes of identifying and controlling hazardous waste materials is generally hard for the average person to undertake. Being human, people assess a risk in relationship to its immediate harmful effects rather than to the long-term (chronic) health effects and subsequent damage to the environment. The assessment of risk in relationship to any kind of a waste disposal operation is not easy, but expenditures for this purpose are being increased by the federal government.

LEGISLATION

The Resource Conservation and Recovery Act (RCRA) was passed by Congress in 1976 to address a problem of enormous magnitude—how to safely dispose of the huge volumes of municipal and industrial solid waste generated nationwide. This law imposes strict controls over the management of hazardous waste throughout its life cycle, from "cradle to grave" (see Figure 19-1). It establishes the statutory framework for a national system that would insure the proper management of hazardous waste by the EPA. The goals set by the RCRA are:

1 To protect human health and the environment.
2 To reduce waste and conserve energy and national resources.

FIGURE 19-1
A system for the tracking of waste from the point of generation to the point of disposal. (*Source: U. S. Environmental Protection Agency, Office of Water and Waste Management, Washington, D.C.*)

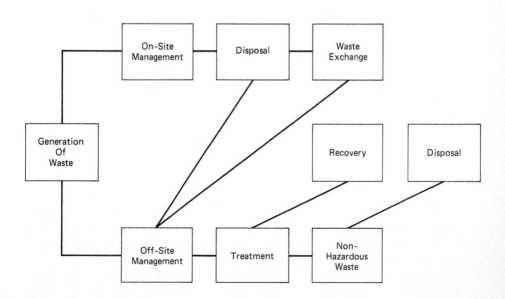

Tracking Hazardous Waste From "Cradle to Grave"

3 To reduce or eliminate the generation of hazardous waste as expeditiously as possible.

This law requires the EPA to:

1 Set standards for those who generate hazardous waste.
2 Set standards for transporters of hazardous waste.
3 Set standards and permit requirements for owners and operators of facilities that treat, store, or dispose of hazardous waste.
4 Create a manifest system for tracing the movement of hazardous waste from the point of generation to the point of disposal.

The law further authorizes states to operate a state hazardous-waste program in lieu of the Federal program if they meet certain requirements under the law.

Along with its authority under RCRA, the EPA also administers the Safe Drinking Water Act, the Clean Water Act, the Toxic Substances Control Act, and the Refuse Act. Under the provisions of certain portions of these acts, EPA was able to create the giant "superfund," the money for enforcing the acts and for the cleanup of abandoned disposal sites.

In June, 1988, Trans World Airlines was fined $100,000 for mishandling hazardous waste near its overhaul base at Kansas City International Airport. Among the violations cited by the EPA were dumping paint wastes near the overhaul base, failure to monitor ground water contamination, failure to have approved plans for closing the waste site, and failure to provide financial assurance for cleanup of the site.

THE RESOURCE CONSERVATION AND RECOVERY ACT

The act is a law which describes the kind of waste management program that Congress wants to establish. The act is divided into nine subtitles, A through I. Subtitles A, B, E, F, G, and H outline general provisions, authorities of the Administrator, duties of the Secretary of Commerce, federal responsibilities, miscellaneous provisions, and research, development, demonstration, and information. Subtitles C, D, and I lay out the framework for the three programs that make up RCRA: the hazardous-waste management program, the solid-waste program, and the underground-storage-tank program.

The Subtitle C program establishes a system to control hazardous waste from generation until ultimate disposal.

The Subtitle D program establishes a system to control solid (primarily nonhazardous) waste, such as household waste.

The subtitle I program regulates toxic substances and petroleum products stored in underground tanks.

The solid-waste program (Subtitle D) is discussed here before the hazardous-waste program (Subtitle C). Although out of alphabetical order, this structure is designed to benefit the reader. The Subtitle D program includes the definition of solid waste, which must be understood before the Subtitle C program can be explained.

Subtitle D of RCRA—Managing Solid Waste

The term "solid waste" is very broad, including not only the traditional non-hazardous solid wastes, such as municipal garbage, but also hazardous solid wastes. The act defines solid waste as:

1 Garbage, that is, milk cartons and coffee grounds, newspapers, and rags.

2 Refuse, that is, metal scrap, wallboard, and empty cartons.

3 Sludge from a waste treatment plant, water supply treatment plant, or pollution control facility, that is, scrubber sludges.

4 Other discarded material including liquids, or contained gaseous material resulting from industrial, commercial, mining, agricultural, and community activities, that is, boiler slag and fly ash.

It should be kept in mind that all solid waste is not solid. As noted above, many solid wastes are liquid, while others are semisolid or gaseous. Although the definition of solid waste includes hazardous waste, the Subtitle D program is concerned primarily with non-hazardous wastes.

Exceptions to the Definition of Solid Waste

The following materials are not considered solid waste under RCRA:

1 Domestic sewage (untreated sanitary wastes that pass through a sewer system)

2 Industrial waste water discharges regulated under the Clean Water Act

3 Irrigation return flows

4 Nuclear materials or by-products, as defined by the Atomic Energy Act of 1954

5 Mining materials that are not removed from the ground during extraction process

The Subtitle D program also regulates the development and implementation of state plans. In addition, it sets standards for solid-waste facilities and identification of open dumps.

State Plans under Subtitle D

States that develop and implement EPA-approved plans are eligible for federal technical and financial assistance. Each plan outlines the steps that a state must take to ensure that the solid waste within its borders is managed in an approved manner. The major components of state plans are to identify poorly managed facilities called "open dumps." Under a state plan, the state must close or upgrade open dumps and must also develop regulatory procedures to ensure that they are operated properly. EPA will provide technical assistance to state and local governments to develop and implement state plans. The federal assistance allotted to the states must be allocated by each state to the state and local agencies carrying out the implementation of the state plan.

Within six months from the date of submittal, the administrator of EPA must either approve or disapprove of the plan. Approval will be granted, if the administrator finds that the plan fulfills all of the minimum requirements set out in the regulations.

Subtitle C of RCRA—Managing Hazardous Waste

The Subtitle C program under RCRA is designed to ensure that mismanagement of hazardous waste does not continue. It does this by creating a Federal "cradle to grave" management system that sets forth statutory and regulatory requirements for:

1 Identifying hazardous waste
2 Regulating generators of hazardous waste
3 Regulating transporters of hazardous waste
4 Regulating owners and operators of facilities that treat, store, or dispose of hazardous waste
5 Issuing operating permits to owners or operators of treatment, storage, and disposal facilities
6 Enforcing the Subtitle C program
7 Transferring the responsibilities of the Subtitle C program from the federal government to the states
8 Requiring public participation in the Subtitle C program

What Is Hazardous Waste?

Hazardous waste comes in all shapes and forms. It may be liquid, solid, or sludge. It may be a by-product of manufacturing processes, or a simple commercial product—such as household cleaning fluid or various acids that have been discarded. To be a hazardous waste, a waste must be a solid, semi-soild, liquid, or contained gas such as garbage, refuse, scrap metal, empty containers, paper and trash, wallboard, and so forth.

Therefore, a hazardous waste is a solid waste or a combination of solid wastes, which because of its quantity, concentration, or physical, chemical, or infectious characteristics may cause, or significantly contribute to, an increase in mortality or cause an increase in serious irreversible illness; or pose a substantial present or potential hazard to human health and the environment when improperly treated, stored, transported, or disposed of, or otherwise managed.

A waste is considered hazardous if it has one or more of the following characteristics: *ignitability, corrosivity, reactivity,* and *toxicity.*

Ignitability Ignitable wastes can create fires under certain conditions. Examples include combustible and flammable liquids, solvents, and paint wastes.

Corrosivity Corrosive wastes include those that are acidic and those that are capable of corroding such metal as tanks, containers, drums, and barrels.

Reactivity If a waste is unstable or undergoes rapid or violent chemical change, it can create explosions and/or toxic fumes, gases, and vapors when mixed with water.

Toxicity When toxic wastes are disposed of on land, contaminated liquid may drain (leach) from the waste and pollute ground water. The EPA can identify toxicity through a laboratory procedure called the "extraction procedure."

EPA regulations require that all waste generators evaluate their wastes to determine if any of the above four characteristics are present.

THE TRACKING SYSTEM

Generators

Generators of hazardous waste are the first link in the cradle-to-grave chain of hazardous-waste management established under RCRA. It is the entity that produces the waste or is the first cause of the waste that becomes subject to RCRA regulations. Generators include industries of all types, small businesses, hospitals, laboratories, and universities.

Generators of more than 100 kilograms of hazardous waste or 1 kilogram of acutely hazardous waste per month must comply with all of the generator regulations under Subtitle C (one 55 gallon barrel = about 200 kilograms of hazardous waste). The regulations for a generator include:

1 Obtaining an EPA identification number.
2 Handling of hazardous waste before transport.
3 Manifesting of hazardous waste.
4 Recording and reporting.

Categories of Hazardous Waste Generators

There are three categories of hazardous waste generators:

1 Generators of no more than 100 kilograms per month of hazardous waste material. They are known as "conditionally exempt small quantity generators."
2 Generators of 100 to 1000 kilograms per month, who are considered "small quantity generators."
3 Generators of 1000 or more kilograms per month.

Transporters

Transporters of hazardous waste are the critical link between the generator and the ultimate off-site treatment, storage, or disposal of hazardous waste. A *transporter* is any person engaged in off-site transportation of waste in the United States by air, rail, highway, or water. Transporters must carry copies of the completed manifests and must also identify the type of waste they are transporting by affixing the proper symbols on the transport vehicle. In case of emergency, these symbols enable firefighters (first responders), police, and other officials to immediately identify the potential hazards. The transporter regulations do not apply to either the on-site transportation of hazardous waste by generators who have their own treatment, storage, and disposal facilities (TSDs) or to TSDs transporting wastes within a facility.

A transporter of hazardous waste must comply with both Department of Transportation (DOT) and EPA regulations. The Subtitle C regulations require a transporter to:

1 Obtain an EPA identification number
2 Comply with the manifest system
3 Deal with hazardous waste discharges

The Manifest

To track transported waste, EPA requires generators to prepare a "Uniform Hazardous Waste Manifest." This one-page form, with carbon copies for participants in the shipment, identifies the type and quantity of waste, the generator, the transporter, and the facility to which the waste is being shipped. Generators must also certify on the manifest that they are minimizing the amount and toxicity of their waste, and that the method of treatment, storage, or disposal they have chosen will minimize the risk to human health and the environment.

The Permitting System

Treatment, storage, and disposal facilities (TSDs) are the last link in the cradle-to-grave hazardous-waste management system. Owners or operators of existing or new facilities that treat, store, or dispose of hazardous waste must obtain an operating permit under Subtitle C.

Treatment facilities use various processes to alter the character or composition of a hazardous waste. Some treatment processes enable waste to be recovered and reused in manufacturing settings, while other treatment processes were designed to reduce the volume of waste to be disposed of. *Storage facilities* hold hazardous waste temporarily until treated or disposed of. Hazardous waste can be stored in 55-gallon drums, tanks, or other containers suitable for the type of waste generated. Hazardous waste may not be stored in tanks if it is likely to cause rupture, leaks, corrosion, or otherwise cause the tank to fail. Historically, most *disposal facilities* buried hazardous waste or piled it on the land. Improper land disposal practices in the past have endangered public health and the environment and continue to pose a threat, particularly to ground water.

Standards for Land Disposal Facilities

RCRA standards for the operation of land disposal facilities have become increasingly stringent. While it is almost impossible to discontinue land disposal of most hazardous wastes (at least for the present time), RCRA has created restrictions that might ensure more thorough protection of the environment, particularly ground water. The most important of these restrictions include the following:

1 Banning liquids from landfills
2 Banning underground injection of hazardous wastes within ¼ mile of a drinking-water well
3 Requiring more stringent design conditions for landfills and surface impoundments
4 Requiring cleanups and/or corrective action if hazardous waste leaks from a facility, including financial assurance that the corrective action can be completed

Subtitle I of RCRA—Managing Underground Storage Tanks

It has been estimated by EPA that as many as 1.5 million underground storage tanks are used in the United States to contain hazardous substances or petroleum products. It has been further estimated that from 100,000 to 300,000 of these tanks are presently leaking and polluting underground water supplies, and more may begin to leak within the coming years.

Subtitle I regulates underground tanks storing petroleum products (including gasoline and crude oil) and any substance defined as hazardous under the Superfund (Superfund is the common name describing the Comprehensive Environmental Response, Compensation and Liability Act). Subtitle I does not regulate above ground tanks storing hazardous wastes as defined by RCRA.

The underground storage tank act has five parts:

1 A ban on unprotected new tanks
2 A notification program by owners of existing or newly installed underground storage tanks to state and local agencies of each tank's age, size, type, location, and use
3 The development of standards for tanks, and regulations concerning leak/detection prevention and corrective action
4 State authorization to carry out the Subtitle I program
5 Inspection and enforcement

ENFORCEMENT

Enforcement may include civil or criminal penalties, orders to correct the violations, fines, and/or imprisonment. For severe or concurrent violations, EPA or the state can levy a penalty of up to $425,000 per day for each day past the abatement date that a facility stays in violation. EPA or the state can also suspend the facility's operating permits or bring criminal action charges against a facility.

Examples of the most serious violations are: transporting waste to a non-permitted facility, transporting waste without a manifest, or omitting—or making false statement in—a label, manifest, report, permit, or compliance document. These acts can carry a penalty up to $50,000 per day or two to five years in jail, per violation.

SUPERFUND

Superfund was created in response to the discovery in the 1970s of a large number of abandoned, leaking, hazardous-waste dumps threatening human health and contaminating the environment. One of the most notorious cases was Love Canal in Buffalo, New York. Because of this situation and because of a frightened public, the President and Congress created legislation to prevent similar occurrences from happening.

Superfund was signed into law in 1980. The Superfund program consists of three functions:

1 To take action in response to the release or threatened releases of hazardous substances, pollutants, or contaminants.

2 To require responsible parties, individuals, or corporations, to take appropriate action, and to oversee their response.

3 To recover monetary expenditures for responses and actions taken by the Federal Government.

Superfund, also called the Substance Response Trust Fund, is funded by taxes on crude oil, petroleum products, certain chemical feedstocks, and appropriation from the Federal Government. Congress gave EPA the power and authority to recover Superfund expenditures by bringing suit against the responsible parties for the cost of the cleanup.

WORD POWER

designated facility A hazardous waste treatment, storage, or disposal facility which has been designated on the manifest by the generator as the facility to which the generator's waste should be delivered.

dechlorination A process for removing chlorine to detoxify chlorinated substances.

EP toxicity A test, called the extraction procedure, that is designed to identify toxic wastes likely to leach into ground water.

EPA identification number The number assigned by EPA to generators or transporters of hazardous waste, and to each treatment, storage, or disposal plant or unit.

federal register A document published daily by the federal government that contains either proposed, or final, regulations.

food-chain crops Crops grown as human food, and crops grown to feed animals whose products are consumed by humans.

ground water Water that naturally flows through, and is stored in, soil and rock bodies beneath the land.

landfill Disposal facilities where hazardous waste is placed in or on land, a surface impoundment, or an injection well.

leachate Any liquid, or suspended components in a liquid, that has drained through from hazardous waste.

resource recovery The recovery of materials or energy from waste.

solid-waste disposal facilities Landfills, surface impoundments, land application facilities, and waste piles.

QUESTIONS AND EXERCISES

1 Does Richard F. Barrett's summarization from the *Wall Street Journal* give you some indication of the immensity of the hazardous waste problem in our country today? Can you see it as a worldwide problem? Explain.

2 Can you cite other situations similar to the Love Canal tragedy that have affected the population of a community or the public in general?

3 How are hazardous-waste or chemical-spill emergencies coordinated in your community? Explain.

4 What is the purpose of ground water monitoring?

5 What is the name given to the first responsible party in the waste cycle?

6 Like generators, transporters of hazardous waste must obtain an EPA identification number. Why?

7 What role does the "Uniform Hazardous Waste Manifest" play in the cradle-to-grave cycle of hazardous wastes?

8 Is your state authorized to operate its own hazardous waste management program? Do some outside research and discuss the program with your class.

9 What is meant by recycling (also referred to as "recovery" and "reuse")?

10 A common practice in some industries was to mix hazardous wastes or used fuel oil with fuels to increase the heating value of the fuels and to reduce the costs of waste disposal. In your opinion, should these practices be more closely regulated or banned completely by EPA? Why?

11 How can the EPA generate more public awareness and interest in the RCRA program?

12 What role can the public play in the success of the RCRA program?

13 If your business generated as little as two 55-gallon drums of hazardous waste per month, would you be required to obtain an EPA identification number?

14 Why should the safety specialist be interested in the waste disposal problem?

15 What are some of the factors that should be considered when deciding whether waste may be discharged in a public sewer?

16 Name some of the products we use that can generate potentially hazardous waste.

17 Explain how the "cradle-to-grave" system of hazardous-waste management is regulated or controlled by RCRA.

18 What are the three major goals set by RCRA?

19 Along with its authority under RCRA, what other laws does EPA administer?

20 What is the purpose of Superfund and how is it funded?

21 What must a state do in order to develop and implement its own statewide hazardous-waste program?

22 What might be some reasons for the public's resistance to hazardous-waste sites in their community?

23 Research some statistics for the purpose of determining how many tons of waste are generated in the United States annually. Include waste generated both by households and industry in your research.

24 The American way of life, as we know it today, depends upon an abundance of manufactured goods, whose manufacture generates hazardous waste as a by-product. If we are to continue to enjoy our present lifestyle, what can we as consumers do to reduce the threat of a hazardous-waste disaster?

BIBLIOGRAPHY

"Environmental Quality," *First Annual Report on the Council on Environmental Quality,* U. S. Congress, August 1970.

"Everybody's Problem: Hazardous Waste," U. S. Environmental Protection Agency. Office of Water and Waste Management, Washington, D.C., SW-826.1980, p. 1.

Federal Register, vol. 45, no. 98, U. S. Environmental Protection Agency, Washington, D.C., May 1980, p. 33071.

RCRA Orientation Manual, U. S. Environmental Protection Agency, Office of Solid Waste Management, Washington, D.C., January 1986, pp. iv–8, 1–5, 11–14, 11–15, 111–114, 111–119, 111–129, 111–134, 135.

"Solving the Hazardous Waste Problem: EPA's RCRA Program," U. S. Environmental Protection Agency, Office of Solid Waste Management, SW-86-037, Washington, D.C., November 1986, pp. 6, 9, 27.

CHAPTER 20

HISTORY AND DEVELOPMENT OF THE SAFETY MOVEMENT

THE INDUSTRIAL REVOLUTION

The history and development of the safety movement in America cannot be accurately described unless its events are coupled with those of the Industrial Revolution.

Between the years 1776, when our nation was founded, and 1850, the United States grew from a small agrarian country to a country of factory owners seeking markets at home in the east, the west, and countries abroad. European immigrants were flocking to the United States by the thousands, seeking freedom and jobs. As American industry began slowly to grow, skilled craftsmen gave way to a new generation of factory workers because the craftsmen were unable to compete with the prices of factory-made goods. An unprecedented industrial explosion of phenomenal growth began to occur in America. Not only did the established cities like New York, Boston, and Philadelphia grow, but new towns sprang up near sources of abundant water power. The textile industry in New England was the first large-scale industry to dominate the dizzying industrial acceleration of making goods faster and cheaper. American business began to grow and prosper. The private enterprise system arose because the founders of this country were dissatisfied with other economic systems, mainly European mercantilism. This new system was designed to give all people the opportunity to achieve material success to the extent that their abilities allowed. It brought significant changes to American business, but had a major fault, its treatment of the worker.

In order to understand the conditions which led to the safety movement in the United States, we should first review what occurred in England just prior to America's Industrial Revolution.

In England, the Industrial Revolution was the transformation of an essentially commercial society into one in which industrial manufacturing became the prevailing mode of economic life. After 1850, the factory was the most important institution shaping politics, social problems, and the character of daily life. The factory provided a new and uncongenial social movement for all who entered its doors. It is little wonder that English laborers, still more used to rural ways than urban ways, feared and hated the advent of "machines." During the late 18th century when the first textile mills were built, workers rose in revolution rather than work in them. Passages in English historical records reveal that workers fearing the advances of machinery actually wrecked and burned entire factory buildings.

HISTORICAL BACKGROUND

Organized Labor

Organized labor, weak in the early years of industrialization, limited its goals to raising wages and reducing long work hours. Eventually, it worked to gain public support for the prevention of accidents in the workplace. Organized labor fought employers and tried to get legislative action for two causes: guarding hazardous machinery and recompense for accident victims and their families, particularly in cases concerning permanent disability or death.

The resistance of management to union demands concerning safety and compensation for serious injuries and death may be explained by the fact that safety became identified with the overall struggle for higher wages, shorter working hours, and better working conditions. Management paid little or no attention to the health and safety of its employees. Working conditions in the mills were deplorable, not only for men and women, but also for the young children who were employed in them. Children as young as 10 years old worked in the cotton and textile mills. It was not until the early 1900s that child labor became a burning issue in this country.

The first organized unions originated as mutual-aid societies, which paid a small sickness and death benefit to its members or their widows. For the most part, these societies had a short life expectancy, the brevity brought on by their decisions to strike for higher wages. Management considered the labor strikes as economic blackmail and resorted to the courts for relief. This form of court action was a highly effective tool for breaking up strikes.

Coal mining and railroading eventually replaced textiles as the leading industries in the late 1800s. Safety-related issues were important here, too; the workers in these industries were intensely concerned about their own personal safety on the job. Mining is a particularly strong example of the situation. The danger in coal mining exceeded that of industry above ground, as indeed it still does today. The too-frequent news of workers trapped in underground coal mines made newspaper headlines. The most dreaded hazard was a gas called "fire-damp," which could explode with the force of dynamite. Toxic or flammable gases, rock falls, and fire were constant threats. In one seven-year period, despite the lack of a major disaster, 566 miners were killed and 1,655 were injured in Schuylkill County, Pennsylvania.

In 1869, a Mining Safety Law—passed by the Pennsylvania legislature—required at least two ways of exiting from each mine. Many authorities believe that this was one of the greatest advances ever made in mine safety. In 1910, the Bureau of Mines was created by the Department of the Interior—an action provoked by the many mine disasters. The purpose of the Bureau of Mines was to investigate the causes of mine accidents, study health hazards in the mines, and find means for making corrections.

In 1877, the bloodiest railroad strike in history broke out. Brakemen in 1871 averaged $1.75 for a 12-hour day. The work was so dangerous that labor historian Robert V. Bruce has said: "A brakeman with both hands and all his fingers was either remarkably skillful, incredibly lucky, or new on the job." Their fellow railroad workers fared little better. Massachusetts, the only state keeping records at the time, counted an average of 42 railroad workers dead each year from job accidents.

Railroading would finally be made safer not by strikes and riots but by a series of inventions that began when George Westinghouse invented the Westinghouse Air Brake in 1869. In 1872, the automatic air brake finally put an entire train under the engineer's control. In the 1880s the automatic coupler, a tremendous step forward, was invented. By the 1890s, the pressure was on Washington to make the automatic coupler mandatory equipment in all American railroads.

How Workers Felt

Carroll D. Wright was director of the pioneer Massachusetts Bureau of Statistics of Labor from 1873 until he became first United States Commissioner of Labor in 1886. One of his many innovations came in 1878, when he surveyed on-the-job working conditions.

Notable among his findings were the following: the average worker missed 13½ days a year because of illness; many regarded their jobs as hazardous; many had surprisingly radical views about capital and labor.

When asked about job hazards, a brass finisher cited dangerous machinery and metallic dusts, a carriage painter was concerned about lead poisoning, file cutters worried about dusts, iron makers about noise, heat, burns, and dusts.

Said an iron molder: "I think that every furnace and rolling mill...any manufacturing establishment where men have to perspire freely...should be obliged to furnish a good washroom and a place where clothes can be dried."

A shoemaker said: "Our shop is not properly ventilated. Each man does his own ventilating by the window before him, ventilating his neighbor...thereby causing him a cold. I think a shop should be ventilated in some other way."

Wright's study also exposed the general dissatisfaction with economic conditions of the 1870s, probably a major cause for the labor violence of the decade. Of 388 respondents to questions about living standards, only 85 said they were living as well or better than they had five years earlier (1873); 303 said they were worse off.

Industrial Hygiene

An occupational disease suddenly started making headlines in 1909, and the public's reaction to it was instantaneous.

"Phossy jaw" is an occupational disease that comes from breathing the fumes of white or yellow phosphorus, which gives off fumes at room temperatures, or from eating with fingers smeared with phosphorus. Match factories were its breeding grounds. British scientists discovered the disease in the 1850s and British industrialistsa took measures to prevent it. American medical scientists said the disease was rare in the United States and concluded that this was due to the more modern, scrupulously clean match factories in this country.

Then in 1908, Carroll Wright of the U. S. Bureau of Labor (then a bureau within the Department of Commerce), while doing a wage study on women and children, spotted phossy jaw among workers in the South. Phossy jaw wasn't all that hard to recognize. The victim's jaws swelled; the pain was intense. Scarring and disfigurement of the face were not uncommon. Sometimes the jaw, or even an eye, had to be removed. The Bureau of Labor expanded its investigation and published its findings in 1909. Meanwhile, a French chemist had come up with a safe substitute for which the Diamond Match Company held the American patent rights. In view of the circumstances, the Diamond Match Company waived its patent rights so that the entire match industry could use the safe substitute, called sequisulphide. Congress then passed the Esch Act which effectively removed white phosphorus matches from the market by taxing them so that they were equal in price to sequisulphide matches.

America's posture toward occupational disease prior to 1910 was one of complacency and unwarranted, even unreasonable optimism, and its industrial hygiene was nonexistent.

The Pittsburgh Survey

America had no statistical evidence of how many workers were being killed and disabled annually in industrial accidents in the United States. There were no laws or regulations requiring any of the Federal agencies to amass such records. The Russell Sage Foundation sought to find out by means of its now historic study, called the Pittsburgh Survey because it was mostly confined to Allegheny County (the greater Pittsburgh area), Pennsylvania, in 1906.

The results were shocking. Sage investigators uncovered 526 on-the-job fatalities and 500 seriously and permanently disabling accidents during 12 months of 1906–1907. The study further revealed that over 50 percent of surviving widows and children of accident victims were left without any death benefits or sources of income. This same report stated that there were 30,000 fatalities from industrial accidents in the United States in 1906. Workmen's compensation benefits did not exist in this country at that time.

Among safety historians, the Pittsburgh Survey is credited with the formation of a safety climate and the advent of organized safety programs in industry. At this particular time in American history (1900–1916), the current of reform, which was a part of the American scene, helped to make the Pittsburgh Survey popular reading by everyone. It was a time when investigative reporting became popular for exposing questionable and corrupt practices by both politicians and big business.

Organized Safety

The first cooperative move toward the formation of an organized safety movement was made in 1912 by the safety committee of the Association of Iron and Steel Engineers in Milwaukee. At a meeting in New York City in the following year, an organization called the National Council of Industrial Safety was formed specifically for the promotion of safety in industry. When the organization met again in 1915, it changed its name to the National Safety Council and broadened its function to include accident prevention—no longer limited to industry, but including safety at home, in the schools, and on the streets and highways. Its membership included representatives from industry, public utilities, labor unions, insurance companies, colleges and universities, and the government sector. It also promoted an annual and well-attended National Safety Congress. By 1927, it had published over 80 pamphlets on safe practices in industry, which included information and tips on safety goggles, good housekeeping, ventilation, and fire extinguishers. Its *Annual Facts,* printed yearly, contains detailed accident statistics on work-related, home, motor vehicle, and public accidents. This publication alone is a credit to the Council and its Statistics Division.

Another organization which contributed heavily to the voluntary safety movement was the National Fire Protection Association which was organized in 1896 as a clearinghouse for information on the subject of fire protection and fire prevention. Some of the technical standards that it issued were accepted by many state and municipal governments as a basis of fire legislation. It also began publishing a wide variety of educational materials in the form of books, pamphlets, fire records, leaflets, posters, and bulletins.

Perhaps most significantly, and probably the first in existence, an in-plant safety department was organized at the Joliet Works of the Illinois State Steel Company in 1892.

Organized industrial safety programs became more prevalent during the 1920s. Companies began competing for safety awards, for working without a lost-time accident. In 1926, Carnegie Steel Company boasted 2,600,000 hours worked without a lost-time accident. Illinois Steel reported 3 million worker-hours without a lost-time accident and Clark Thread Company worked over 10 million hours without a lost-time accident.

Federal Intervention

In this climate, the Federal government commenced expressing interest in public safety—an interest which began with the Pure Food and Drug Act of 1906 and culminated with the Occupational Safety and Health Act of 1970. For example, the nation's first workmen's compensation law was passed in 1908 covering federal employees, and the United States Bureau of Mines was established in 1910. The United States Department of Labor was established in 1913. Certainly, occupational safety and health was making some headway. For those injured on the job, workmen's compensation now provided some relief. By 1921, 46 states had some form of what is now properly called workers' compensation. The cost of work accidents was recognized as a cost of doing business. On the average, American workers today are manifestly far better off than their counterparts of some 75 years ago.

In 1970, the Occupational Safety and Health Act became law. It was probably one of the most important pieces of legislation, from the standpoint of 55 million people covered by it, ever passed by the Congress of the United States. This act, under the Secretary of Labor, empowered OSHA to enter the workplace, inspect for and cite violations, and set abatement deadlines. It has made a considerable impact on the occupational safety and health movement in the United States. It became inevitable that companies which had never had a systematized approach to accident prevention began formulating one and that companies with perfunctory programs were upgrading them.

Concern for the harm to human health and the environment of this country also became a prime issue, eliciting Federal intervention out of concern for the citizenry. In 1976, Congress passed the Resource Conservation and Recovery Act, followed by the Clean Water Act, the Safe Drinking Water Act, the Toxic Substances Control Act, and the Refuse Act. Of special importance in the Resource Conservation and Recovery Act was a provision stating that "public participation in the implementation and enforcement of this Act is encouraged and assisted." The goal of these acts was to control hazardous wastes and to prevent environmentally unsound disposal practices before they caused harm to human health and to the environment in which we live.

Women in the Workplace

The Equal Employment Opportunity Commission (EEOC) enforces Title VII of the Civil Rights Act that prohibits any discrimination based on sex, race, color, religion, or national origin. This law gives women the right to be considered for jobs on an equal rights basis with men. This law also allows them to be exposed to the same hazards in industry as men.

Women were once told their place is in the home. This is obviously no longer true. In 1976, according to the U. S. Department of Labor, 49 of every 100 women were in the labor force. In 1979, about 45 percent of the 100 million females in the United States were in the labor force. Between 1980 and 1984, women accounted for 48 percent of the work force, and by 1987 the figure rose to around 50 percent. Because of seasonal work and plant closings, this figure may vary. By 1990, working women are expected to number 54 million.

Most of these women will not be "working wives," but heads of households. The increasing number of single-parent homes in this country will increase the demand for jobs. Also, birthrates are dropping, and as old attitudes and stereotypes are being cast aside, these trends encourage many married women to work for a second family income.

This dramatic shift from home to the workplace has elicited concern in industry for safety. The questions of concern are applicable to the great majority of women in the work force. Does a normal woman with a normal pregnancy and a normal fetus confront any hazards in the workplace? Another question: What bearing do toxic exposures in the workplace have on a fetus? To date, limited information is available about physical and chemical influences upon the developing fetus. Obviously enough, the limited scientific data on chemical substances which can harm reproductive capabilities frustrate safety and health officials and workers alike.

The Environment

The *environment,* both physical and social, is also changing. Today's worker is exposed to various air and water pollutants over an extended period of time; to food additives and preservatives; to complex laundry and cleaning compounds; and to many innocuous "work-savers" in the home. Industrial workers come into contact daily with many new substances utilized in their work processes. Many workers, both men and women, come to the workplace with long family histories of sensitivities to all kinds of synthetic materials and pharmaceuticals and also with physiological and psychological tendencies to certain types of health problems. Two important examples having to do with the social environment are drugs and alcohol. Many workers are emotionally or physiologically dependent on certain drugs, and some combine drugs with alcoholic beverages—thus compounding the original problem. Many employees come to work every day with alcohol in their body systems in order to try to control their inner conflicts and problems. Management, with the aid of labor, must design assistance programs to help such employees through periods of stress. The problem of alcohol will not disappear on its own; it must be dealt with in order to protect industry's most important asset—its employees.

Will trends such as these create changes resulting in accidents? Will the number of industrial accidents decline, remain the same, or rise? If the accident trend does rise or fall, will it do so in the form of a straight or delayed curve? Naturally, only time will tell. It is possible that trends will change based on both the physical and social environment. But even if present trends continue, years will elapse before definite changes in accident patterns will actually appear. Although something is always happening in the "safety world," radical changes will not and cannot take place overnight.

But whatever trends there are, and whatever their effects may be on accidents, safety on the job can be improved if employers will treat the subject of safety as a humanitarian effort rather than a *work station* requirement. Management's thinking about the safety of its employees should not begin and stop in the workplace, but should take off-the-job safety into consideration as well.

As the history of labor shows, until the late 1800s agriculture was the principal industry in the United States; then manufacturing became the primary source of jobs. Now, the trend is towards the *service* industries, which means an increase in the number and proportions of service workers.

LOOKING AHEAD

The management of safety in the workplace is a continually evolving process, reflected in research and technology development, new regulations and amendments to the existing Occupational Safety and Health Act, and a supportive management unobstructed by labor difficulties over safety and health issues.

Since the Occupational Safety and Health Act was passed, we have seen substantial progress in solving complex safety problems in the workplace, both above ground and underground. The climate for a productive, accident-free future looks more promising than in the early days of this century. Public interest in the problems associated with safety in the workplace has steadily increased over the past several years and is likely to remain high. Knowledge related to safety and health issues and the technical capabilities to cope with them will continue to grow with time as our experience and expertise increase. Technological changes in industry, population expansion, and economic growth will present new challenges in safety and health. The cooperation and participation of industry, labor, government, and the public will ensure that these challenges are met.

The foregoing description in this chapter of the development of the safety movement in industry gives only the highlights—and even these are in brief form. Any reasonably complete description would, in itself, be the basis of a historical treatment of the subject long enough to fill the pages of this textbook.

QUESTIONS AND EXERCISES

1 Using outside sources, explain the major theory advanced by Karl Marx and Friedrich Engels about the spread of the Industrial Revolution worldwide.
2 What part did the labor movement play in reducing accidents on the job during the Industrial Revolution in America?
3 In your opinion, how might the accident statistics compiled by OSHA today fare against those of the early 1800s?
4 You may have heard it said that a manufacturing environment is safer than the

home environment. What is this idea based on? What is your personal opinion of this theory?

5 Do you feel that there is a correlation between the use of alcohol and industrial accidents? Between alcohol and highway accidents?

6 Do you feel that the use of drugs has more or less influence than alcohol on industrial accidents?

7 What possible changes do you foresee in industrial safety programs as a result of the increasing number of women entering the workplace annually?

8 What were the demands made by organized labor from about 1850 and onward? Were safety issues among them?

9 In your opinion, did the workers during the Industrial Revolution in England have working conditions any better than their counterparts in America during the same time period? Explain.

10 Do you feel that if management and labor had joined hands in the formative years of the safety movement, and had worked together toward a common goal, the outcome would have been different? Explain.

11 Perhaps the greatest hope of reducing injury and death tolls in the workplace lies in the interest and participation in safety activities of organized labor. How do you feel that organized labor goes about fulfilling this role?

12 How did the Pittsburgh Survey help to bring about the passage of workers' compensation laws in the United States?

13 What events led to the development of the Occupational Safety and Health Act?

14 What is the nature and function of the National Safety Council?

15 Can you name any other private organizations that have made contributions to safety ideals and methodology?

BIBLIOGRAPHY

Kalis, David B., Carl Musacchio, Peter J. Sheridan, Jonathan Foreman, and Ronald E. Lucas: "The Evolution of America's Industrial Safety Movement," *Occupational Hazards,* Parts I, II, III, IV, Cleveland, Ohio, September 1975, pp. 56, 58, 59, 60, 67, 71.

SAFETY AND HEALTH
MILESTONES

1866	National Board of Fire Underwriters is founded.
1869	Railroad air brake is invented by George Westinghouse.
1869	Pennsylvania passes the first state law for coal mine inspections.
1869	Massachusetts organizes the first state Bureau of Statistics of Labor.
1872	Automatic air brake is invented.
1877	Massachusetts passes the first State Factory Inspection Law.
1878	The Knights of Labor, the nation's largest union, demands that Congress adopt measures for health and safety in mining, manufacturing, and construction.
1885	American Public Health Association is founded.
1886	The first United States Commissioner of Labor is appointed.
1892	America's first industrial Safety Department is organized at the Joliet Works of the Illinois Steel Company.
1894	Underwriter's Laboratories is organized.
1895	State legislators begin to change common law affecting employer's liability.
1896	National Fire Protection Association is founded.
1902	United States Public Health Service is established.
1906	Pure Food and Drug Act is established.
1906	The Pittsburgh Survey reveals the frequency and seriousness of industrial accidents.
1908	Nation's first federal workers' compensation law covering federal employees is enacted by Congress.

1909	United States Bureau of Labor publishes its findings on phossy jaw among workers in match factories.
1910	United States Bureau of Mines is established.
1911	New York passes the first state workers' compensation law.
1911	American Society of Safety Engineers is established.
1913	United States Department of Labor is established.
1913	National Safety Council is established.
1916	American Occupational Medical Association is chartered.
1919	E. W. Bullard designs the first protective safety helmet for industry.
1920	F. R. Davis introduces the first industrial first-aid kit.
1926	Carnegie Steel Company records 2.6 million worker-hours worked without a lost-time accident; Illinois Steel reported 3 million hours; and Clark Thread Company reported over 10 million hours.
1931	United States Public Health Service establishes the Dermatoses Investigation Branch.
1933	National Recovery Act (NRA) safety codes are formulated.
1934	Division of Labor Standards is organized.
1936	Walsh-Healy Public Contracts Act is passed.
1938	Federal Interdepartmental Safety Council is organized.
1938	The Center for Safety Education of New York University is founded.
1954	The Atomic Energy Act becomes law.
1960	Safety standards for government contracts under the Walsh-Healy Act of 1936 are passed.
1961	Accident frequency rates "bottom out" and begin to climb steadily.
1962	President Lyndon Johnson convenes the President's Conference on Occupational Safety.
1965	Ralph Nader publishes *Unsafe at Any Speed.*
1965	Congress enacts the Water Quality Act.
1966	Congress enacts the Federal Metal and Nonmetal Mines Safety Act.
1967	Congress enacts the Air Quality Act.
1967	Congress enacts the Fire Research and Safety Act.
1969	Congress enacts the Construction Safety Act.
1969	Congress enacts the National Environmental Policy Act.
1970	The Occupational Safety and Health Act (OSHA) becomes law.
1971	National Institute for Occupational Safety and Health (NIOSH) is established.
1972	The Consumer Product Safety Act becomes law.
1976	Congress enacts the Toxic Substances Control Act.

Note: Included in this chronology are all the dates considered important to the safety movement. There may be others that have been unknowingly omitted.

CASE STUDIES

Studying actual cases is a good way to become acquainted with the area of accident prevention. Accordingly, this appendix presents ten cases; they contain a minimum of detail because there is no mystery about how accidents in general occur, or about the everyday kinds of accidents that are common in the workplace.

Almost 90 percent of all work-related accidents are due to negligence on the part of the injured employee; the cases here follow that pattern. In almost every case, the preventive action ultimately recommended or taken indicates that training of the employee by management is the most important method that can be used to reduce accidents. In fact, training is *always* important and it should not be neglected at any point, from the preemployment orientation until the end of an employee's career.

CASE STUDY 1

Extent of Injury Fatality.

Description of Accident An employee was struck by a 5-ton riveting machine which fell while he was moving it with a wall-mounted electrically controlled crane. The machine fell because the lifting eyebolt assembly had failed: the nut holding the machine to the lifting eyebolt came off.

Cause of Accident The nut holding the eyebolt had loosened, causing the eyebolt threads to become stripped.

Corrective Action Required (1) Make periodic inspections of lifting equipment in the plant. (2) Design an inspection checklist for this program, making sure that eyebolts and lifting lugs are included. (3) Make it a plant procedure that employees visually inspect lifting components before using equipment.

CASE STUDY 2

Extent of Injury Amputation of left thumb.

Description of Accident An axle packager in a metals plant placed an axle on a layout table and sprayed it with a rust preventative before packing it for shipment. The spraying was done with a gun operated at a pressure of 40 pounds per square inch (psi). The gun clogged while the packager was spraying, and he wiped the spray tip with his left thumb. At that instant the gun discharged and the employee received an injection of rust preventative in his left thumb.

Cause of Accident Cleaning a clogged spray gun with a bare finger.

Corrective Action Required Remove pressure from hose line before attempting repair work on a spray gun or applicator.

Follow-up Action (1) Caution employees never to point a gun or applicator under pressure toward any part of the body. (2) Warn employees of the possibility of dangerous substances penetrating the skin.

CASE STUDY 3

Extent of Injury Amputation of left little finger.

Description of Accident A drill-press operator was operating a drill press while wearing gloves. She attempted to change the tooling while the machine was running. Her glove caught on the revolving drill, and her left little finger was amputated.

Cause of Accident (1) Not shutting off the drill press before changing the tooling. (2) Wearing gloves around revolving machinery.

Corrective Action Required (1) Instruct employees to shut off machinery when tooling changes must be made. (2) Instruct employees not to wear gloves around revolving machinery.

Follow-up Action Employees should be further instructed not to wear rings, bracelets, jewelry, or long sleeves around rotating machinery; shirts and blouses should be tucked into trousers; and employees with long hair should be required to wear approved hairnets or snoods.

CASE STUDY 4

Description of Injury Fracture of skull.

Description of Accident A die setter and a co-worker had each rigged a chain around one end of a 5-ton die to move it by crane to a press line. The die setter did not double-check his rigging. As he turned to walk away, his co-worker signaled the crane operator to take up the slack in the chain. The chain which the die setter had rigged was against the keeper pin instead of the die notch. The sudden pressure from the chain caused the keeper pin to shear off; it struck the die setter across the back of the head, knocking him unconscious.

Cause of Accident Improper rigging of die.

Corrective Action Required Reinstruct die setters to check die rigging before allowing anyone to signal a crane operator to lift the die off the floor.

CASE STUDY 5

Extent of Injury Slight redness of the face and a case of the "shakes."

Description of Accident A machine operator moving a tub of material noted a 440-volt extension cord lying on the floor in her way. Instead of moving it by hand, she kicked it with her foot, causing the plug to explode with a blinding flash.

Cause of Accident Investigation revealed that the electrical cord had a bare wire protruding from its casing and had apparently "shorted" out when it struck a piece of metal on the floor.

Corrective Action Required Handle electrical cords and electrical plugs with extreme caution.

Note An accident such as this could have been a fatality.

CASE STUDY 6

Extent of Injury   Loss of left eye.

Description of Accident   A carpenter was nailing a 2- by 4-inch wood strip into place on the floor of a boxcar. He hit the head of the nail off center, causing it to glance upward, striking him in the left eye.

Cause of Accident   The carpenter was not required to wear safety glasses.

Corrective Action Required   Set up safety glass requirements for the department doing carpentry; the program should involve both management and employees.

Follow-up Action Required   Study other departments of the plant for similar hazards and include them in the safety glass procedure if necessary. Furthermore, instigate an educational program on a plantwide basis to stimulate interest in eye protection.

CASE STUDY 7

Extent of Injury   Deep laceration of the forehead requiring seven sutures.

Description of Accident   In a machine shop, a machinist was using a cold chisel with a mushroomed head and striking it with a hammer on a piece of tool steel. As he struck the chisel with his hammer, a chip of the mushroomed head flew upward and struck him in the forehead.

Cause of Accident   Mushroomed chisel head.

Corrective Action Required   Reinstruct all machinists in the proper procedure for inspecting chisels for mushroomed heads. Furthermore, have the machine shop supervisor inspect all of employees' tools for the purpose of finding any other hazards.

CASE STUDY 8

Extent of Injury   Minor first-aid cases.

Description of Accident   One of the oldest (in seniority), and most experienced, toolmakers in an automobile assembly plant suddenly developed the habit of fumbling and dropping tools and other pieces of work on the floor. He spoiled work that formerly had been accurately done and suffered five first-aid cases in three weeks.

Cause of Accident At the advice of his supervisor, the toolmaker had a thorough physical examination which included an eye examination and a hearing test, without, however, finding a solution of his difficulty. There were no changes in the shop's environment and working conditions remained the same as before. Apparently a condition other than one directly related to the work at the plant impaired the ability of the employee to work safely.

An analysis of the situation by the supervisor revealed that the types of accidents the employee was involved in occurred with "struck-by" tools and materials, and the unsafe act was "lack of skill," or, more accurately perhaps, temporary failure to exercise good judgment and skill in work performance.

Corrective Action Closer supervision by the supervisor, consisting of observation, more frequent safety contacts, and making the necessary corrections when necessary.

CASE STUDY 9

You are the plant manager of a small manufacturing plant employing 150 employees. Because of the small size of your plant, you have delegated safety responsibilities to your personnel manager. Although your plant has never been thoroughly studied for hazards, your manufacturing experience keeps you well aware that your plant is not as safe as you would like to see it and that some of the operations require considerable care on the part of your supervisors and employees. You have had an average number of minor injuries in the past few years; however, you have escaped without a lost-time injury, and feel very lucky for having done so.

You discover that 12 of your employees in one department have been working with a new compound, identified only by a trade name, which you suspect may be a potential health hazard.

Describe what steps you would take to identify the material and to help your employees deal safely with this condition. Would you take into consideration the formation of a hazardous materials committee in your plant? Would you have your personnel manager design a training program for such conditions? Describe how you would proceed to establish this program.

CASE STUDY 10

Extent of Injury   Compound fracture of right leg.

Property Damage $3,100—Broken shaft and belts on a large press and broken guard on conveyor belt.

Description of Accident   Millwright was reaching out to make an adjustment on a flywheel chain on press while standing on a 20-foot ladder. In doing so, he lost his balance and fell onto the shaft and then struck a conveyor and fell to the floor, approximately 15 feet below.

Cause of Accident

1 Ladder was improperly positioned (too far out for support)
2 Ladder not tied off
3 Ladder did not have safety treads

Corrective Action Required

1 Instruct employees to position ladders properly and to tie them off when in use.
2 Instruct employees to make a visual inspection of ladders before each use.
3 Instigate a monthly physical inspection of all ladders by the maintenance department and correct deficiencies.

Note   You are the safety manager of the plant where the above accident occurred. You have read the supervisor's accident report and recommendations for corrective action. Are you satisfied with these recommendations? If not, what other requirements would you recommend to prevent recurrence of this accident?

THREE

DIRECTORY OF USEFUL MATERIALS AND HANDBOOKS CONTAINING ENGINEERING TABLES, FORMULAS, AND OTHER DATA

The following listing of engineering handbooks, manuals, and technical materials is often revised; no attempt has been made to list the latest information available. Therefore, when corresponding or ordering information from this listing, the latest or most recent edition(s) should be requested.

American Conference of Governmental Industrial Hygienists, Committee on Industrial Ventilation, P. O. Box 1937, Cincinnati, O. 45201. *Air Sampling Instruments Manual, Industrial Ventilation: A Manual of Recommended Practice.*

American Industrial Hygiene Assn., 475 Wolf Ledges Parkway, Akron, O. 44311. *Industrial Noise Manual,* 3d ed., 1975.

American Mutual Insurance Alliance, 20 North Wacker Drive, Chicago, Ill. 60606. "Handbook of Organic Industrial Solvents," Technical Guide no. 6, 1980.

American Society for Testing and Materials, 1916 Race St., Philadelphia, Pa. 19103. "Standard Metric Practice Guide," E380 (ANSI 2210.1).

American Welding Society, 2501 NW 7th St., Miami, Fl. 33125. *Welding Handbook.*

Baumeister, Theodore, (ed.): *Standard Handbook for Mechanical Engineers,* McGraw-Hill Book Co., New York, N.Y. 10020.

A. M. Best Company, Oldwick, N.J. 08858. *Best's Safety Directory* (2 vols.).

Bolz, Harold A. (ed.); *Materials Handling Handbook,* The Ronald Press Co., John Wiley & Sons, Inc., New York, N.Y. 10016.

Car, Clifford C.: *Croft's American Electricians' Handbook,* McGraw-Hill Book Co., New York, N.Y. 10020.

Carmichael, Colin, and J. K. Salisbury: *Kent's Mechanical Engineers Handbook,* John Wiley & Sons, Inc., New York, N.Y. 10016.

De Garmo, E. Paul: *Materials and Processes in Manufacturing,* 4th ed., 1974, The Macmillan Co., New York, N.Y. 10022.

Eshbach, Ovid W.: *Handbook of Engineering Fundamentals,* John Wiley & Sons, Inc., New York, N.Y. 10016.

Factory Mutual Engineering Corp., Norwood, Mass. 02062. *Loss Prevention Data,* 1973.

Hackman, J. R. and Oldham, G. R.: *Work Redesign,* 1980, Addison-Wesley Press, Reading, Mass.

Harris, Cyril M.: *Handbook of Noise Control,* 1957, McGraw-Hill Book Co., New York, N.Y. 10020.

Hosey, A.D., and Powell C.H. (eds.): *Industrial Noise—A Guide to Its Evaluation and Control,* 1967, U. S. Department of Health, Education, and Welfare. (Available through U. S. Government Printing Office, Washington, D.C. 20402.)

How to Dispose of Toxic Waste Substances and Industrial Wastes, Noyes Data Corporation, Park Ridge, N.J. 07656.

Industrial Safety Products News (published monthly), Philadelphia, Pa. 19120.

Knowlton, A. E. (ed.): *Standard Handbook for Electrical Engineers,* McGraw-Hill Book Co., New York, N.Y. 10020.

LaLonde, William S., Jr., and Milo F. Janes: *Concrete Engineers Handbook,* McGraw-Hill Book Co., New York, N.Y. 10020.

National Association of Women in Construction, Greater Phoenix (Az.) Chapter #98. *Construction Dictionary,* 1973.

National Fire Protection Association, Batterymarch Park, Quincy, Mass. 02110. *Fire Protection Handbook,* 15th ed., 1981.

National Safety Council, *Accident Prevention Manual for Industrial Operations,* 8th ed., 1980, Chicago, Ill. 60611.

NIOSH, *Pocket Guide to Chemical Hazards,* Superintendent of Documents, U. S. Government Printing Office, Washington, D.C. 20402

Occupational Safety and Health Guidance Manual for Hazardous Waste Site Activities, Superintendent of Documents, U. S. Government Printing Office, Washington, D.C. 20402.

Rizzi, E. A.: *Design and Estimating for Heating, Ventilation, and Air Conditioning,* 1980, Van Nostrand Reinhold, New York, N.Y.

Robb, Dean A., and Harry M. Philo: *Lawyers Desk Reference: A Source Guide to Safety Information, What to Find, How to Find It,* The Lawyers Cooperative Publishing Co., Rochester, N.Y. 14603.

Rossnagel, W. E.: *Handbook of Rigging,* McGraw-Hill Book Co., New York, N.Y. 10020.

Standard Operating Safety Guidelines, Environmental Response Branch, Hazardous Response Support Division, Office of Emergency and Remedial Response, U. S. Environmental Protection Agency, Washington, D.C. 20460.

Tarrants, W. E.: *The Measurement of Safety Performance,* 1980, Garland STPM Press, New York, N.Y.

U. S. Department of Health, Education, and Welfare, *Mechanical Power Press Safety Engineering Guide,* 1976, Superintendent of Documents, U. S. Government Printing Office, Washington, D.C. 20402.

Index

Page references in *italic* refer to illustrations; page references in **boldface** refer to tables.